Varieties of Integration

© 2015 by
The Mathematical Association of America (Incorporated)
Library of Congress Catalog Card Number 2015944846
Print Edition ISBN 978-0-88385-359-7
Electronic Edition ISBN 978-1-61444-217-2
Printed in the United States of America
Current Printing (last digit):
10 9 8 7 6 5 4 3 2

The Dolciani Mathematical Expositions

NUMBER FIFTY-ONE

Varieties of Integration

C. Ray Rosentrater
Westmont College

Published and Distributed by
The Mathematical Association of America

The DOLCIANI MATHEMATICAL EXPOSITIONS series of the Mathematical Association of America was established through a generous gift to the Association from Mary P. Dolciani, Professor of Mathematics at Hunter College of the City University of New York. In making the gift, Professor Dolciani, herself an exceptionally talented and successful expositor of mathematics, had the purpose of furthering the ideal of excellence in mathematical exposition.

The Association, for its part, was delighted to accept the gracious gesture initiating the revolving fund for this series from one who has served the Association with distinction, both as a member of the Committee on Publications and as a member of the Board of Governors. It was with genuine pleasure that the Board chose to name the series in her honor.

The books in the series are selected for their lucid expository style and stimulating mathematical content. Typically, they contain an ample supply of exercises, many with accompanying solutions. They are intended to be sufficiently elementary for the undergraduate and even the mathematically inclined high-school student to understand and enjoy, but also to be interesting and sometimes challenging to the more advanced mathematician.

MAA Service Center
P.O. Box 91112
Washington, DC 20090-1112
1-800-331-1MAA   FAX: 1-301-206-9789

Preface

While the primary audience for this book is an advanced undergraduate mathematics student, the contents will appeal to any mathematician who has wondered how the integrals introduced in elementary calculus and in real analysis courses fit together. By the time a young mathematician has completed the first year of graduate school, she will have encountered three versions of the integral: Riemann, introduced in elementary calculus; Darboux, studied in a first real analysis course and often still called a Riemann integral; and Lebesgue, developed in an advanced analysis course. Most often, these integrals are studied in isolation and with very little connection or comparison made between the different definitions. This book provides a comparative study of four approaches to integration over an interval $[a, b]$: Riemann, Darboux, Lebesgue, and gauge.

In addition to serving as a reference, this book can serve as a text for a second course in real analysis. Indeed, this manuscript is written with such users particularly in mind. The prerequisite first course should include the standard topics of supremum, infimum, compactness, the mean value theorem, and sequences of functions. The reader should also be familiar with using the formal ε-δ definitions of limit and continuity in proofs. A series of appendices containing statements of the most relevant definitions and results from a first real analysis course is provided for readers who have encountered the requisite ideas but may need to refresh their memories. In addition, readers may find the notational index found at the beginning of the index helpful.

While the most celebrated milestone in the development of calculus comes from the late 17th century work of Newton and Leibniz, questions and ideas that lie at the heart of integral calculus were introduced by Eudoxus (4th century BCE) and Archimedes (3rd century BCE). The ideas of the differential and integral calculus (brought together by Newton and Leibniz) were powerful forces in the advancement of science. But cracks in the foundations of the subject (identified early on by Bishop Berkeley) became increasingly apparent toward the end of the 18th century. By the end of the 19th century, Cauchy, Riemann, and Darboux had addressed these foundational issues and had provided solid foundations for the integral calculus. But the newly grounded integral calculus was challenged again by a new set of issues arising from applications in differential equations. Lebesgue, Henstock, Kurzweil, and others developed new approaches to meet these challenges. Subsequently, Lebesgue's ideas have been extended into other domains in which both the regions of integration and the types of values produced by the integrals are greatly generalized.

This book explores the critical contributions by Riemann, Darboux, Lebesgue, Henstock, and Kurzweil and provides a glimpse of more recent variations of the integral. Though historical background is useful in motivating and framing the questions to be addressed, the primary focus of this book is not historical. The primary goals are (1) to provide you with an understanding of and an appreciation for the work done in formalizing and extending the ideas of the integral calculus, (2) to help you to think like a mathematician, and (3) to provide you with an opportunity to develop into a reader of professional mathematics. While the first goal is self-explanatory, the other two deserve some explanation.

Most often mathematical texts present material in a final, polished form. This is good. However, the budding mathematician is often left with the impression that ideas spring full-grown from the ground and that they could never have been otherwise. Since you are reading this book, you already know that mathematical knowledge is hard-won, but you may not have the sense for how many different paths have been tried and abandoned in the process of developing the theories that we have today. The general structure of this book, as it moves through the development of various integrals, encourages you to think about how mathematics develops and the implications of different approaches. This book also addresses the issue in a more localized fashion. The exercises will continually prod you to think about why certain choices in a proof were made or why an alternate, apparently simpler or more obvious approach was not used. The intended effect is that you, the reader, will come away with a greater flexibility in the way you see mathematics.

This text also seeks to improve your mathematics reading ability. Proof-oriented mathematics texts try to include all the details and background used in a proof. Student writing tends to reflect this practice by including long sequences of algebraic derivations. This happens even (particularly?) when a student's solution or proof fails to address a critical logical step or two. Mathematics found in professional journals takes a different approach. The writing is terse. The reader is expected to be familiar with a significant breadth of background information and to understand (or at least appreciate) the motivation for the work. The professional proof then provides navigational landmarks that the reader is expected to use to construct a path from the hypothesis to the conclusion. In particular, the reader is expected to fill in much of the algebra. To use another metaphor, the version of a proof that appears in a professional journal serves as a skeleton and the reader must fill in many details to "flesh out" the proof. This type of writing requires significant engagement on the part of its readers since, before understanding a proof, the reader must identify the places where details must be filled in and must construct the bridging argument or computation. Consequently, it is not uncommon to spend an hour or even a couple of days working to digest a proof.

You may well ask why this difference in writing styles exists. This is an excellent question to which I offer only tentative, partial answers. One factor is the mathematical culture. This is the way mathematicians have written for quite some time and the practice is not likely to change soon. Of course, this answer does nothing to explain how things came to be this way. One significant pressure in this direction is the cost of publishing or, moving further into the past, the effort required to make papyrus sheets. Until fairly recently, it was customary to charge authors or their institutions a fee for preparing a paper for publication. These charges could be as high as several hundred dollars per page. In addition, most journals have limits on the length of the papers they will publish. Both of these practices exert pressure on authors to compress their writing.

With the widespread use of TEX, papers no longer need to be retyped into a form suitable for publication so the assessment of page charges has largely disappeared. Online publishing has the potential to eliminate the constraints on article length. In this new environment, perhaps you will help create a cultural shift so that professional mathematics papers will include more of the details. In the meantime, it is important for developing mathematicians to learn to read journals as they are written rather than as one might wish they were written.

This book attempts to help you become more adept at this task. The proofs at the beginning of the book will call attention to the places where details must be filled in. Usually this takes the form of a reference to an exercise that often (but not always) includes suggestions about how to approach the problem. These suggestions should be taken for what they are: suggestions. You should feel free to approach the problem a different way or have another type of insight. Some exercises will ask you to fill in minor steps in a proof and will not require a great deal of work. Other exercises will ask you to modify a proof to fill in the missing details signaled by the phrase "similarly, . . . " that appears so frequently in mathematical writing. These exercises are generally grouped under the classification of "Filling the Gaps." Other exercises ask you to reflect on the structure of definitions and proofs. Still other exercises are independent of the proofs and are designed to help you gain a deeper understanding of the material or greater facility working with the ideas. The latter types of exercise are labeled as "Deeper Reflection" and may demand more significant effort. Occasionally, a separate section labeled "Related Ideas: Deeper Reflections" is included at the end of a chapter. The problems in this section build on but are not directly related to the ideas in the chapter. Since they play such an important role in this text, you should at least read all of the "Filling the Gaps" exercises even if you do not intend to complete any of them. It is highly recommended that you read the "Deeper Reflections" exercises as well.

As the book progresses, the exercises will remain, but they will be referenced less frequently in the main body. At this point, you will be well served by keeping a finger or bookmark in the exercises when reading a proof. Note those places where you have identified details that should be verified. Then read the exercises to see if there are any significant details that you have missed.

While the proofs use a style comparable to that used in a professional journal, the transitional material is more conversational, informal, and reflective. The transitions provide a time to discuss the historical reasons for the track of investigation, to suggest motivational questions, to outline the general arc of the work, and to compare the ways that the different approaches to integration affect the way that proofs are constructed.

Thank you for engaging this book. Please address any error notices, comments, and suggestions to rosentr@westmont.edu.

I would like to thank Professor Russell Howell and students Tyler Brannan and Daniel Ray for their careful reading of the book and for their many corrections and suggestions. I also owe a huge debt of gratitude to the reviewers and editors of the Dolciani series whose subsequent suggestions for additions and modifications greatly strengthened the book and whose careful reading identified many needed corrections.

Contents

CHAPTER 1

Historical Introduction

The Riemann integral is usually introduced in elementary calculus classes via a problem with roots in early Greek mathematics—the problem of finding the area of a region R. Modern readers of mathematics expect the area to be expressed as a number. But what does this number mean? One way of interpreting the number is to think of it as representing the length of one side of a rectangle whose other side has unit length. Then the areas of the region R and the rectangle are the same.

1.1 Greek foreshadowing

Following this line of thought, a Greek solution to an area problem consisted of finding a square (or equivalently a rectangle or triangle) having the same area as the region in question.

The lunes of Hippocrates of Chios (c. 470–410 BCE) provide an early example of this type of problem. Hippocrates' work was done in an attempt to find a square or triangle with the same area as a given circle. Referring to Figure 1.1, Hippocrates was able to prove that the area of a portion of the circle with diameter AD has the same area as a triangle. Specifically, the crescent $AEDF$ has the same area as the triangle ACD. (See exercise 1.) Hippocrates' result reduced the task of finding a square to represent the area of a circle to the problem of constructing a square to represent the area of the region $AGDE$. About two millennia later, Ferdinand von Lindemann would show that this task is impossible by demonstrating that π is a transcendental number. (See [Baker] and [Lindemann].)

Archimedes of Syracuse (287–212 BCE) proved that the area of a circle is equal to the area of a triangle whose base has the same length as the circumference of the circle and whose height is the radius of the circle. (See exercises 7 and 8.) In modern notation, the area of the triangle, and so of the circle, is $\frac{1}{2} \cdot (2\pi r) \cdot r$ or πr^2.

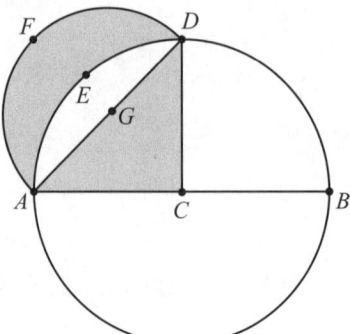

Figure 1.1. Lune of Hippocrates

While Archimedes' formulation of the result may seem odd to us, we need to appreciate that the first glimmers of algebraic notation did not appear until the 3rd-century CE work of Diophantus of Alexandria and that algebraic notation did not reach a state of relative maturity until the 17th century in the work of Descartes and Fermat. Consequently, the statement that the area of a circle is πr^2 would have been far more foreign to the Greeks than Archimedes' statement relating the area of the circle and a triangle is to us.

While the development of algebra is critical to the development of modern calculus, Archimedes' work on the area of a circle engaged in significant calculus-like thinking without algebra. Here we review a result of Archimedes that has an even stronger calculus flavor: the quadrature[1] of the parabola.

Suppose that points A and B are the endpoints of a section of a parabola as illustrated in Figure 1.2. Let C be the intersection between the parabola and the line l constructed through the midpoint M of AB and parallel to the axis of the parabola. (In terms of coordinate geometry, C is the point on the parabola whose x-coordinate is midway between x-coordinates of the two endpoints.) Then the area of the parabolic section between A and B is equal to $\frac{4}{3}$ the area of the inscribed triangle ABC.

Notice that the section of the parabola determined by A and B consists of three parts: the triangle ABC, the parabolic section determined by A and C, and the parabolic section determined by C and B. Expand the triangle ABC to become an inscribed polygon P_2 by attaching the inscribed triangle for each smaller parabolic section. In this first stage, the added triangles use the points C_1 and C_2 located at the intersections of the parabola with the lines

---

[1] The problem of quadratures is the problem of finding the area of an object. The word *quadrature* refers to finding a square of the same area.

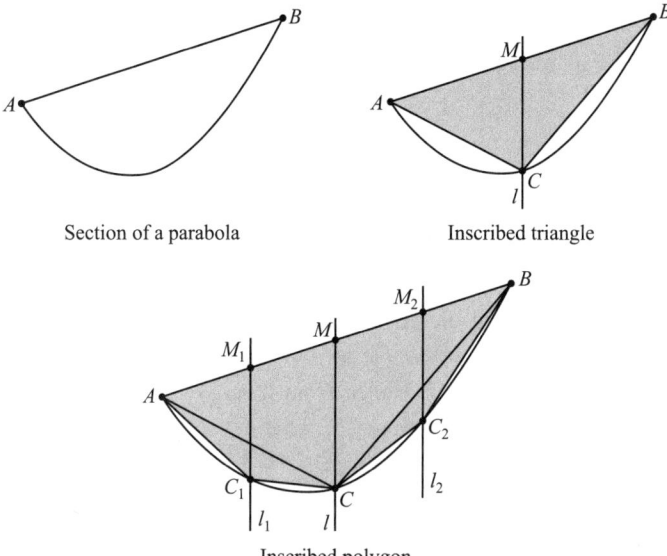

Section of a parabola   Inscribed triangle

Inscribed polygon

Figure 1.2. Quadrature of the Parabola

l_1 and l_2 parallel to the axis of the parabola and through the midpoints of AC and CB respectively. (A similar triangles argument shows that the lines l_1 and l_2 could also be defined as passing through the midpoints of AM and MB.) This process can be repeated indefinitely introducing more, but smaller, triangles and parabolic sections. Each succeeding stage produces a polygon that better fits the original parabolic section. At each stage, the combined area of the added triangles is $\frac{1}{4}$ the area of triangles on which they are based. Exercise 2 outlines a proof of this fact.

Let T be the area of ABC, let P be the area of the parabolic section between A and B, and let P_n be the area of the polygon that results from n iterations of the process described in the preceding paragraph. Using the results of the exercises,

$$P_0 = T$$
$$P_1 = T + \frac{1}{4}T = \frac{1}{3}\left(4 - \frac{1}{4}\right)T$$
$$P_2 = T + \frac{1}{4}T + \frac{1}{4^2}T = \frac{1}{3}\left(4 - \frac{1}{4^2}\right)T$$
$$P_3 = T + \frac{1}{4}T + \frac{1}{4^2}T + \frac{1}{4^3}T = \frac{1}{3}\left(4 - \frac{1}{4^3}\right)T$$
$$\cdots$$
$$P_n = T + \frac{1}{4}T + \frac{1}{4^2}T + \cdots + \frac{1}{4^n}T = \frac{1}{3}\left(4 - \frac{1}{4^n}\right)T.$$

A modern student with some background in calculus would conclude that

$$P = \sum_{k=0}^{\infty} \frac{T}{4^k} = \frac{T}{1 - \frac{1}{4}} = \frac{4}{3}T$$

or

$$P = \lim_{n\to\infty} \frac{1}{3}T\left(4 - \frac{1}{4^n}\right) = \frac{4}{3}T.$$

Not only did Archimedes lack the tools of calculus, but the Greek mind would have recoiled at the very idea of the completed infinity implicit in $\sum_{k=0}^{\infty} \frac{T}{4^k}$. Instead, Archimedes argued that P could not be greater than $\frac{4}{3}T$ nor could it be less than $\frac{4}{3}T$. Hence P must be $\frac{4}{3}T$.

Suppose that P were less than $\frac{4}{3}T$. Then for a sufficiently large value of n, we would find that $P < \frac{4}{3}T - \frac{1}{3\cdot4^n}T = \frac{1}{3}T\left(4 - \frac{1}{4^n}\right) = P_n$. But this is impossible since the polygon is contained in the section of the parabola. On the other hand, if P were greater than $\frac{4}{3}T$, we could keep inscribing triangles until $P_n > \frac{4}{3}T$. But $P_n = \frac{1}{3}T\left(4 - \frac{1}{4^n}\right)$ is manifestly less than $\frac{4}{3}T$. Thus by a double *reductio ad absurdum* argument, $P = \frac{4}{3}T$.

1.2   Newton and Leibniz

Although many of the foundational ideas were previously known in a vague form, the 17th and 18th centuries witnessed the birth of what we now call calculus. Isaac Newton (1642–1727) and Gottfried Leibniz (1646–1716) realized the power of the antiderivative and introduced its use. For Newton and Leibniz the integral *was* the antiderivative. Using modern notation, given a function f on $[a, b]$,

$$^{N}\!\!\int_{a}^{b} f\,(x)\,\mathrm{d}x = F\,(b) - F\,(a)\,, \text{ where } F' = f.$$

(Here the N reflects the fact that we are referring to Newton's integral.) In this context, the fundamental theorem of calculus and the definition of the integral are one and the same. There is nothing to prove.

Of course the preceding statement is a gross oversimplification of the situation. For Newton, the *fluxion* and the *fluent* were inverse operations in that the *fluxion* represented the ratio of the change in a quantity over an infinitely small time interval and a *fluent* captured the movement of an object that had a given ratio of change to time. Similarly, for Leibniz the integral accumulated the results of an infinite number of infinitesimals in the form of differentials—mathematical objects smaller than any real number but having some of the properties of real numbers. So instead of starting with a definition of the integral based on the area under the curve and showing that it can be computed using an antiderivative, Newton and Leibniz started with an integral defined by an antiderivative and needed to show that the integral represents the area under a curve.

Figure 1.3 is a drawing used by Leibniz to support a geometric proof of the connection between the antiderivative and the area under a curve.

> I shall now show that the general problem of quadratures can be reduced to the finding of a [curve] that has a given law of tangency.
>
> —Leibniz [Struik, pg 282]

In other words, the problem of finding the area under a given curve ($AH(H)$ in the diagram) is to be solved by finding a second curve ($C(C)$ in the diagram) whose tangent line at any point has a slope that is equal to the distance of the curve $AH(H)$ from the line $AF(F)$. In more modern terminology, the problem of finding the area under the curve $y = f(x)$ is to be solved by finding another function F whose derivative is f.

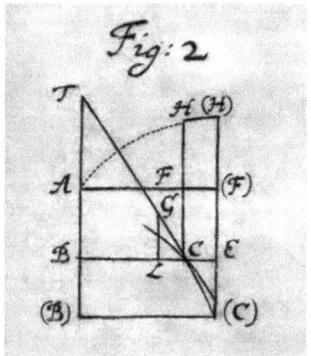

Figure 1.3. Liebniz diagram

In the diagram, the line $AF(F)$ is essentially what we would call the x-axis and the curves $AH(H)$ and $C(C)$ are viewed from the perspective of this axis. However, $C(C)$ does not represent a negative function. Instead, the viewer should concentrate on the distance between the curve and the "x-axis". Similarly, while the slope[2] of the tangent line TC appears to be negative according to modern conventions, the distance between $AF(F)$ and

---

[2] Note that Leibniz never mentions slope.

TC is increasing near C so the slope is actually positive. The segments GL, LC, CE, and the curve between C and (C) are all infinitesimals and have no values—*unassignable* in Leibniz's terminology. However, the segments can be used in ratios. Leibniz used the similarity of TBC, GLC, and CEC' (where C' is the unmarked point where TC intersects $E(C)$) together with the fact that $E(C)$ and EC' are essentially equal to draw his conclusion.[3] Leibniz assumed, without verification, that the desired curve $C(C)$ exists.

The Newton-Leibniz integral proved to be surprisingly useful for solving both physical and mathematical problems. As is illustrated in the following example, the previous development of solid algebraic notation and procedures was critical to the integral's usefulness.

Independently of his work on the calculus, Newton developed a generalized binomial formula. By solving for leading coefficients in special cases, Newton concluded that

$$(1 + x)^\alpha = \sum_{k=1}^{\infty} \binom{\alpha}{k} x^k$$

where

$$\binom{\alpha}{k} = \frac{\alpha (\alpha - 1) (\alpha - 2) \cdots (\alpha - k + 1)}{k!}.$$

When α is a positive integer, $\binom{\alpha}{k}$ agrees with the usual definition of the binomial coefficient for $0 \le k \le \alpha$ and, since $\alpha - (\alpha + 1) + 1 = 0$, $\binom{\alpha}{k} = 0$ for $k \ge \alpha + 1$. This means that Newton's formula agrees with the standard binomial theorem for positive integer values of α.

Newton did not prove his formula for general values of α. That would have to wait until Lagrange and Cauchy provided error bounds for Taylor series. Instead, Newton convinced himself of the correctness of his expression for $(1 + x)^\alpha$ by considering special cases. Using $\alpha = \frac{1}{2}$ for example, Newton truncated the series after a manageable number of terms and verified that, when squared, the truncated series matched $1 + x$ through the same number of terms. (See exercise 11.)

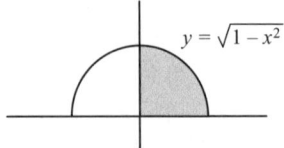

$y = \sqrt{1 - x^2}$

Figure 1.4. Area of $\pi/4$

[3] For a detailed explanation, see [Nauenberg].

In Newton's capable hands, the generalized binomial expansion provided a tool for approximating π by computing the area under $y = \sqrt{1 - x^2}$ from $x = 0$ to $x = 1$.

$$
\begin{aligned}
\frac{\pi}{4} &= {}^N\!\int_0^1 \sqrt{1 - x^2}\, dx \\
&= {}^N\!\int_0^1 \sum_{k=0}^\infty \binom{1/2}{k} (-1)^k\, x^{2k}\, dx \\
&= \sum_{k=0}^\infty \binom{1/2}{k} (-1)^k\; {}^N\!\int_0^1 x^{2k}\, dx \\
&= \sum_{k=0}^\infty (-1)^k \binom{1/2}{k} \left. \frac{x^{2k+1}}{2k+1} \right|_0^1 \\
&= \sum_{k=0}^\infty (-1)^k \binom{1/2}{k} \frac{1}{2k+1} \\
&= 1 - \frac{1}{6} - \sum_{k=2}^\infty \frac{1 \cdot 3 \cdots (2k-3)}{2^k k!\, (2k+1)}.
\end{aligned}
$$

This expression for π is not particularly elegant or efficient; quite a few terms are needed to obtain much accuracy. But there are more basic concerns here. No justification is provided for interchanging the infinite sum with the integral. It happens to work in this case, but the interchange can produce erroneous results. In fact, Newton did not even prove that the sum converged. The ratio test, typically used to show that $\sum_{k=2}^\infty \frac{1 \cdot 3 \cdots (2k-3)}{2^k k!(2k+1)}$ converges (see exercise 12), was developed by Jean-Baptiste le Rond d'Alembert who was only nine years old when Newton died.

1.3   Cauchy, Riemann, and Darboux

While the Newton-Leibniz integral was useful, its foundations were not solid. Philosopher Bishop George Berkeley strongly criticized mathematicians for using techniques with such flimsy groundings.

It must, indeed, be acknowledged, that [Newton] used Fluxions, like the Scaffold of a building, as things to be laid aside or got rid of, as soon as finite Lines were found proportional to them. But then these finite Exponents are found by the help of Fluxions. Whatever therefore is got by such Exponents and Proportions is to be ascribed to Fluxions:

which must therefore be previously understood. And what are these Fluxions? The Velocities of evanescent Increments? And what are these same evanescent Increments? They are neither finite Quantities nor Quantities infinitely small, nor yet nothing. May we not call them the Ghosts of departed Quantities?—Bishop Berkeley in *The Analyst*

Foundational questions about fluxions, differentials, and infinitesimals might be ignored, but difficulties related to convergence of infinite series were even more apparent and more difficult to disregard. Newton, Leibniz, and their contemporaries made frequent use of infinite series. But what does an infinite series mean? How can one contemplate adding an infinitude of terms? Particular series such as $\sum_{n=0}^{\infty} (-1)^n$ were recognized as being problematic quite early on. By regrouping, the series seems to converge to 1 or 0.

$$1 + (-1 + 1) + (-1 + 1) + (-1 + 1) + \cdots = 1,$$

but

$$(1 - 1) + (1 - 1) + (1 - 1) + (1 - 1) + \cdots = 0.$$

Concern over the legitimate handling of the infinitely large and infinitely small rose to a level where, in 1784, the Berlin Academy offered a prize to any mathematician who could provide a "clear and precise theory of what is called the infinite." Even though the prize was awarded in 1786, the mathematical community still was not satisfied that it had a satisfactory accounting of the infinite. That would have to wait another 35 years until Cauchy published *Cours d'Analyse* in 1821.[4] In this text, Cauchy introduced the ε-δ definition of convergence and eliminated the need to deal with infinitesimals.

Cauchy used the ε-δ paradigm not only to define the limit of a function but also to provide a clear statement of the meaning of the integral. Cauchy's definition takes a left-endpoint approach.

Definition 1. *Let f be a function defined on $[a, b]$. Then f is **integrable** over $[a, b]$ if there is a value A such that for every $\varepsilon > 0$ one can find a value $\delta > 0$ such that*

$$\left| \sum_{i=1}^{n} f(x_{i-1})(x_i - x_{i-1}) - A \right| < \varepsilon$$

---

[4] See [Bradley and Sandifer] for an annotated English translation.

for any choice of values $a = x_0 < x_1 < x_2 < \cdots < x_n = b$ *satisfying* $x_i - x_{i-1} < \delta$ *for* $1 \leq i \leq n$. *The value of the integral of* f *over* $[a, b]$, *denoted by* ${}^C\!\int_a^b f$, *is* A.[5]

While Cauchy's definition can be applied more generally, he only applied the definition to continuous (or at least piecewise continuous) functions. In the context of continuous functions, the question of integrability is moot. Cauchy provided an argument that all continuous functions are integrable (i.e., the value A required by his definition always exists). In addition, Cauchy proved that if F has a continuous derivative on $[a, b]$, then

$$ {}^C\!\int_a^b F' = F(b) - F(a). $$

You will immediately recognize this statement as a form of the fundamental theorem of calculus. Spend a few minutes comparing Cauchy's definition and theorem with those of Newton and Leibniz. Notice the shift from geometric conceptualization to a more algebraic presentation.

Even before Cauchy published *Cours d'Analyse*, mathematical developments were occurring for which Cauchy's definition proved to be awkward and inadequate. In 1807, Joseph Fourier submitted a memoir entitled *Theory of the propagation of heat in solid bodies*. In his paper, Fourier proposed using series of trigonometric polynomials (what we now call Fourier series) to solve the differential equation that models the steady-state temperature in a lamina or thin plate. Fourier's paper was rejected. Powerful mathematicians of the day, including Laplace, Lagrange, and Poisson, objected to Fourier's approach in large part because it violated their sense of what a function should be. But while Fourier's methods were viewed as problematic, his approach seemed to produce solutions that agreed with experimental results.

In 1810, the Institut de France announced that the next Grand Prize in Mathematics would be on the subject of "the propagation of heat in solid bodies." Fourier rewrote his paper and submitted it to the competition. In 1812, Fourier was awarded the substantial prize associated with the competition, but his prize-winning paper was not immediately published. While disciplinary politics certainly played a role, the delay was also a result of the view that, while Fourier had the right equations, his methods were problematic.

---

[5] The symbol ${}^C\!\int$ is used to denote the Cauchy integral. It does not refer to a multiple of an integral. Similar notation will be used throughout the text to identify particular types of integrals.

In addition to requiring a modified view of a function, Fourier's methods raised questions related to convergence of series and the interactions
between integration and series.[6] Ten years after winning the prize, Fourier's
work was published as *The analytical theory of heat*. Later (after Fourier was
elected secretary of the Institut de France), his original paper was published
in two parts in the Memoirs of the Académie des Sciences. The second part
appeared in 1826, the year of Bernhard Riemann's birth.

Riemann's first university lectures, given in 1854, introduced what would
later become Riemannian geometry and opened up the idea of using higher
dimensions (beyond three or four) to describe physical reality. In this same
year, Riemann submitted a paper entitled *On the representability of a function by a trigonometric series* in support of his certification as lecturer at the
University of Göttingen. This paper, not published until 1868, introduced a
general theory of trigonometric series that extended the ideas behind Fourier
series. Riemann's paper also included a major milestone in the development
of integral calculus. Section 4: *On the concept of a definite integral and the
extent of its validity* modified Cauchy's definition of the integral to the one
now found in elementary calculus courses. Specifically, Riemann's definition
required that

$$\left| \sum_{i=1}^{n} f(t_i)(x_i - x_{i-1}) - A \right| < \varepsilon$$

not just when t_i is the left endpoint of $[x_{i-1}, x_i]$ but for any choice of
$t_i \in [x_{i-1}, x_i]$. While this change has the appearance (and reality) of com-
plicating the definition, it makes many of the proofs involving the integral
significantly simpler.

Since the functions produced by trigonometric series are not necessarily
continuous, Riemann was interested in when the integral could be applied
to discontinuous functions. Although the result was not proved until well
after Riemann's death, any bounded function that is continuous almost everywhere (a concept to be defined in later chapters) is Riemann integrable.
Any function with a finite or even a countably infinite number of points of
discontinuity is included in the set of Riemann integrable functions. As a
result, the fundamental theorem of calculus can be extended to conclude that

$$^R\!\!\int_a^b F' = F(b) - F(a)$$

as long as F' is bounded and continuous almost everywhere on $[a, b]$.

---

[6] See Section 2 on trigonometric series in Chapter 4 for more information on Fourier's method
and its problems.

The last major player of this era was Gaston Darboux (1842–1917). Darboux edited and republished Fourier's *The analytical theory of heat* in 1888. Darboux also reformulated Riemann's definition of the integral by replacing the value of the function at a single point, $f(t_i)$, with the infimum and supremum of the values of f over the interval $[x_{i-1}, x_i]$. Intuitively, a function is Darboux integrable over $[a, b]$ if the sums

$$\sum_{i=1}^{n} \inf_{[x_{i-1}, x_i]} f \cdot (x_i - x_{i-1}) \quad \text{and} \quad \sum_{i=1}^{n} \sup_{[x_{i-1}, x_i]} f \cdot (x_i - x_{i-1})$$

are essentially the same. (Here $\inf_{[x_{i-1}, x_i]} f = \inf\{f(x) : x \in [x_{i-1}, x_i]\}$ is the infimum of the set of values of f over the input set $[x_{i-1}, x_i]$. The supremum is to be understood similarly.) If one considers the integral geometrically for positive functions, Darboux used inscribed and circumscribed rectangles instead of using general approximating rectangles. This formulation of the integral, while again appearing to be more complicated, had the effect of greatly simplifying the proofs associated with the Darboux integral.

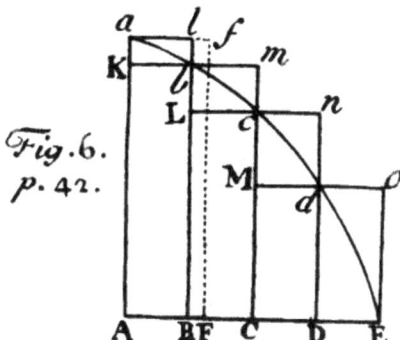

Figure 1.5. Newton's diagram

Interestingly, Darboux's definition comes full circle to one of Newton's results. In Book I, Section I of *The mathematical principles of natural philosophy*, Newton's lemma II states that, for decreasing functions, the sums of the areas of the inscribed rectangles and the sums of the areas of the circumscribed rectangles converge to the area under the original curve. (See Figure 1.5.)

... Then if the breadth of those parallelograms be supposed to be diminished and their number augmented *in infinitum*: I say that the ultimate ratios which the inscribed figure AKbLcMdD, the circumscribed figure AalbmcndoE, and the curvilinear figure AabcdE, will have to one another, are ratios of equality.[7]

1.4 Lebesgue

While Riemann and Darboux provided precise and workable definitions of the integral and developed criteria that guaranteed that a function is integrable, the state of the theory remained unsatisfactory from the perspective of Fourier series (a particular type of trigonometric series which will be introduced in Chapter 4). Niels Henrik Abel (1802–1829) and Johann Peter Gustav Lejeune Dirichlet (1805–1859) developed theories and techniques that could be used to prove that Fourier series converged, but the interaction between Fourier series and integrals remained murky. One major difficulty was that, while the convergence of a Fourier series could be established, the resulting function might not be Riemann integrable. Cauchy had proved that when a sequence (or series) of continuous functions converges uniformly, the limit is continuous. In a more general context, the uniform limit of a sequence (or series) of integrable functions is again integrable. Unfortunately, some of the most interesting problems (including one of Fourier's original examples) involve discontinuous functions. In such cases, the Fourier series cannot converge uniformly.

By approaching integration from a different perspective, Henri Lebesgue (1875–1941) found a way to address this problem. Instead of partitioning $[a, b]$, the domain of f, Lebesgue partitioned f's range. In some sense, the Lebesgue approach slices the region under a function horizontally instead of vertically. Whereas a term in a Riemann sum is $f(t_i)(x_i - x_{i-1})$ where $[x_{i-1}, x_i]$ is a subinterval of $[a, b]$ and $t_i \in [x_{i-1}, x_i]$, a Lebesgue term is $y_i \lambda(A_i)$ where $[y_{i-1}, y_i]$ is a subinterval of the range of f, A_i is the set of all $x \in [a, b]$ for which $y_{i-1} < f(x) \le y_i$, and $\lambda(A_i)$ is the "length" of A_i. (See Figure 1.6.) The sums used to define the Lebesgue integral have the form

$$\sum_{i=1}^{n} y_i \lambda(A_i).$$

A function is Lebesgue integrable if these sums converge in an appropriate sense.

---

[7] You can read an English translation of Newton's lemma and proof on Google books at books.google.com/books?id=Tm0FAAAAQAAJ. See page 42 and the unnumbered page after 44.

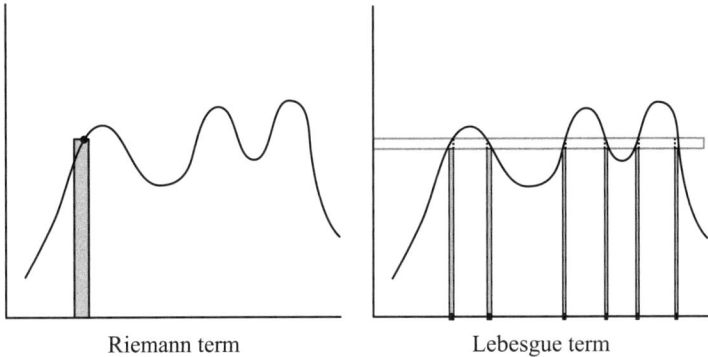

Riemann term   Lebesgue term

Figure 1.6. Comparison of Riemann and Lebesgue terms

In addition to being visually more complicated, Lebesgue's approach requires us to make sense of the "length" of rather arbitrary sets. This is not an easy task, but the effort to generalize length to a broader class of sets is richly rewarded. The Fourier series of a bounded function will always converge and it will do so in a way that allows termwise integration. Lebesgue's results, published in a 1902 paper, finally resolved the issues generated by Fourier's work over 90 years earlier.

1.5 Henstock and Kurzweil

Lebesgue's formulation of the integral is not the only way to address the questions flowing from Fourier's work. The gauge integral, developed primarily by Ralph Henstock (1923–2007) and Jaroslav Kurzweil (1926–) around 1960, retains the Riemann integral's focus on the domain of the function to be integrated but works more locally than does the Riemann integral.

The Riemann integral requires that

$$\left| \sum_{i=1}^{n} f(t_i)(x_i - x_{i-1}) - A \right| < \varepsilon$$

for any choice of $a = x_0 < x_1 < x_2 < \cdots < x_n = b$ satisfying $x_i - x_{i-1} < \delta$ and $t_i \in [x_{i-1}, x_i]$ for $1 \leq i \leq n$. The gauge integral exerts a more localized control on the size of the subintervals by specifying that $x_i - x_{i-1} < \delta(t_i)$ for $1 \leq i \leq n$. In this formulation, $\delta(\cdot)$ is a positive function on $[a, b]$ called a gauge.

For example, if

$$f(x) = \begin{cases} \frac{1}{\sqrt{x}}, & x > 0 \\ 0, & x \le 0 \end{cases}$$

then f is unbounded and so cannot be Riemann integrable over $[0, 1]$. However if we choose the gauge

$$\delta(t) = \begin{cases} t/2, & t > 0 \\ 1, & t \le 0 \end{cases}$$

then any choice of points satisfying $0 = x_0 < x_1 < x_2 < \cdots < x_n = 1$, $t_i \in [x_{i-1}, x_i]$, and $x_i - x_{i-1} < \delta(t_i)$ for $1 \le i \le n$ must take $t_1 = 0$. No other choice for t_1 allows the left endpoint of the first interval to be 0. This has the effect of ensuring that the corresponding gauge sum will remain bounded. (See exercise 17.)

Of course, the boundedness of the gauge sums does not guarantee that f is gauge integrable. However, given a positive ε it is possible to construct a gauge that will ensure that $\sum_{i=1}^{n} f(t_i)(x_i - x_{i-1})$ is within ε of 2. Thus $^{g}\!\int_0^1 f = 2$. Since the process of constructing such a gauge is too involved for our current purposes, we will not do so here. But we will return to this example in Chapter 7.

As is the case for Lebesgue integrable functions, the limit of any sequence of gauge integrable functions will itself be gauge integrable. This means that all the manipulations used by Fourier to construct trigonometric series to solve differential equations are justified in the context of the gauge integral.

Of all the standard integrals on $\mathbb{R}$, the gauge integral has the strongest version of the fundamental theorem of calculus. As long as F is continuous on $[a, b]$ and the derivative of F exists at all but a countable number of points in $[a, b]$,

$$^{g}\!\int_a^b F' = F(b) - F(a).$$

1.6   Extensions

Not all modifications of the Riemann integral were made for the purpose of solving the difficulties associated with Fourier's work or extending the fundamental theorem of calculus. In 1894, Thomas Joannes Stieltjes (1856–1894) published the definition of a new type of integral that did not treat all intervals of the same length equally. Given functions f and g, Stieltjes considered sums of the form

$$\sum_{i=1}^{n} f\left(t_{i}\right)\left[g\left(x_{i}\right) - g\left(x_{i-1}\right)\right]$$

where $a = x_0 < x_1 < x_2 < \cdots < x_n = b$ and $t_i \in [x_{i-1}, x_i]$. When the sums converge to a finite value in an appropriate sense, the value, denoted by $^{R\text{-}S}\int_a^b f \, dg$, is called the Riemann-Stieltjes integral of f over $[a, b]$ with respect to g. The Riemann-Stieltjes integral is a generalization of the Riemann integral since Stieltjes' integral is Riemann's integral when $g(x) = x$.

While Stieltjes' work was motivated by the problem of calculating the moment of inertia of an object with varying density, the integral has much broader application. In the moment of inertia problem, $g(x_i) - g(x_{i-1})$ reflects the mass between x_{i-1} and x_i. Alternatively, if you wish to model travel over terrain of unequal difficulty, the value of $g(x_i) - g(x_{i-1})$ can be interpreted as the effort required to travel from x_{i-1} to x_i.

From a mathematical point of view, probably the greatest value of the Riemann-Stieltjes integral is that it provides a single treatment of all types of probability on the real line. There are two principal types of probability spaces: discrete and continuous. Discrete spaces model situations where the set of possible outcomes forms a finite or countable set. Counting the number of phone calls arriving in an hour or counting the number of flips of a coin required to see the first heads are typical examples. In such cases, computations related to probability and expected value involve sums. On the other hand, continuous distributions used to model phenomena such as the height of an object or the time until the next arrival of an airplane use the Riemann integral for calculation. As a result, the typical probability or statistics text written for upper-division mathematics students states many if not most theorems twice: once for the discrete case and once for the continuous case. And while texts typically present only one type of proof and indicate that the other proof is similar, most theorems require two different proofs for complete justification.[8]

The Riemann-Stieltjes integral captures both types of distributions as well as distributions that are mixtures of the two types. Consequently only one statement and one proof is needed when probability and statistics is done using the Riemann-Stieltjes integral. Moreover, the proof is generally just as easy or hard as the proof of one of the individual cases in the more common

---

[8] Actually, complete proofs would be even more complicated since a distribution could mix types or otherwise not fit neatly into one of the two principle types of distributions.

approach. Why then is the Riemann-Stieltjes integral not used in probability and statistics courses? Among other reasons is the fact that text writers
cannot assume that students have prior knowledge of the Riemann-Stieltjes
integral.

The notion that different intervals should have varying weights has a natural affinity with Lebesgue's idea of working with the "lengths" of various sets. Instead of thinking of the length of a set, we should consider its
"measure." The idea of measure encompasses many applications such as the
weight of an object over the set A, the total difficulty in traveling over the
set A, or the probability of A. To integrate a function relative to the measure
μ, we use Lebesgue's approach and partition the range of f. The partitions
generate sums of the form

$$\sum_{i=1}^{n} y_i \mu (A_i)$$

where $[y_{i-1}, y_i]$ is a subinterval of the range of f, A_i is the set of all $x \in$
$[a, b]$ for which $y_{i-1} < f(x) \le y_i$, and $\mu (A_i)$ is the "measure" of A_i. The
function is Lebesgue-Stieltjes integrable with respect to μ if the sums converge appropriately. In this case, the integral is written as ${}^{M}\!\int_a^b f \, d\mu$.[9] The
Riemann-Stieltjes and Lebesgue-Stieltjes integrals are connected by building
μ from the foundational assumption that $\mu ((x_{i-1}, x_i]) = g(x_i) - g(x_{i-1})$.

When we step back from particular applications and consider what could
happen with arbitrary choices of g, things can appear rather odd. Suppose
that g is decreasing on the interval $[x_{i-1}, x_i]$. Then $g(x_i) - g(x_{i-1})$ will
be negative. This makes no sense if the associated measure represents probability. But other types of interpretations might make sense. If g reflects
the resistance along a path, then a negative value of $g(x_i) - g(x_{i-1})$ can
be understood as a section of the path that puts energy back. Perhaps the
section is downhill. A negative value in a moment of inertia application
might reflect a certain type of anti-matter or a repelling rather than attracting
force.

Once such possibilities are contemplated, a whole different type of mathematical universe opens up. Measures can be put on sets much more general
than $\mathbb{R}$ or $\mathbb{R}^n$. Moreover, measures need not be restricted to taking on only
positive, real, or even complex values. In particular, projection-valued measures defined on subsets of the complex numbers are used in operator theory,
a mathematical subject that plays a central role in quantum physics.

---

[9] M is for measure.

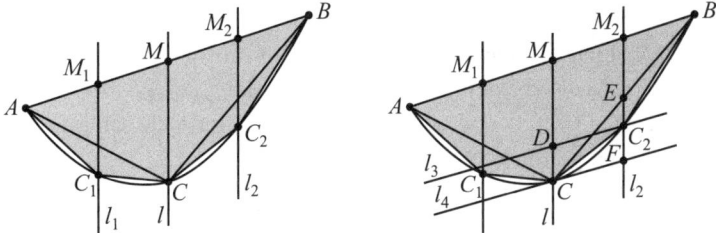

Figure 1.7. Quadrature of the parabola

Mathematicians and physicists have extended the concept of the integral in a myriad of other ways that we will not explore in this text. In the remaining chapters, you will be introduced to the ideas and proofs that undergird the ideas sketched out in this historical overview, and you should be prepared to study an even broader range of interpretations of integration.

1.7 Exercises

1.1 Greek foreshadowing: filling the gaps

1. In Figure 1.1, C is the midpoint of AB and thus the center of the large circle. Prove that the triangle ACD and the crescent $AEDF$ have the same areas by providing explanations of the following. (Your write-up will be stronger if you blend the individual explanations into a single proof rather than answering each part separately.)

   (a) How are the areas of the circles with diameters AB and AD related? Why?

   (b) How are the areas of the semicircle $GDFA$ and the quarter circle $CDEA$ related?

   (c) Why do the crescent $AEDF$ and the triangle ACD have the same area?

2. Give a geometric proof of the claim in Archimedes' proof that the combined area of triangles ACC_1 and BCC_2 is $\frac{1}{4}$ that of ABC. (Use the following outline to construct a proof.)

   (a) Add lines l_3 and l_4 parallel to AB through C_2 and C respectively and use D, E, and F to label the intersection points as indicated in the parabola on the right of Figure 1.7.

   (b) Explain why triangles EC_2B and EC_2C have the same area.

(c) How is the area of M_2BE related to the area of MBC?

(d) Archimedes knew from Euclid's *Elements* that $\frac{CD}{CM} = \frac{MM_2^2}{MB^2}$. Use this fact to prove that M_2E is twice as long as C_2E.

(e) Explain why triangles M_2BE and CBC_2 have the same area.

(f) Argue that the combined area of triangles ACC_1 and BCC_2 is $\frac{1}{4}$ that of ABC.

1.1 Greek foreshadowing: deeper reflections

3. (Quadrature of two squares) Given two squares, explain how to construct a single square with the same area as the combined areas of the two original squares.

4. Given a square, how can one construct a second square with half its area?

5. (Quadrature of a rectangle) Given a rectangle, explain how to construct a square of the same area. (Work with a circle whose diameter is the sum of the lengths of the two sides of the rectangle.)

6. (Quadrature of a triangle) Given a triangle, how can one construct a square of the same area? (Use the previous exercises.)

Exercises 7 and 8 outline Archimedes' proof that the area of a circle is equal to the area of a triangle with height equal to the radius of the circle and with base length equal to its circumference. Let C be a circle and let T be a triangle with height equal to the radius of C and with base length equal to the circumference of C. Let A_C and A_T be the corresponding areas.

7. Prove that $A_C \not> A_T$. (Combine the steps into a flowing exposition.)

(a) Suppose that $D = A_C - A_T > 0$.

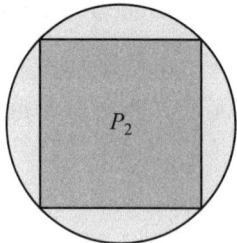 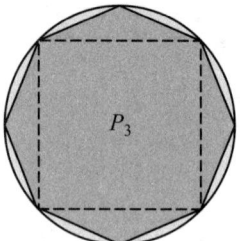

polygons P_n of 2^n sides

(b) Construct polygons P_n of 2^n sides contained in C is the following manner.

    i. P_2 is an inscribed square.

    ii. P_{n+1} is constructed from P_n by adding an isosceles triangle on each edge whose third vertex is on C.

(c) Explain why more than half of the area between C and P_n is removed to make P_{n+1}.

(d) Give a simple description of the area of P_n in terms of its circumference.

(e) Explain how this produces a contradiction.

8. Prove that $A_C \not< A_T$ using circumscribed polygons in an argument similar to that of the previous exercise.

9. Give a calculus-based proof of the claim that the combined area of triangles ACC_1 and BCC_2 of exercise 2 is $\frac{1}{4}$ that of ABC. (Use integrals to compute the areas between lines.)

10. Use calculus to prove directly that the area of the region bounded by $y = sx^2$ and the line through $A = (a, sa^2)$ and $B = (b, sb^2)$ is $\frac{4}{3}$ times the area of the triangle with vertices at A, B, and $C = (c, sc^2)$ where $c = \frac{1}{2}(a + b)$.

1.2 Newton and Leibniz: deeper reflections

11. According to Newton's formula, $\sqrt{1 - x^2} = \sum_{k=0}^{\infty} \binom{1/2}{k} (-1)^k x^{2k}$.

(a) Expand $\sum_{k=0}^{5} \binom{1/2}{k} (-1)^k x^{2k}$.

(b) Verify that $\left(1 - \frac{1}{2}x^2 - \frac{1}{8}x^4 - \frac{1}{16}x^6\right)^2 = 1 - x^2 + O\left(x^8\right)$.

(c) Verify that $\left(1 - \frac{1}{2}x^2 - \frac{1}{8}x^4 - \frac{1}{16}x^6 - \frac{5}{128}x^8\right)^2 = 1 - x^2 + O\left(x^{10}\right)$.

Here $O(x^n)$ is used to indicate that the power of x in the remaining terms is n or greater.

12. Use the ratio test for series to verify that

$$\sum_{k=2}^{\infty} \frac{1 \cdot 3 \cdots (2k - 3)}{2^k k! (2k + 1)}$$

converges.

1.3 Cauchy, Riemann, and Darboux: deeper reflections

13. Suppose that the function

$$f(x) = \begin{cases} 0, & x \in [0,1) \\ 3, & x \in [1,5] \end{cases}$$

represents the current flowing through a line if a switch is closed at time $t = 1$. If the battery to which the line is connected maintains a voltage of 2 volts, then the power consumed over the time period $[0,t]$, where $0 \le t \le 5$, is given by $\int_0^t 2f(x)\,dx$.

   (a) Explain why this integral does not exist in the Newton-Leibniz sense.
   (b) Explain why the Riemann integral $^R\!\int_1^t f(x)\,dx$ does exist for $1 \le t \le 5$.

14. Let

$$F(x) = \begin{cases} x^2 \sin \frac{\pi}{x}, & x \ne 0 \\ 0, & x = 0. \end{cases}$$

   (a) Compute the derivative F' of F.
   (b) Explain why Cauchy's version of the fundamental theorem of calculus cannot be used to conclude that $^C\!\int_0^1 F' = 0$.
   (c) Explain why Riemann's version of the fundamental theorem of calculus can be used to conclude that $^R\!\int_0^1 F' = 0$.

1.4 Lebesgue: deeper reflections

15. Many of the more interesting questions related to Fourier series involve discontinuous functions. Explain why the Fourier series of a discontinuous function cannot converge uniformly.

16. One of the hurdles in defining the Lebesgue integral is the problem of making sense of the "length" of sets that might arise as inverse images of an interval. For the purpose of thinking about how to construct a general definition, what is your intuitive sense for an appropriate value for the "length" of the following sets? How do you arrive at your value?

   (a) $[0,1]$
   (b) $[0.2, 0.3] \cup (0.4, 0.5) \cup [0.7, 0.9]$
   (c) $[0.2, 0.4] \cup (0.3, 0.5) \cup [0.7, 0.9]$
   (d) $\{\frac{1}{2}, \frac{1}{3}, \frac{1}{4}\}$
   (e) $\{\frac{1}{n} : n \in \mathbb{N}\}$

(f) the rational numbers in $[0, 1]$

(g) the irrational numbers in $[0, 1]$

What properties do you expect "length" to have?

1.5 Henstock and Kurzweil: deeper reflections

17. Define
$$f(x) = \begin{cases} \frac{1}{\sqrt{x}}, & x > 0 \\ 0, & x \le 0. \end{cases}$$

(a) Given $0 = x_0 < x_1 < x_2 < \cdots < x_n = 1$, explain how the evaluation points can be chosen so that the corresponding Riemann sum is larger than any predetermined value v. Explain why you can conclude that f is not Riemann integrable over $[0, 1]$.

(b) Set
$$\delta(t) = \begin{cases} t/2, & t > 0 \\ 1, & t \le 0 \end{cases}$$

and suppose that we have values satisfying $0 = x_0 < x_1 < x_2 < \cdots < x_n = 1$, $t_i \in [x_{i-1}, x_i]$, and $x_i - x_{i-1} < \delta(t_i)$ for $1 \le i \le n$.

Prove that $\sum_{i=1}^{n} f(t_i)(x_i - x_{i-1}) < 3$. (Combine the following hints into a well presented argument.)

i. Explain why $t_1 = 0$. (What is the minimal value of x_0 if $0 < t_1$?)

ii. Prove that $g(x) = 3\left(\sqrt{x} - \sqrt{a}\right) - \frac{1}{\sqrt{a}}(x - a)$ is increasing for $x \in [a, 2a]$.

iii. Use (ii) to prove that $\sum_{i=1}^{n} f(t_i)(x_i - x_{i-1}) < 3 \sum_{i=2}^{n} \left(\sqrt{x_i} - \sqrt{x_{i-1}}\right)$.

1.6 Extensions: deeper reflections

18. Suppose that the random variable X has a $\frac{2}{3}$ probability of being 0 and $\frac{1}{3}$ probability of being 1. Define a function $g : \mathbb{R} \to [0, 1]$ so that $g(x_i) - g(x_{i-1})$ is the probability that the value of X is in the interval $(x_{i-1}, x_i]$.

19. Suppose that the random variable X has equal likelihood of taking on any value in the interval $[0, 1]$ and never takes on a value outside of $[0, 1]$. Define a function $g : \mathbb{R} \to [0, 1]$ so that $g(x_i) - g(x_{i-1})$ is the probability that the value of X is in the interval $(x_{i-1}, x_i]$.

20. Let
$$g(x) = \begin{cases} 0, & x < 0 \\ 3x, & 0 \le x \le 2 \\ 8, & 2 < x. \end{cases}$$

Compute

(a) $^{R\text{-}S}\!\int_0^1 1\,dg.$

(b) $^{R\text{-}S}\!\int_{-2}^5 1\,dg.$

(c) $^{R\text{-}S}\!\int_0^2 x\,dg.$

(For (c), write out the Riemann-Stieltjes sums and then reinterpret them as Riemann sums.)

21. Suppose that

$$\mu\,(A) = \begin{cases} 0, & 1,2 \notin A \\ 1, & 1 \in A,\ 2 \notin A \\ 2, & 1 \notin A,\ 2 \in A \\ 3, & 1,2 \in A. \end{cases}$$

Compute

(a) $^{M}\!\int_0^2 1\,d\mu.$

(b) $^{M}\!\int_0^1 x\,d\mu.$

(c) $^{M}\!\int_0^2 x\,d\mu.$

(Partition the interval and consider which terms contribute to the sum.)

22. Let

$$P_2 = \frac{1}{2}\begin{bmatrix} 1 & 1 \\ 1 & 1 \end{bmatrix}$$

and

$$P_4 = \frac{1}{2}\begin{bmatrix} 1 & -1 \\ -1 & 1 \end{bmatrix}.$$

Then P_2 and P_4 are projections onto the spaces spanned by $\left\{\begin{bmatrix} 1 \\ 1 \end{bmatrix}\right\}$ and $\left\{\begin{bmatrix} 1 \\ -1 \end{bmatrix}\right\}$ respectively. Suppose that

$$\mu\,(A) = \begin{cases} 0, & 2,4 \notin A \\ P_2, & 2 \in A,\ 4 \notin A \\ P_4, & 2 \notin A,\ 4 \in A \\ P_2 + P_4, & 2,4 \in A. \end{cases}$$

(a) Use the ideas from the previous exercise and the definitions of matrix operations to determine the matrix values of

i. $^{M}\!\int_0^3 1\,d\mu.$

ii. $^{M}\!\int_0^5 1 \, d\mu$.

iii. $A = \, ^{M}\!\int_0^5 x \, d\mu$.

iv. $B = \, ^{M}\!\int_0^5 x^2 \, d\mu$.

(b) Compute A^2.

(c) Write an integral expression that you think is likely to represent $\sqrt{A}$.

(d) Evaluate your integral from part (c) and compute its square.

1.8 References

Baker, Alan (1975). *Transcendental Number Theory*, Cambridge University Press.

Berkeley, G. (1734). *The Analyst*. Edited by David Wilkins (2002) http://www.maths.tcd.ie/pub/HistMath/People/Berkeley/Analyst/

Bradley, R.E. and C.E. Sandifer, C.E. (2009). *Cauchy's Cours d'analyse, An Annotated Translation*. Sources and Studies in the History of Mathematics and Physical Sciences. Springer.

Bressoud, D.M. (2006). *A Radical Approach to Real Analysis* (2nd ed.). Mathematical Association of America.

Burk, F.E. (2007). *A Garden of Integrals*. Mathematical Association of America.

Edwards Jr., C.H. (1994). *The Historical Development of the Calculus* (3rd ed.). Springer.

Fourier, J. (1822). *The Analytical Theory of Heat*. Translated by Alexander Freeman (1878, re-released 2003). Dover Publications. books.google.com/books?id=No8IAAAAMAAJ.

Hawkins, T. (1975). *Lebesgue's Theory of Integration* (2nd ed.). Chelsea.

Heath, T.L. (2011). *The Works of Archimedes* (2nd ed.). CreateSpace.

Henstock, R. (1968). A Riemann type integral of Lebesgue power, *Canadian Journal of Math.*, **20**: 79–87.

Lebesgue, H. (1903). Sur une condition de convergence des séries de Fourier, *C.R. Acad. Sci, Paris* **140**: 229–242.

Lebesgue, H. (1903). Recherches sur la convergence des séries de Fourier, *Math. Ann.* **61**: 251–280.

Lindemann, F. (1882). Über die Zahl π, *Mathematische Annalen* **20**: pp. 213–225.

Nauenberg, M. (Forthcoming) Barrow and Leibniz on the Fundamental Theorem of the Calculus, *Annals of Science* arXiv:1111.6145 [math.HO].

Newton, I. (1687). *The Mathematical Principles of Natural Philosophy*, Translated by B. Motte (1724). Middle-Temple-Gate, Fleetstreet. books.google.com/books?id=Tm0FAAAAQAAJ.

Riemann, B. (1868). Über die Darstellbarkeit einer Function durch eine trigonometrische Reihe, H. Weber (ed.), *B. Riemann's Gesammelte Mathematische Werke*, Dover, reprint (1953) pp. 227–271 (Original: Göttinger Akad. Abh. **13**).

Stein, S.K. (1999). *Archimedes: What Did He Do Besides Cry Eureka?* Mathematical Association of America.

Stillwell, J. (2004). *Mathematics and its History* (2nd ed.). Springer.

Struik, D.J. (1969). *A Source Book in Mathematics, 1200–1800*. Harvard University Press.

Swain, G. and D. Thomas (1998). Archimedes' quadrature of the parabola revisited, *Mathematics Magazine* **71** (2): 123–30. JSTOR 2691014.

Truesdell, C.A. (1980). *The Tragicomical History of Thermodynamics, 1822–1854*. Springer.

CHAPTER 2

The Riemann Integral

Modern students of mathematics are typically introduced to integration through the Riemann integral, so you are no doubt familiar with its definition. But, since the definition requires a significant amount of supporting notation that may vary from text to text, we will provide a review. Moreover, for ease of transition and to make later comparisons more straightforward, we will use a unified terminology and notation throughout this text. While Riemann did not use this notation and it may be unfamiliar initially, the ideas and approaches behind the terminology remain Riemann's.

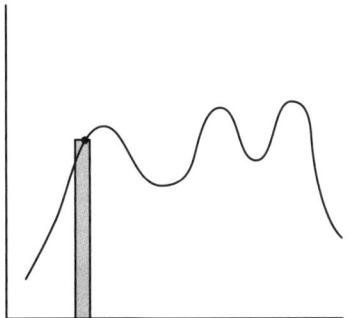

Figure 2.1. A Riemann Term

Given an interval $[a, b]$, a **partition** of $[a, b]$ is a finite set of contiguous intervals $[x_{k-1}, x_k]$ with $a = x_0 < x_1 < x_3 < \cdots < x_n = b$. The endpoints of the intervals interior to $[a, b]$, $\{x_i\}_{i=1}^{n-1}$, are called the **division points** of the partition. A **tagged partition** is a partition together with a set of **tags** $\{t_k\}_{k=1}^{n}$ where $t_k \in [x_{k-1}, x_k]$. For example, $\left\{ \left[\frac{k-1}{n}, \frac{k}{n} \right] \right\}_{k=1}^{n}$ is a partition of $[0, 1]$ for any value of $n \in \mathbb{N}$ and

$$\mathcal{P}_1 = \left\{ \left(0, \left[0, \tfrac{1}{n} \right] \right), \left(\tfrac{1}{n}, \left[\tfrac{1}{n}, \tfrac{2}{n} \right] \right), \left(\tfrac{2}{n}, \left[\tfrac{2}{n}, \tfrac{3}{n} \right] \right), \ \dots \ , \left(\tfrac{n-1}{n}, \left[\tfrac{n-1}{n}, 1 \right] \right) \right\}$$

$$\mathcal{P}_2 = \left\{ \left(\tfrac{1}{n}, \left[0, \tfrac{1}{n} \right] \right), \left(\tfrac{2}{n}, \left[\tfrac{1}{n}, \tfrac{2}{n} \right] \right), \left(\tfrac{3}{n}, \left[\tfrac{2}{n}, \tfrac{3}{n} \right] \right), \ \dots \ , \left(1, \left[\tfrac{n-1}{n}, 1 \right] \right) \right\}, \text{ and}$$

$$\mathcal{P}_3 = \left\{ \left(\tfrac{1}{2n}, \left[0, \tfrac{1}{n} \right] \right), \left(\tfrac{3}{2n}, \left[\tfrac{1}{n}, \tfrac{2}{n} \right] \right), \left(\tfrac{5}{2n}, \left[\tfrac{2}{n}, \tfrac{3}{n} \right] \right), \ \dots \ , \left(1 - \tfrac{1}{2n}, \left[\tfrac{n-1}{n}, 1 \right] \right) \right\}$$

are tagged partitions where the tags are respectively the left endpoints, right endpoints, and midpoints of the subintervals. These three tagged partitions use equal-length subintervals, but that is not necessary.

$$\mathcal{P}_4 = \left\{ \left(0, \left[0, \tfrac{1}{2} \right] \right), \left(\tfrac{1}{2}, \left[\tfrac{1}{2}, \tfrac{2}{3} \right] \right), \left(\tfrac{2}{3}, \left[\tfrac{2}{3}, \tfrac{3}{4} \right] \right), \left(\tfrac{3}{4}, \left[\tfrac{3}{4}, 1 \right] \right) \right\}$$

is also a tagged partition of $[0, 1]$. On the other hand, the set

$$\left\{ \left(0, \left[0, \tfrac{1}{2} \right] \right), \left(\tfrac{1}{2}, \left[\tfrac{1}{2}, \tfrac{2}{3} \right] \right), \left(\tfrac{2}{3}, \left[\tfrac{2}{3}, \tfrac{3}{4} \right] \right), \dots \right\}$$

is not a tagged partition of $[0, 1]$ because the set is not finite, nor does it terminate with a right endpoint of 1.

Given a partition $\mathcal{P} = \{[x_{k-1}, x_k]\}_{k=1}^{n}$ of $[a, b]$, we denote the width of the kth subinterval by Δx_k and the **mesh** of $\mathcal{P}$ by

$$\|\mathcal{P}\| = \max_{1 \le k \le n} \Delta x_k = \max_{1 \le k \le n} (x_k - x_{k-1}).$$

The tagged partitions $\mathcal{P}_1$, $\mathcal{P}_2$, and $\mathcal{P}_3$ all have a mesh of $\tfrac{1}{n}$. The mesh of $\mathcal{P}_4$ is $\|\mathcal{P}_4\| = \tfrac{1}{2}$.

Given a real-valued function f defined on $[a, b]$, we compute the **Riemann sum** for a tagged partition $\mathcal{P} = \left\{ \left(t_k, [x_{k-1}, x_k] \right) \right\}_{k=1}^{n} = \{(t_k, I_k)\}$ of $[a, b]$ as

$$S_R(f, \mathcal{P}) = \sum_{k=1}^{n} f(t_k)(x_k - x_{k-1}) = \sum_{k=1}^{n} f(t_k) \Delta x_k = \sum_{\mathcal{P}} f \, \Delta x.$$

Notice the varying levels of detail in the notation introduced above. Going forward, we will use the most compact notation that is unambiguous in the context. For example, we will use $\{(t_k, I_k)\}$ rather than $\left\{ \left(t_k, [x_{k-1}, x_k] \right) \right\}_{k=1}^{n}$ when we do not need to reference the endpoints of the intervals and $\sum_{\mathcal{P}} f \, \Delta x$ rather than $\sum_{k=1}^{n} f(t_k) \Delta x_k$ when the particular tags are not important. This will help us reason at higher levels of abstraction.

For future reference, note that $\sum_{\mathcal{P}} \Delta x = \sum_{k=1}^{n} (x_k - x_{k-1})$ which telescopes to $x_n - x_0 = b - a$. This fact will be used frequently and without reference.

Definition 2 (Riemann Integral). *Let f be a real-valued function defined on $[a, b]$. Then f is **Riemann integrable** over $[a, b]$ if there is a real number A such that for any $\varepsilon > 0$ we can find a $\delta > 0$ so that*

$$|S_R(f, \mathcal{P}) - A| < \varepsilon \tag{2.1}$$

*whenever $\mathcal{P}$ is a tagged partition of $[a, b]$ with $\|\mathcal{P}\| < \delta$. In this case, A is the **integral** of f over $[a, b]$ and we write $^R\!\int_a^b f = {}^R\!\int_a^b f(x) \, dx = A$.*

Note that (2.1) must hold for all partitions with mesh $\|\mathcal{P}\| < \delta$ (evenly spaced or not) and for all choices of the tags (left, right, center, or otherwise).

Though stated here for the Riemann integral, we will address the following three questions for all the integrals we will study:

1. What functions are Riemann integrable?

2. When is it true that $\frac{d}{dx} {}^R\!\int_a^x f = f$ and $^R\!\int_a^b f' = f(b) - f(a)$?

3. Under what circumstances is $^R\!\int_a^b \sum_n f_n = \sum_n {}^R\!\int_a^b f_n$ or, equivalently, under what circumstances is $^R\!\int_a^b \lim_{n \to \infty} f_n = \lim_{n \to \infty} {}^R\!\int_a^b f_n$?

We address each question in turn, obtaining only partial answers initially.

2.1 Riemann integrability

The first thing to note about the Riemann integral is that any Riemann integrable function must be bounded. You are asked to provide a proof of this fact in exercise 2.

Before addressing the general questions about Riemann integrable functions, we present several examples that will be considered for each new type of integral. Pay attention to both the results and the means by which the results are obtained.

Example 1 (Constant functions). Let $f(x) = c$ on $[a, b]$. Let $\mathcal{P} = \{(t_k, I_k)\}$ be any tagged partition of $[a, b]$. Since $f(t_k) = c$, we find that

$$S_R(f, \mathcal{P}) = \sum_{\mathcal{P}} c \Delta x = c \sum_{\mathcal{P}} \Delta x = c(b - a).$$

Hence we see that

$$|S_R(f, \mathcal{P}) - c(b - a)| = 0 < \varepsilon$$

for any $\varepsilon > 0$ and any tagged partition $\mathcal{P}$ of $[a, b]$. Thus the constant function $f(x) = c$ is integrable over $[a, b]$ with

$$R\!\int_a^b f = c(b-a).$$

Example 2 (Dirichlet). The Dirichlet function

$$d(x) = \begin{cases} 1, & x \in \mathbb{Q} \\ 0, & x \notin \mathbb{Q}, \end{cases}$$

where $\mathbb{Q}$ is the set of rational numbers, is not Riemann integrable over any non-degenerate interval $[a, b]$. To see why, suppose that $\mathcal{P}$ is a partition of $[a, b]$. Create two tagged partitions from $\mathcal{P}$. Choose rational numbers for the tags of $\mathcal{P}_1$ and irrational tags for $\mathcal{P}_2$. Then, irrespective of the mesh of $\mathcal{P}$, we see that

$$S_R(d, \mathcal{P}_1) = \sum_{\mathcal{P}_1} 1\, \Delta x = b - a$$

and

$$S_R(d, \mathcal{P}_2) = \sum_{\mathcal{P}_2} 0\, \Delta x = 0.$$

If $d(x)$ were integrable with a value of A, we would be able to find a $\delta > 0$ so that any tagged partition $\mathcal{Q}$ with a mesh less than δ would satisfy $|S_R(d, \mathcal{Q}) - A| < \frac{b-a}{4}$. In that case we could begin with a partition $\mathcal{P}$ with mesh $\|\mathcal{P}\| < \delta$ and use the triangle inequality to show that

$$b - a = |S_R(d, \mathcal{P}_1) - S_R(d, \mathcal{P}_2)|$$
$$< |S_R(d, \mathcal{P}_1) - A| + |A - S_R(d, \mathcal{P}_2)| < \frac{b-a}{2}.$$

This clearly false conclusion implies that $d(x)$ is not Riemann integrable over $[a, b]$.

Example 3 (Identity function). Let f be the identity function $f(x) = x$ on $[0, 2]$ and take $\mathcal{P}_n = \left\{ \left(\frac{2k}{n}, \left[\frac{2k-2}{n}, \frac{2k}{n} \right] \right) \right\}_{k=1}^n$. Then $\|\mathcal{P}_n\| = \frac{2}{n}$ and

$$S_R(f, \mathcal{P}_n) = \sum_{k=1}^n \frac{2k}{n} \frac{2}{n} = \frac{4}{n^2} \sum_{k=1}^n k = \frac{4}{n^2} \frac{n(n+1)}{2} = 2 + \frac{2}{n}.$$

As n gets larger and the mesh decreases, $S_R(f, \mathcal{P}_n)$ decreases to 2. We conclude that **if** f is Riemann integrable, then $R\!\int_0^2 f = 2$.

How can we show that f is Riemann integrable? This task is rather more difficult than identifying 2 as the potential value of the integral. We will show f to be Riemann integrable by connecting the Riemann sums associated with arbitrary tagged partitions to the Riemann sums associated with $\mathcal{P}_n$ of example 3. To that end, begin with an arbitrary tagged partition $\mathcal{P} = \{(t_k, [x_{k-1}, x_k])\}$ and create a new (untagged) partition $\mathcal{Q}_n$ that has $\{x_0, x_1, x_2, \ldots, x_n, \frac{2}{n}, \frac{4}{n}, \frac{6}{n}, \ldots, \frac{2n}{n}\}$ as its division points. Place these division points in order, relabel them as $\{y_0, y_1, y_2, \ldots, y_m\}$.

This type of construction occurs frequently enough to merit its own notation. Recognizing that the set of division points of $\mathcal{Q}_n$ is the union of the division points of $\mathcal{P}_n$ and $\mathcal{P}$, we use the notation $\mathcal{Q}_n = \mathcal{P}_n \cup \mathcal{P}$. To avoid possible confusion, even when the two partitions are tagged, the resulting partition is not. The union of two partitions (tagged or untagged) is always untagged. One natural way to attach tags to an untagged partition is to use one of the endpoints of each subinterval. We will use subscripts of L and R to designate the tagged partition created by using the left and right endpoints.

We claim that

1. $S_R(f, \mathcal{P}_L) \leq S_R(f, \mathcal{P}) \leq S_R(f, \mathcal{P}_R)$,

2. $S_R(f, \mathcal{P}_L) \leq S_R(f, \mathcal{Q}_{n,L}) \leq S_R(f, \mathcal{Q}_{n,R}) \leq S_R(f, \mathcal{P}_R)$,

3. $S_R(f, \mathcal{P}_{n,L}) \leq S_R(f, \mathcal{Q}_{n,L}) \leq S_R(f, \mathcal{Q}_{n,R}) \leq S_R(f, \mathcal{P}_{n,R})$,

4. $0 \leq S_R(f, \mathcal{P}_R) - S_R(f, \mathcal{P}_L) \leq 2 \|\mathcal{P}\|$, and

5. $0 \leq S_R(f, \mathcal{P}_{n,R}) - S_R(f, \mathcal{P}_{n,L}) \leq 2 \|\mathcal{P}_n\| = \frac{4}{n}$.

Assuming these results for the moment, note that $\mathcal{P}_n = \mathcal{P}_{n,R}$ so that (3) and (5) imply that

$$|S_R(f, \mathcal{Q}_{n,R}) - S_R(f, \mathcal{P}_n)| \leq 2 \|\mathcal{P}_n\| = \frac{4}{n}.$$

With a bit more effort, we observe that (1), (2), and (4) imply

$$|S_R(f, \mathcal{P}) - S_R(f, \mathcal{Q}_{n,R})| \leq 2 \|\mathcal{P}\|.$$

This is most easily seen by considering the values of $S_R(f, \mathcal{P}_L)$ and $S_R(f, \mathcal{P}_R)$ to be endpoints of an interval. Points known from (1) and (2) to be in that interval cannot be further from each other than the length of the interval.

Now suppose that we are given an $\varepsilon > 0$. Choose n so that $6/n < \varepsilon/2$ and suppose that our tagged partition $\mathcal{P}$ has a mesh $\|\mathcal{P}\| < \delta = \varepsilon/4$. Then

$$
\begin{aligned}
&|S_R(f,\mathcal{P}) - 2| \\
&\leq |S_R(f,\mathcal{P}) - S_R(f,\mathcal{Q}_{n,R})| \\
&\quad + |S_R(f,\mathcal{Q}_{n,R}) - S_R(f,\mathcal{P}_n)| + |S_R(f,\mathcal{P}_n) - 2| \\
&< \varepsilon/2 + 4/n + 2/n = \varepsilon/2 + 6/n < \varepsilon/2 + \varepsilon/2 = \varepsilon.
\end{aligned}
$$

This means that the identity function is integrable over $[0,2]$ with $^R\!\int_0^2 f = 2$.

Now what about the five claims? There are some general principles operating behind these statements that apply more broadly. So instead of verifying the claims for the identity function, we will prove the more general results.

Lemma 1. *Suppose that f is an increasing[1] function on the interval $[a,b]$. Then $S_R(f,\mathcal{P}_L) \leq S_R(f,\mathcal{P}) \leq S_R(f,\mathcal{P}_R)$ for any tagged partition $\mathcal{P}$ of $[a,b]$.*

Proof. The proof of this lemma is left as an exercise (exercise 3). □

Lemma 1 establishes claim (1) and the inner inequalities of claims (2) and (3). Before turning to the next lemma, note that the critical feature of $\mathcal{Q}_n$ is that its set of division points includes the division points of both $\mathcal{P}$ and $\mathcal{P}_n$. The following definition places this idea in a more general setting.

Definition 3 (Refinement). *Let $\mathcal{P}$ and $\mathcal{Q}$ be partitions of an interval $[a,b]$. We say that $\mathcal{Q}$ is a **refinement of** $\mathcal{P}$ if the division points of $\mathcal{P}$ are included in the division points of $\mathcal{Q}$. Alternatively, $\mathcal{Q}$ is a **refinement of** $\mathcal{P}$ if each subinterval of $\mathcal{Q}$ is contained in a subinterval of $\mathcal{P}$.*

Lemma 2. *Suppose that f is an increasing function on the interval $[a,b]$, that $\mathcal{P}$ is a partition of $[a,b]$, and that $\mathcal{Q}$ is a refinement of $\mathcal{P}$. Then $S_R(f,\mathcal{P}_L) \leq S_R(f,\mathcal{Q}_L) \leq S_R(f,\mathcal{Q}_R) \leq S_R(f,\mathcal{P}_R)$.*

Proof. The inner inequality is a consequence of lemma 1.

To prove the first inequality, assume that $\mathcal{Q}$ is created from $\mathcal{P}$ by the addition of a single division point, y, that falls in the interval $[x_{j-1}, x_j]$ of

---

[1] We say that a function with domain X is **increasing** if whenever $x, y \in X$ with $x \leq y$, we have $f(x) \leq f(y)$. We will use **strictly increasing** when $x < y$ implies $f(x) < f(y)$. **Decreasing** is defined analogously.

$\mathcal{P}$. Since all the other subintervals of $\mathcal{P}_L$ and $\mathcal{Q}_L$ are equal, their corresponding terms will cancel when the Riemann sums are subtracted. The non-cancelling terms are generated from the intervals $[x_{j-1}, y]$ and $[y, x_j]$ from $\mathcal{Q}$ and $[x_{j-1}, x_j]$ from $\mathcal{P}$. Since f is increasing,

$$
\begin{aligned}
S_R(f, \mathcal{Q}_L) &- S_R(f, \mathcal{P}_L) \\
&= \left[f\left(x_{j-1}\right)\left(y - x_{j-1}\right) + f\left(y\right)\left(x_j - y\right) \right] - f\left(x_{j-1}\right)\left(x_j - x_{j-1}\right) \\
&\geq \left[f\left(x_{j-1}\right)\left(y - x_{j-1}\right) + f\left(x_{j-1}\right)\left(x_j - y\right) \right] - f\left(x_{j-1}\right)\left(x_j - x_{j-1}\right) \\
&= 0.
\end{aligned}
$$

The general result now follows by induction. The right inequality is left as an exercise.   □

Claims (4) and (5) are consequences of the next lemma.

Lemma 3. *Suppose that f is an increasing function on the interval $[a, b]$. Then $0 \leq S_R(f, \mathcal{P}_R) - S_R(f, \mathcal{P}_L) \leq (f(b) - f(a)) \|\mathcal{P}\|$ for any tagged partition $\mathcal{P}$ of $[a, b]$.*

Proof. The left inequality follows from lemma 1. For the right-hand side, note that $0 \leq f(x_k) - f(x_{k-1})$ so that

$$
\begin{aligned}
S_R(f, \mathcal{P}_R) - S_R(f, \mathcal{P}_L) &= \sum_{k=1}^{n} f(x_k)\, \Delta x_k - \sum_{k=1}^{n} f(x_{k-1})\, \Delta x_k \\
&= \sum_{k=1}^{n} \left(f(x_k) - f(x_{k-1}) \right) \Delta x_k \\
&\leq \sum_{k=1}^{n} \left(f(x_k) - f(x_{k-1}) \right) \|\mathcal{P}\| \\
&= (f(b) - f(a)) \|\mathcal{P}\|.
\end{aligned}
$$
                                                                       □

The five claims on page 29 are thus established.

What we have just done properly strikes most students as a tremendous amount of work to verify that $f(x) = x$ is integrable over $[0, 2]$. Newton and Leibniz did not worry about such concerns. For them, the integral obviously existed and the only question was how to evaluate it. As a cautionary note, consider the Dirichlet function. Had we simply evaluated the Riemann sums for the partitions $\mathcal{P}_n = \left\{ \left[\frac{k-1}{n}, \frac{k}{n} \right] \right\}$ using left-, right-, or midpoints, we would have easily, but incorrectly, deduced that the value of the integral of the Dirichlet function over $[0, 1]$ is 1.

The need for the careful analysis we have just completed arose from considerations of trigonometric series. It is not at all clear that such functions will be better behaved than the Dirichlet function. In fact, in Chapter 4 we will show how trigonometric series can produce rather ill-behaved functions. One might be tempted to ignore the issues raised by the Dirichlet function as issuing from an annoying, but irrelevant, pathological example. However, trigonometric series functions arise as solutions to differential equations and cannot be so ignored. We do, indeed, need to expend the effort to develop criteria for integrability for poorly-behaved functions rather than assuming that any naturally arising function is integrable.

Happily, the effort we devoted to verifying that $^R\!\int_0^2 x = 2$ is rewarded by providing us with insights into ways of modifying the definition of the integral that allow us to apply similar techniques to a wider range of functions. We will do this in the next chapter, but we do not need to wait until then to see some return on our effort. We can get additional mileage from our work right now by considering general increasing functions. If you reflect on the structure of example 3, you will see that we have most of the tools needed to prove that increasing functions are integrable. What we lack is a method to identify a value for the integral.

Theorem 4. *If f is an increasing function defined on $[a, b]$ then f is integrable over $[a, b]$.*

Proof. To identify the value of the integral, we will use the fact that any bounded increasing sequence of real numbers converges. Consider the partition $\mathcal{P}_n$ generated by the division points $\{a + \frac{k}{2^n}(b - a)\}_{k=1}^{2^n}$. Since $\mathcal{P}_{n+1}$ is a refinement of $\mathcal{P}_n$, we can use lemma 2 (page 30) to conclude that

$$S_R(f, \mathcal{P}_{1,L}) \le S_R(f, \mathcal{P}_{2,L}) \le S_R(f, \mathcal{P}_{3,L}) \le \cdots \le S_R(f, \mathcal{P}_{1,R}).$$

Let A be the limit of the bounded, increasing sequence $\{S_R(f, \mathcal{P}_{n,L})\}$. The proof can now be completed in a manner similar to example 3 (page 28). (See exercise 6b.)   □

There is another general class of functions whose members are integrable: continuous functions.

Theorem 5. *Suppose that f is continuous on the interval $[a, b]$. Then f is integrable over $[a, b]$.*

Proof. Since $[a, b]$ is a compact set, f is bounded and uniformly continuous[2] on $[a, b]$. In particular, there is a real number B so that $|f(x)| < B$ for all $x \in [a, b]$. Take $\Delta x = (b - a)/n$ and let $\mathcal{P}_n$ be the partition defined by the division points $\{a + k\Delta x\}_{k=1}^{n-1}$. Since

$$-B(b - a) \leq S_R(f, \mathcal{P}_{n,L}) \leq B(b - a)$$

by exercise 7, the sequence $\{S_R(f, \mathcal{P}_{n,L})\}_{n=1}^{\infty}$ must have a cluster point A. We claim that $^R\!\int_a^b f = A$.

Let $\varepsilon > 0$ and use the uniform continuity of f to choose $\delta > 0$ so that $|f(t) - f(s)| < \varepsilon$ whenever $s, t \in [a, b]$ with $|s - t| < \delta$. Choose n so that $|S_R(f, \mathcal{P}_{n,L}) - A| < \varepsilon$ and $\|\mathcal{P}_n\| = \frac{b-a}{n} < \delta$. Now let $\mathcal{P} = \{t_k, [x_{k-1}, x_k]\}_{k=1}^{n}$ be a tagged partition of $[a, b]$ with $\|\mathcal{P}\| < \delta$. Finally, define $\mathcal{Q} = \mathcal{P}_n \cup \mathcal{P}$ and write $\mathcal{Q} = \{[y_{j-1}, y_j]\}_{j=1}^{m}$.

Focus for the moment on a single subinterval $[x_{i-1}, x_i]$ of $\mathcal{P}$. Since $\mathcal{Q}$ is a refinement of $\mathcal{P}$, there are values j and k so that $x_{i-1} = y_{j-1} < \cdots < y_k = x_i$. In other words, $[x_{i-1}, x_i]$ is made up of one or more subintervals of $\mathcal{Q}$. Since $t_i, y_j, y_{j+1}, \ldots, y_k$ are all in the interval $[x_{i-1}, x_i]$ and $|x_i - x_{i-1}| < \delta$, we can conclude that $|f(t_i) - f(y_p)| < \varepsilon$ for $j \leq p \leq k$.

Now consider the difference of the terms in the Riemann sums corresponding to these intervals.

$$\left| f(t_i)(x_i - x_{i-1}) - \left[f(y_j)(y_j - y_{j-1}) + \cdots + f(y_k)(y_k - y_{k-1}) \right] \right|$$
$$= \left| \left[f(t_i)(y_j - y_{j-1}) + \cdots + f(t_i)(y_k - y_{k-1}) \right] \right.$$
$$\left. - \left[f(y_j)(y_j - y_{j-1}) + \cdots + f(y_k)(y_k - y_{k-1}) \right] \right|$$
$$\leq |f(t_i) - f(y_j)|(y_j - y_{j-1}) + \cdots + |f(t_i) - f(y_k)|(y_k - y_{k-1})$$
$$< \varepsilon(y_j - y_{j-1}) + \cdots + \varepsilon(y_k - y_{k-1})$$
$$= \varepsilon(y_k - y_{j-1}) = \varepsilon(x_i - x_{i-1}).$$

Since the same analysis applies to all the subintervals of $\mathcal{P}$ and every subinterval of $\mathcal{Q}$ is accounted for this way, we see that

$$|S_R(f, \mathcal{P}) - S_R(f, \mathcal{Q}_R)|$$
$$< \varepsilon(x_1 - x_0) + \varepsilon(x_2 - x_1) + \cdots + \varepsilon(x_n - x_{n-1}) = \varepsilon(b - a).$$

---

[2] Statements of definitions and important results from real analysis can be found in the appendices.

Because $\mathcal{Q}$ is also a refinement of $\mathcal{P}_n$ the same analysis proves that

$$|S_R(f,\mathcal{P}_{n,L}) - S_R(f,\mathcal{Q}_R)| < \varepsilon\,(b-a)$$

as well. Pulling the various pieces together, we see that

$$|S_R(f,\mathcal{P}) - A|$$
$$\leq |S_R(f,\mathcal{P}) - S_R(f,\mathcal{Q}_R)|$$
$$+ |S_R(f,\mathcal{Q}_R) - S_R(f,\mathcal{P}_{n,L})| + |S_R(f,\mathcal{P}_{n,L}) - A|$$
$$< 2\varepsilon\,(b-a) + \varepsilon.$$

This is sufficient to conclude that f is integrable with ${}^R\!\int_a^b f = A$. (See exercise 8.) $\qquad\square$

A function need not be continuous everywhere to guarantee Riemann integrability. A function f is also Riemann integrable when it has only isolated discontinuities.

Theorem 6. *Suppose that f is bounded and has a finite number of discontinuities in $[a,b]$. Then f is Riemann integrable over $[a,b]$.*

Proof. Suppose that f is continuous on $[a,b]$ except at $x = a$. Since f is bounded, there is a B so that $|f(x)| < B$ for all $x \in [a,b]$. We know from theorem 5 that the integral $s_n = {}^R\!\int_{a+\frac{1}{n}}^b f$ is defined whenever $n > \frac{1}{b-a}$. Moreover, when $m > n$,

$$|s_m - s_n| = \left|{}^R\!\int_{a+\frac{1}{m}}^{a+\frac{1}{n}} f\right| \leq {}^R\!\int_{a+\frac{1}{m}}^{a+\frac{1}{n}} |f| \leq {}^R\!\int_{a+\frac{1}{m}}^{a+\frac{1}{n}} B = B\left(\frac{1}{n} - \frac{1}{m}\right) < \frac{B}{n}$$

so that $\{s_n\} = \left\{{}^R\!\int_{a+\frac{1}{n}}^b f\right\}$ is a Cauchy sequence. Let $A = \lim_{n\to\infty} s_n$.

Now let $\varepsilon > 0$ and choose n_0 so that $|s_{n_0} - A| < \varepsilon/6$, $\frac{1}{n_0} < \varepsilon/3B$, and $n_0 > \frac{1}{b-a}$. Set $a_0 = a + \frac{1}{n_0}$. Since f is integrable over $[a_0, b]$, we can find a δ so that $|S_R(f,\mathcal{P}) - s_{n_0}| < \varepsilon/6$ for any tagged partition $\mathcal{P}$ of $[a_0, b]$ with $\|\mathcal{P}\| < \delta$.

Set $\delta' = \min\left\{\delta, \frac{\varepsilon}{6B}\right\}$ and suppose that P is a tagged partition of $[a,b]$ with mesh $\|\mathcal{P}\| < \delta'$. If a_0 is not a division point of $\mathcal{P}$, create a new tagged partition $\mathcal{P}^*$ by inserting a_0 as a division point and using the left endpoints as the tags of the newly created subintervals. By construction, $\mathcal{P}$ and $\mathcal{P}^*$ will agree except for the subintervals from $\mathcal{P}$ and $\mathcal{P}^*$ containing a_0. Since the values of f at any pair of tags for a given interval can differ by at most $2B$, we can use the techniques of lemma 3 (page 31) to conclude that

$$|S_R(f, \mathcal{P}) - S_R(f, \mathcal{P}^*)| \leq 2B \cdot \delta < \varepsilon/3.$$

(See exercise 9b.) If a_0 is a division point of $\mathcal{P}$, then take $\mathcal{P}^* = \mathcal{P}$ so the difference of the Riemann sums is zero.

Split $\mathcal{P}^*$ into two tagged partitions $\mathcal{P}_1^*$ and $\mathcal{P}_2^*$ of $[a, a+\frac{1}{n_0}]$ and $[a+\frac{1}{n_0}, b]$ respectively. Then

$$|S_R(f, \mathcal{P}_1^*)| = \left| \sum_{\mathcal{P}_1^*} f \, \Delta x \right| \leq \sum_{\mathcal{P}_1^*} B \, \Delta x = B \cdot \frac{1}{n_0} < \varepsilon/3$$

and, since the mesh of $\mathcal{P}_2^*$ is less than δ,

$$|S_R(f, \mathcal{P}_2^*) - A| \leq |S_R(f, \mathcal{P}_2^*) - s_{n_0}| + |s_{n_0} - A| < \varepsilon/3.$$

Thus

$$
\begin{aligned}
&|S_R(f, \mathcal{P}) - A| \\
&\leq |S_R(f, \mathcal{P}) - S_R(f, \mathcal{P}^*)| + |S_R(f, \mathcal{P}^*) - A| \\
&\leq |S_R(f, \mathcal{P}) - S_R(f, \mathcal{P}^*)| + |S_R(f, \mathcal{P}_1^*)| + |S_R(f, \mathcal{P}_2^*) - A| \\
&< \varepsilon.
\end{aligned}
$$

We conclude that f is integrable over $[a, b]$.

The case where the right endpoint is a point of discontinuity is treated similarly. (See exercise 9d.) The proof extends to a finite number of discontinuities located anywhere in $[a, b]$ by using the results of the next section to break the original interval into a finite number of subintervals. (exercise 9e.)   □

One might be tempted to conclude, on the basis of the Dirichlet function, that any function with an infinite number of discontinuities fails to be Riemann integrable. This conclusion would be false as illustrated by the following two examples.

Example 4. Define

$$f(x) = \begin{cases} 1, & x = \frac{1}{n}, n \in \mathbb{N} \\ 0, & \text{otherwise.} \end{cases}$$

Suppose that $\mathcal{P}$ is a tagged partition of $[0, 1]$ with $\|\mathcal{P}\| < \frac{1}{N^2}$. The subintervals of $\mathcal{P}$ contained in $\left[0, \frac{1}{N}\right]$ can contribute at most $\frac{1}{N}$ to $S_R(f, \mathcal{P})$. The intervals intersecting $\left[\frac{1}{N}, 1\right]$ contribute at most N nonzero terms. Since the width of these subintervals is less than $\frac{1}{N^2}$, $0 \leq S_R(f, \mathcal{P}) < \frac{1}{N} + N\frac{1}{N^2} = \frac{2}{N}$. Thus f is Riemann integrable over $[0, 1]$ with $^R\!\int_0^1 f = 0$.

Not only can a Riemann-integrable function have an infinite set of discontinuities, the set of discontinuities can be dense.

Example 5. Define

$$s_c(x) = \begin{cases} 0, & x < c \\ 1, & c \leq x. \end{cases}$$

(See Figure 2.2.) Let $\{r_i\}_{i=1}^{\infty}$ be an enumeration of the rational numbers in the interval $[0, 1]$ and define f by

$$f(x) = \sum_{n=1}^{\infty} \frac{1}{2^n} s_{r_n}(x).$$

The function f is discontinuous on the dense set of rational points of $[0, 1]$ (exercise 10) but is Riemann integrable since f is increasing.

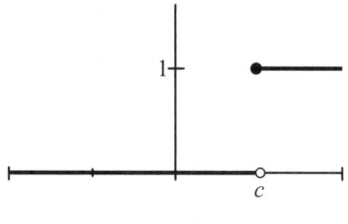

Figure 2.2. $s_c(x)$

These examples make evident the fact that we do not yet have a clearly demarcated boundary between functions that are Riemann integrable and those that are not. We will set that question aside temporarily and take it up again in the next chapter.

Reflections.   Before proceeding, it is worthwhile spending some time reflecting on the proofs. There are several common features that will reappear in our subsequent work. Moreover, some of the techniques will illuminate the advantages of alternative definitions of the integral.

Probably the most obvious observation is that the proofs thus far have generally consisted of two parts: (1) identifying a potential value for the integral and (2) verifying that the Riemann sums corresponding to partitions of small mesh are close to this value. The first phase is generally accomplished by examining a specific sequence of Riemann sums. One verifies that the corresponding sequence of values is bounded or Cauchy and uses the properties of sequences to identify a potential value for the integral.

The second phase of the proof typically consists of three parts. The first critical component is to find a way to relate the values of Riemann sums in the special sequence to the values of Riemann sums for arbitrary partitions with small mesh. This is accomplished using refinements. At this point, the techniques seem rather ad hoc; the approaches used for increasing functions, continuous functions, and functions with a single discontinuity are rather different. The need to connect the values of the Riemann sums of a partition and of one of its refinements is nonetheless present.

The second common feature can be described as bound-and-telescope. This technique is used to show that two Riemann sums are close in value by considering the difference of the two sums. In the case of increasing functions (lemma 3, page 31), the length of the subintervals was bounded and the value of the function was telescoped. For continuous functions (theorem 5, page 32), the difference in the values of the function at the tags is bounded and the subintervals are telescoped. Later, we will see proofs that combine both types of bound-and-telescope.

The third ingredient is the triangle inequality. A member of the special sequence of Riemann sums is selected to be close to the potential value of the integral. Then a generic partition with sufficiently small mesh is chosen. A refinement of the two partitions is constructed and its sum is shown to be close to the Riemann sums of both of the other two partitions. An application of the triangle inequality then completes the proof.

You are strongly encouraged to review the proofs in this section. Classify the steps in each proof according to the taxonomy just outlined. Going forward, both in this and subsequent chapters, watch for these moves. This discipline will help you make sense of the proofs.

2.2 Subintervals

Suppose f is integrable over $[a, b]$ and $[b, c]$. Is f integrable over $[a, c]$? If so, is ${}^R\!\int_a^c f = {}^R\!\int_a^b f + {}^R\!\int_b^c f$? While this may seem obvious, the conclusion fails for the Riemann-Stieltjes integral introduced in Chapter 8. The result is true for the Riemann integral.

Theorem 7. *Suppose f is Riemann integrable over $[a, b]$ and $[b, c]$. Then f is integrable over $[a, c]$ and ${}^R\!\int_a^c f = {}^R\!\int_a^b f + {}^R\!\int_b^c f$.*

Proof. For notational efficiency set $A_1 = {}^R\!\int_a^b f$, $A_2 = {}^R\!\int_b^c f$, and $A = A_1 + A_2$.

Now suppose that $\varepsilon > 0$. Since f is integrable over $[a, b]$ and $[b, c]$, we can conclude that there is a value B such that $|f(x)| < B$ for all $x \in [a, c]$. Additionally, we can find a value $\delta_1 > 0$ so that whenever $\mathcal{P}_1$ is a tagged partition of $[a, b]$ with $\|\mathcal{P}\| < \delta_1$ then $|S_R(f, \mathcal{P}_1) - A_1| < \varepsilon/4$. We also can find a corresponding $\delta_2 > 0$ for the interval $[b, c]$. Set $\delta = \min\{\delta_1, \delta_2, \varepsilon/4B\}$ and suppose that $\mathcal{P} = \{(t_i, [x_{i-1}, x_i])\}$ is a tagged partition of $[a, c]$ with $\|\mathcal{P}\| < \delta$. In exercise 20 you are asked to prove that if b is a division point of $\mathcal{P}$ then $|S_R(f, \mathcal{P}) - A| < \varepsilon/2$.

If b is not a division point of $\mathcal{P}$, create a new tagged partition $\mathcal{P}^*$ by adding b as a division point and taking b as the tag for both newly created subintervals. Now $x_{i-1} < b < x_i$ for some i between 1 and n. Since all the other tagged intervals are the same in $\mathcal{P}$ and $\mathcal{P}^*$, their corresponding terms in the Riemann sums cancel leaving

$$
\begin{aligned}
\left| S_R(f, \mathcal{P}) - S_R(f, \mathcal{P}^*) \right| &= |f(t_i)(x_i - x_{i-1}) - (f(b)(b - x_{i-1}) \\
&\quad + f(b)(x_i - b))| \\
&= |(f(t_i) - f(b))(x_i - x_{i-1})| \\
&\leq 2B(x_i - x_{i-1}) < 2B\delta < \varepsilon/2.
\end{aligned}
$$

To complete the proof, apply the result of exercise 20 (the case when b is a division point) to $\mathcal{P}^*$ and conclude that

$$
|S_R(f, \mathcal{P}) - A| \leq \left| S_R(f, \mathcal{P}) - S_R(f, \mathcal{P}^*) \right| + \left| S_R(f, \mathcal{P}^*) - A \right| < \varepsilon. \qquad \square
$$

What about the other direction? If f is integrable over $[a, b]$ and $[c, d] \subset [a, b]$ must f be integrable over $[c, d]$? Yes!

Theorem 8. *If f is integrable over $[a, b]$ and $[c, d] \subset [a, b]$, then f is integrable over $[c, d]$.*

Proof. As in the proof of theorem 5 (page 32), for each $n \in \mathbb{N}$ create a partition $\mathcal{P}_n$ of $[c, d]$ using the division points $\{c + \frac{k}{n}(c - d)\}_{k=1}^{n-1}$ and let A be a cluster point of $\{S_R(f, \mathcal{P}_n)\}_{n=1}^{\infty}$. Let $\varepsilon > 0$ be given and choose n_0 so that $|S_R(f, \mathcal{P}_{n_0}) - A| < \varepsilon/2$. Also choose a $\delta > 0$ so that $\left| S_R(f, \mathcal{P}^*) - {}^R\!\int_a^b f \right| < \varepsilon/4$ whenever $\mathcal{P}^*$ is a tagged partition of $[a, b]$ with $\|\mathcal{P}^*\| < \delta$. Then if $\mathcal{P}^*$ and $\mathcal{Q}^*$ are tagged partitions of $[a, b]$ with meshes less than δ, $|S_R(f, \mathcal{P}^*) - S_R(f, \mathcal{Q}^*)| < \varepsilon/2$.

Let $\mathcal{R}_a$, $\mathcal{P}$, and $\mathcal{R}_b$ be tagged partitions of $[a, c]$, $[c, d]$, and $[d, b]$ respectively with meshes less than δ. ($\mathcal{R}_a$ or $\mathcal{R}_b$ will be empty if $[a, c]$ or $[d, b]$ is degenerate.) The partitions $\mathcal{R}_a$, $\mathcal{P}$, and $\mathcal{R}_b$ combine to create a partition $\mathcal{P}^*$ of $[a,b]$ with $\|\mathcal{P}^*\| < \delta$. Similarly, $\mathcal{R}_a$, $\mathcal{P}_{n_0}$, and $\mathcal{R}$ create a partition $\mathcal{P}_{n_0}^*$. Since $\mathcal{P}^*$ and $\mathcal{P}_{n_0}^*$ share $\mathcal{R}_a$ and $\mathcal{R}_b$ in common,

$$\left| S_R\left(f,\mathcal{P}\right) - S_R\left(f,\mathcal{P}_{n_0}\right) \right| = \left| S_R\left(f,\mathcal{P}^*\right) - S_R\left(f,\mathcal{P}_{n_0}^*\right) \right| < \varepsilon/2.$$

Hence

$$\left| S_R\left(f,\mathcal{P}\right) - A \right| \le \left| S_R\left(f,\mathcal{P}\right) - S_R\left(f,\mathcal{P}_{n_0}\right) \right| + \left| S_R\left(f,\mathcal{P}_{n_0}\right) - A \right| < \varepsilon.$$

Thus f is Riemann integrable over $[c,d]$ with ${}^R\!\int_c^d f = A$. $\qquad\square$

2.3   The fundamental theorems

One of the important ideas of elementary calculus is the fundamental theorem of calculus that connects the concepts of the Newton-Leibniz and Riemann integrals. Intuitively, the fundamental theorem of calculus says that integration and differentiation are inverse operations. As we shall see, the relationship is not quite that simple. We begin with statements and proofs of the two forms of the fundamental theorem.

Theorem 9 (FTC-1). *If F is a differentiable function on the interval $[a,b]$ and F' is continuous on (a,b) then*

1. F' is Riemann integrable on $[a,b]$ and

2. ${}^R\!\int_a^x F' = F(x) - F(a)$ for all $x \in [a,b]$.

Proof. The first statement follows from theorem 5 (page 32).

To verify the value of the integral, let $\mathcal{P} = \{[x_{i-1}, x_i]\}_{i=1}^n$ be any partition of $[a,x]$. Applying the mean value theorem to F on the interval $[x_{i-1}, x_i]$, we can select tags $\{t_i\}_{i=1}^n$ so that $F'(t_i)(x_i - x_{i-1}) = F(x_i) - F(x_{i-1})$. Then, by telescoping the sum,

$$S_R\left(F',\mathcal{P}\right) = \sum_{k=1}^n F'(t_k)(x_k - x_{k-1})$$

$$= \sum_{k=1}^n (F(x_k) - F(x_{k-1})) = F(x) - F(a).$$

Since we know that F' is integrable over $[a,x]$, we can conclude that ${}^R\!\int_a^x F' = F(x) - F(a)$ for all $x \in [a,b]$. (See exercise 23b.) $\qquad\square$

The requirement that F' be continuous is somewhat annoying. If, as Poisson believed, differentiation and integration truly are inverse operations, then we should have $^R\!\int_a^x F' = F(x) - F(a)$ whenever F is differentiable on $[a, b]$ and $x \in [a, b]$. But the theorem fails when the assumption of continuity is removed.

Example 6. Let

$$f(x) = \begin{cases} x^2 \sin \frac{1}{x^2}, & x \neq 0 \\ 0, & x = 0. \end{cases}$$

Then f is differentiable everywhere, but f' is not bounded. Thus f' is not Riemann integrable. (See exercise 24.)

Unbounded derivatives are not the only barrier to a generalized version of FTC-1. In Chapter 4 we will construct a function with a bounded derivative for which FTC-1 fails.

If you review the proof of FTC-1, you will see that the role played by the assumption that F' is continuous is to ensure that F' is Riemann integrable. In light of this fact, the push to determine exactly which functions are Riemann integrable (and, if possible, to extend this set) takes on a heightened importance.

On the other hand, if there is a single point at which F fails to be differentiable, the conclusion of the fundamental theorem may be false even when the point where F is not differentiable is a removable discontinuity of F'.

Example 7. Define

$$F(x) = \begin{cases} 0, & x \in [0, 1/2) \\ 1, & x \in [1/2, 1]. \end{cases}$$

Then F is differentiable on $[0, 1]$ except at $x = 1/2$. Since F' is zero except at $x = 1/2$, the discontinuity of F' at $x = 1/2$ is removable. Then $^R\!\int_0^1 F' = 0$, but $F(1) - F(0) = 1$.

There is, of course, a second form of the fundamental theorem of calculus.

Theorem 10 (FTC-2). *Suppose that f is Riemann integrable on $[a, b]$. Define F on $[a, b]$ by $F(x) = {}^R\!\int_a^x f$.*

1. Then F is continuous on $[a, b]$.

2. If f is continuous at $x_0 \in (a, b)$, then F is differentiable at x_0 and $F'(x_0) = f(x_0)$.

Proof. First note that, in order to be Riemann integrable, f must be bounded by some value B. Thus if $x < y$, we can use theorems 7 and 8 and exercise 1 to show that

$$|F(y) - F(x)| = \left| {}^R\!\!\int_x^y f \right| \leq {}^R\!\!\int_x^y B = B(y - x).$$

The continuity of F follows easily. (See exercises 25b and 25c.)

Now suppose that f is continuous at x_0. For any $\varepsilon > 0$ we can find a $\delta > 0$ so that $f(x_0) - \varepsilon < f(x) < f(x_0) + \varepsilon$ for any $x \in [x_0, x_0 + \delta] \cap [a, b]$. Since constants are Riemann integrable, monotonicity of the Riemann integral (exercise 1c) implies that

$$ {}^R\!\!\int_{x_0}^x (f(x_0) - \varepsilon) \leq {}^R\!\!\int_{x_0}^x f \leq {}^R\!\!\int_{x_0}^x (f(x_0) + \varepsilon)$$

so that

$$(f(x_0) - \varepsilon)(x - x_0) \leq F(x) - F(x_0) \leq (f(x_0) + \varepsilon)(x - x_0).$$

Rearranging,

$$-\varepsilon \leq \frac{F(x) - F(x_0)}{x - x_0} - f(x_0) \leq \varepsilon.$$

A similar argument applies to values to the left of x_0 so that F is differentiable at x_0 and $F'(x_0) = f(x_0)$. □

2.4 Convergence theorems

We now take up the driving question behind the intense historical investigation of integration. How does the Riemann integral interact with limits? We begin with a set of examples that illustrate some of the issues.

Example 8. Let $\{r_n\}_{n=1}^\infty$ be an enumeration of the rational numbers in $[0, 1]$. Define

$$f_n(x) = \begin{cases} 1, & x \in \{r_1, r_2, \ldots, r_n\} \\ 0, & \text{otherwise.} \end{cases}$$

Then f_n has a finite number of discontinuities and thus is Riemann integrable. With a bit more work, one can verify that ${}^R\!\!\int_0^1 f_n = 0$ (exercise 26). On the other hand, if $x \in [0, 1]$ then $\lim_{n \to \infty} f_n(x) = d(x)$, the Dirichlet function of example 2 (page 28). Since the Dirichlet function is not Riemann integrable, ${}^R\!\!\int_0^1 \lim_{n \to \infty} f_n \neq \lim_{n \to \infty} {}^R\!\!\int_0^1 f_n$. The left-hand side does not exist.

Example 9. Define f_n on $[0, 1]$ by

$$f_n(x) = \begin{cases} n, & x \in (0, 1/n) \\ 0, & \text{otherwise.} \end{cases}$$

Then f_n is Riemann integrable with ${}^R\!\int_0^1 f_n = 1$. In this case, the limit function, $f(x) = \lim_{n \to \infty} f_n(x) = 0$, is Riemann integrable, but ${}^R\!\int_0^1 \lim_{n \to \infty} f_n \neq \lim_{n \to \infty} {}^R\!\int_0^1 f_n$. Both sides exist, but they do not agree.

Example 9 is the more troubling of the two examples. Since both ${}^R\!\int_0^1 \lim_{n \to \infty} f_n$ and $\lim_{n \to \infty} {}^R\!\int_0^1 f_n$ exist, there is less of an indication that something may be amiss. Of course, often things work out "right" in the sense that the equation ${}^R\!\int_0^1 \lim_{n \to \infty} f_n = \lim_{n \to \infty} {}^R\!\int_0^1 f_n$ is true. When is this the case? The problem in the two examples is that, while $f_n(x_0) \to f(x_0)$ for any $x_0 \in [0, 1]$, no matter how large the value of n, there remain other values of $x \in [0, 1]$ for which $f_n(x)$ is far from $f(x)$. The key to getting the Riemann integral to interact "nicely" with limits is uniform convergence.

Definition 4 (Uniform Convergence). *A sequence of functions $\{f_n\}$ converges **uniformly** to a function f on a set S if for any $\varepsilon > 0$, it is possible to find an $N \in \mathbb{N}$ such that $|f_n(x) - f(x)| < \varepsilon$ for all $n > N$ and all $x \in S$.*

The modifier "uniformly" refers to the fact that the value of N can be chosen without regard for the value of x. One value of N will serve for all $x \in S$. The sequence of functions in example 9 converges (pointwise) but does not converge uniformly.

Theorem 11. *Suppose that $\{f_n\}$ is a sequence of functions that are Riemann integrable over $[a, b]$ and converge uniformly on $[a, b]$ to f. Then f is Riemann integrable over $[a, b]$ and*

$$^R\!\int_a^b \lim_{n \to \infty} f_n = {}^R\!\int_a^b f = \lim_{n \to \infty} {}^R\!\int_a^b f_n.$$

Proof. Since f_n is Riemann integrable on $[a, b]$, f_n is bounded on $[a, b]$. Because $\{f_n\}$ converges uniformly, we can find a value B so that $|f_n(x)| < B$ for all $x \in [a, b]$ and all $n \in \mathbb{N}$. (See exercise 27.) Hence

$$-B(b - a) \leq {}^R\!\int_a^b f_n \leq B(b - a).$$

Consequently, $\{ {}^R\!\int_a^b f_n \}$, being a bounded sequence, must have a cluster point A.

We will show that f is integrable with ${}^R\!\int_a^b f = A$. To that end, suppose that $\varepsilon > 0$ and select n so that $|\,{}^R\!\int_a^b f_n - A| < \varepsilon/3$ and $|f_n(x) - f(x)| < \varepsilon/3\,(b-a)$ for all $x \in [a, b]$. (Why does such an n exist?) Then for any tagged partition $\mathcal{P} = \{(t_k, [x_{k-1}, x_k])\}_{k=1}^m$,

$$
|S_R(f, \mathcal{P}) - S_R(f_n, \mathcal{P})| = \left| \sum_{k=1}^m (f(t_k) - f_n(t_k)) \Delta_k \right|
$$

$$
\leq \sum_{k=1}^m \frac{\varepsilon}{3\,(b-a)} \Delta_k = \varepsilon/3.
$$

As f_n is integrable over $[a, b]$, we can find a $\delta > 0$ so that $|S_R(f_n, \mathcal{P}) - {}^R\!\int_a^b f_n| < \varepsilon/3$ for any tagged partition $\mathcal{P}$ of $[a, b]$ with mesh $\|\mathcal{P}\| < \delta$. But then for any such tagged partition $\mathcal{P}$ of $[a, b]$,

$$
|S_R(f, \mathcal{P}) - A|
$$

$$
\leq \left| S_R(f, \mathcal{P}) - S_R(f_n, \mathcal{P}) \right| + \left| S_R(f_n, \mathcal{P}) - {}^R\!\int_a^b f_n \right| + \left| {}^R\!\int_a^b f_n - A \right| < \varepsilon.
$$

We conclude that f is integrable over $[a, b]$ with ${}^R\!\int_a^b f = A = \lim_{n \to \infty} {}^R\!\int_a^b f_n$.

To complete the proof, we need to verify that $\left\{ {}^R\!\int_a^b f_n \right\}$ converges to A. By the uniqueness of the integral (exercise 1) the sequence can have only one cluster point. Thus $\left\{ {}^R\!\int_a^b f_n \right\}$ must converge to A (exercise 29).   □

We close out this chapter with a final example showing that, while uniform convergence is sufficient to ensure that ${}^R\!\int_a^b \lim_{n \to \infty} f_n = \lim_{n \to \infty} {}^R\!\int_a^b f_n$, uniform convergence is not necessary.

Example 10. Let $f_n(x) = x^n$ for $x \in [0, 1]$. Then $\{f_n\}$ converges pointwise, but not uniformly, to

$$
f(x) = \begin{cases} 1, & x = 1 \\ 0, & x \in [0, 1). \end{cases}
$$

Nevertheless, ${}^R\!\int_0^1 f_n = \frac{1}{n+1}$ converges to $0 = {}^R\!\int_0^1 f$.

2.5 Exercises

2.1 Riemann integrability: filling the gaps

1. Verify the standard integral properties for the Riemann integral. These properties will be used without comment in the text.

   (a) **Uniqueness.** The value of the Riemann integral is unique (if it exists).

   (b) **Linearity.** Let $c \in \mathbb{R}$. If f and g are Riemann integrable over the interval $[a, b]$, then so are $f + g$ and cf. Moreover, $^R\!\int_a^b (f + g) = {}^R\!\int_a^b f + {}^R\!\int_a^b g$ and $^R\!\int_a^b cf = c\, {}^R\!\int_a^b f$.

   (c) **Monotonicity.** If f and g are Riemann integrable over the interval $[a, b]$ with $f(x) \leq g(x)$ for $x \in [a, b]$, then $^R\!\int_a^b f \leq {}^R\!\int_a^b g$.

   (d) **Triangle inequality.** If f and $|f|$ are Riemann integrable over $[a, b]$, then $\left| {}^R\!\int_a^b f \right| \leq {}^R\!\int_a^b |f|$.

2. Show that if f is unbounded over the interval $[a, b]$, then f is not integrable over $[a, b]$. (Show that if f is not bounded above, given any partition $\mathcal{P}$ and any potential value A for the integral, tags can be chosen so that $|S_R(f, \mathcal{P})| > A + 1$.)

3. Let f be an increasing function defined on $[a, b]$ and let $\mathcal{P}$ be a tagged partition $\mathcal{P}$ of $[a, b]$. Prove that $S_R(f, \mathcal{P}_L) \leq S_R(f, \mathcal{P}) \leq S_R(f, \mathcal{P}_R)$.

4. Prove that the two definitions of refinement given in Definition 3 (page 30) are equivalent.

5. In the proof of lemma 2 (page 30),

   (a) Complete the induction proof of the left inequality.

   (b) Prove the right inequality.

   (c) Where and how would the proof fail if f were not increasing.

6. In the proof of theorem 4 (page 32)

   (a) Why is $\mathcal{P}_{n+1}$ a refinement of $\mathcal{P}_n$?

   (b) Use example 3 (page 28) as a guide to complete the proof of theorem 4 (page 32).

7. Suppose that f is bounded by B on $[a, b]$; in other words, $|f(x)| \leq B$ for all $x \in [a, b]$. Show that $-B(b - a) \leq S_R(f, \mathcal{P}) \leq B(b - a)$ for any tagged partition $\mathcal{P}$ of $[a, b]$.

8. Suppose that for any $\varepsilon > 0$ you can select a $\delta > 0$ so that $|S_R(f, \mathcal{P}) - A| < 15\varepsilon$ for any partition $\mathcal{P}$ with mesh $\|\mathcal{P}\| < \delta$. Explain how this fact can be used to prove that for any $\varepsilon > 0$ it is possible to select a $\delta > 0$ so that for any partition $\mathcal{P}$ with $\|\mathcal{P}\| < \delta$ the condition $|S_R(f, \mathcal{P}) - A| < \varepsilon$ is satisfied.

9. In the proof of theorem 6 (page 34)

   (a) The proof that f is Riemann integrable over $[a, b]$ if f is bounded and continuous except at the left endpoint, a, implicitly used theorem 7. Identify where and how this theorem was used.
   (b) Explain why $|S_R(f, \mathcal{P}) - S_R(f, \mathcal{P}^*)| \le 2B \cdot \delta$. (Most terms cancel.)
   (c) Why is $|\sum_{\mathcal{P}_1^*} f \, \Delta x| \le B \cdot \frac{1}{n_0}$?
   (d) Provide a proof that f is Riemann integrable over $[a, b]$ if f is bounded and continuous except at the right endpoint, b.
   (e) Use theorem 7 (page 37) to prove that if f is bounded on the interval $[a, b]$ and continuous except for a finite number of points, then f is Riemann integrable over $[a, b]$.

10. Define
$$s_c(x) = \begin{cases} 0, & x < c, \\ 1, & c \le x. \end{cases}$$

Let $\{r_i\}_{i=1}^{\infty}$ be an enumeration of the rational numbers in the interval $[0, 1]$ and define f by $f(x) = \sum_{n=1}^{\infty} \frac{1}{2^n} s_{r_n}(x)$.

   (a) Why is f defined? In other words, how do you know that $\sum_{n=1}^{\infty} \frac{1}{2^n} s_{r_n}(x)$ converges for $x \in [0, 1]$?
   (b) Prove that f is strictly increasing.
      Let $c \in [0, 1]$.
   (c) Prove that $\lim_{x \to c+} f(x) = \sum_{r_k \le c} \frac{1}{2^k}$.
   (d) Prove that $\lim_{x \to c-} f(x) = \sum_{r_k < c} \frac{1}{2^k}$.
   (e) Conclude that f is continuous at c if and only if c is an irrational number.
   (f) Prove that $^R\!\int_0^1 \sum_{n=1}^{\infty} \frac{1}{2^n} s_{r_n} = \sum_{n=1}^{\infty} \frac{1}{2^n} \, ^R\!\int_0^1 s_{r_n}$.
   (g) Express the value of $^R\!\int_0^1 \sum_{n=1}^{\infty} \frac{1}{2^n} s_{r_n}$ without using an integral.

11. Use theorems 7 and 8 (page 37) to explain why $^R\!\int_a^{c-\frac{1}{m}} f - {}^R\!\int_a^{c-\frac{1}{n}} f = {}^R\!\int_{c-\frac{1}{n}}^{c-\frac{1}{m}} f$ whenever $m > n > \frac{1}{c-a}$. (See theorem 6 on page 37.)

2.1 Riemann integrability: deeper reflections

12. Give an example of a function f that is not Riemann integrable but for which $|f|$ is Riemann integrable.

13. Give an alternate proof that $f(x) = x$ is Riemann integrable over $[0, 2]$ by verifying and using the following facts about a partition $\mathcal{P} = \{(t_k, [x_{k-1}, x_k])\}_{k=1}^{n}$ of $[0, 2]$.

    (a) $\sum_{k=1}^{n} \left(x_k^2 - x_{k-1}^2 \right) = 4$.
    (b) $t_k \approx \frac{x_k + x_{k-1}}{2}$ when $\|\mathcal{P}\|$ is small.

14. Prove that if f is a decreasing function on $[a, b]$ then f is Riemann integrable. (Work smart, not hard.)

15. Prove that if f is Riemann integrable over $[a, b]$ and g agrees with f except at $c \in [a, b]$, then g is also Riemann integrable and $^R\!\!\int_a^b g = {}^R\!\!\int_a^b f$.

16. Suppose that for any $\varepsilon > 0$ you can select a $\delta > 0$ so that $|S_R(f, \mathcal{P}) - A| \le \varepsilon$ for any partition $\mathcal{P}$ with $\|\mathcal{P}\| < \delta$. Explain how this fact can be used to prove that for any $\varepsilon > 0$ it is possible to select a $\delta > 0$ so that for any partition $\mathcal{P}$ with mesh $\|\mathcal{P}\| < \delta$ we are assured that $|S_R(f, \mathcal{P}) - A| < \varepsilon$.

17. Explain how the proof of theorem 5 (page 32) can be modified to arrive at the conclusion that $|S_R(f, \mathcal{P}) - A| < \varepsilon$. Why do you think this was not done in the provided proof?

18. Prove that

$$f(x) = \begin{cases} \sin \frac{1}{x}, & x \neq 0 \\ 0, & x = 0 \end{cases}$$

    is Riemann integrable over $[0, 1]$.

19. Prove that if f is Riemann integrable over $[a, b]$ then so is $|f|$. (Hint: Use the fact that for any x and y in the domain of f, $\big| |f(x)| - |f(y)| \big| \le |f(x) - f(y)|$.)

2.2 Subintervals: filling the gaps

20. Fill in the gap in the proof of theorem 7 (page 37) by showing that if b is a division point of $\mathcal{P}$, then $|S_R(f, \mathcal{P}) - A| < \varepsilon/2$. (Hint: Split $\mathcal{P}$ into tagged partitions of $[a, b]$ and $[b, c]$.)

21. In the proof of theorem 8 (page 38)

    (a) Why must $\{S_R(f, \mathcal{P}_n)\}_{n=1}^{\infty}$ have a cluster point?

(b) Why must we have $|S_R(f, \mathcal{P}^*) - S_R(f, \mathcal{Q}^*)| < \varepsilon/2$?
(c) Explain how $\mathcal{R}_a$, $\mathcal{P}$, and $\mathcal{R}_b$ can be combined to create a partition $\mathcal{P}^*$ of $[a, b]$.

2.2 Subintervals: deeper reflections

22. Suppose that f is bounded on $[a, b]$, that f is Riemann integrable over $[t, b]$ for all $t \in (a, b)$, and that $\lim_{t \to a+} {}^R\!\int_t^b f$ exists. Prove that f is Riemann integrable over $[a, b]$ with ${}^R\!\int_a^b f = \lim_{t \to a+} {}^R\!\int_t^b f$.

2.3 Fundamental theorems: filling the gaps

23. In the proof of FTC-1

(a) Explain how the mean value theorem is used to select the tags.
(b) Complete the final piece of the proof of FTC-1 by using the definition of integrability to show that if $A = {}^R\!\int_a^x F'$ then $A = F(x) - F(a)$.

24. Let

$$f(x) = \begin{cases} x^2 \sin \frac{1}{x^2}, & x \neq 0 \\ 0, & x = 0. \end{cases}$$

(a) Show that

$$f'(x) = \begin{cases} 2x \sin \frac{1}{x^2} - \frac{2}{x} \cos \frac{1}{x^2}, & x \neq 0 \\ 0, & x = 0. \end{cases}$$

(Be sure to address the $x = 0$ case.)
(b) Explain why FTC-1 fails for this function.

25. In the proof of theorem 10 (page 40)

(a) Why is $F(x)$ defined for $x \in [a, b]$?
(b) Explain how theorems 7 and 8 (page 37) and exercise 1 can be used to show that $|F(x) - F(y)| = \left| {}^R\!\int_x^y f \right| \leq B(y - x)$ when $x < y$.
(c) Explain how $|F(x) - F(y)| \leq B|y - x|$ implies that F is continuous.
(d) Explain how

$$(f(x_0) - \varepsilon)(x - x_0) < F(x) - F(x_0) < (f(x_0) + \varepsilon)(x - x_0)$$

follows from

$${}^R\!\int_{x_0}^x (f(x_0) - \varepsilon) < {}^R\!\int_{x_0}^x f < {}^R\!\int_{x_0}^x (f(x_0) + \varepsilon).$$

(e) Supply the details for the proof of the differentiability of F in FTC-2 when x is to the left of x_0.

2.4 Convergence theorems: filling the gaps

26. Prove that $g(x) = \begin{cases} 1, & x \in \{r_1, r_2, \ldots, r_n\} \\ 0, & \text{otherwise} \end{cases}$ is Riemann integrable with ${}^R\!\int_a^b g = 0$.

27. Prove that if $\{f_n\}$ is a sequence of bounded functions that converges to f uniformly on a set S, then $\{f_n\}$ is uniformly bounded on S. In other words, there is a value B so that $|f_n(x)| < B$ for all $n \in \mathbb{N}$ and $x \in S$.

28. In the proof of theorem 11 (page 42)

    (a) Why can we choose m and N so that $\left| {}^R\!\int_a^b f_m - A \right| < \varepsilon/2$ and $|f_n(x) - f_m(x)| < \varepsilon/2 (b-a)$ for all $x \in [a,b]$ and $n > N$?
    (b) Why can we select n so that $\left| {}^R\!\int_a^b f_n - A \right| < \varepsilon/3$ and $|f_n(x) - f(x)| < \varepsilon/2 (b-a)$ for all $x \in [a,b]$?
    (c) How does the uniqueness of the integral imply that $\left\{ {}^R\!\int_a^b f_n \right\}$ can have only one cluster point?

29. Prove that any bounded sequence with a unique cluster point converges to that cluster point.

30. Prove that the sequence $\{f_n\}$ of example 10 (page 43) converges pointwise but not uniformly to f.

31. Define

$$f_n(x) = \begin{cases} 1, & x \in \left[0, \frac{1}{n}\right] \\ 0, & x \in \left(\frac{1}{n}, 1\right]. \end{cases}$$

    Show that $\{f_n\}$ does not converge uniformly. Nevertheless, ${}^R\!\int_a^b \lim_{n\to\infty} f_n = \lim_{n\to\infty} {}^R\!\int_a^b f_n$.

2.4 Convergence theorems: deeper reflections

32. Prove that

$$r(x) = \begin{cases} \frac{1}{n}, & x = \frac{m}{n} \text{ with } \frac{m}{n} \text{ in lowest terms} \\ 0, & \text{otherwise} \end{cases}$$

    is Riemann integrable with ${}^R\!\int_0^1 r = 0$.

33. Give a counterexample to the conclusion of exercise 27 if $\{f_n\}$ only converges to f pointwise rather than uniformly. In other words, find a convergent sequence of bounded functions that is not uniformly bounded.

34. Provide an alternate proof that $\lim_{n \to \infty} {}^R\!\int_a^b f_n$ exists in theorem 11 (page 42) by showing that $\left\{ {}^R\!\int_a^b f_n \right\}$ is a Cauchy sequence.

35. Modify exercise 31 to give an example of a sequence of Riemann-integrable functions that is not uniformly bounded but for which ${}^R\!\int_a^b \lim_{n \to \infty} f_n = \lim_{n \to \infty} {}^R\!\int_a^b f_n$.

2.5 Related ideas: deeper reflections

36. Let $\mathcal{P}$ be a partition of $[a, b]$ and let $\varepsilon > 0$. Explain how one could select tags $\{r_k\}_{k=1}^n$ and $\{s_k\}_{k=1}^n$ so that if $\mathcal{P}_r$ and $\mathcal{P}_s$ are the tagged partitions using those tags, then $S_R\,(f, \mathcal{P}_r) - \varepsilon/4 < S_R\,(f, \mathcal{P}_t) < S_R\,(f, \mathcal{P}_s) + \varepsilon/4$ for any other tagged partition $\mathcal{P}_t$ also based on $\mathcal{P}$.

37. Prove the following generalization of exercise 15. If f is Riemann integrable over $[a, b]$ and $g\,(x) = f\,(x)$ except for finitely many points in $[a, b]$, then g is Riemann integrable over $[a, b]$ and ${}^R\!\int_0^1 g = {}^R\!\int_0^1 f$.

Prior to Riemann's work, Cauchy introduced a definition of the integral. In Cauchy's definition, only left endpoints are used.

Definition (Cauchy Integral) Let f be a real-valued function defined on $[a, b]$. Then f is **Cauchy Integrable** over $[a, b]$ if there is a number A such that for any $\varepsilon > 0$ we can find a $\delta > 0$ so that whenever $\mathcal{P}$ is a partition of I with $\|\mathcal{P}\| < \delta$,

$$|S_R(f, \mathcal{P}_L) - A| < \varepsilon.$$

In this case, A is the **Cauchy integral** of f over $[a, b]$ and we write ${}^C\!\int_a^b f = {}^C\!\int_a^b f\,(x)\,dx = A$.

38. Prove that if f is Riemann integrable over $[a, b]$, then f is Cauchy integrable over $[a, b]$ and the two integrals agree.

39. Prove that the Dirichlet function is not Cauchy integrable over $[0, 1]$.

40. Prove that every function that is Cauchy integrable over $[a, b]$ is bounded on $[a, b]$.

2.6   References

Bressoud, D.M. (2006). *A Radical Approach to Real Analysis* (2nd ed.). Mathematical Association of America.

Burk, F.E. (2007). *A Garden of Integrals*. Mathematical Association of America.

DePree, J. and C. Swartz (1988). *Introduction to Real Analysis*. John Wiley & Sons.

Gelbaum, B.R. and J.M.H. Olmsted (2003). *Counterexamples in Analysis*. Dover.

Riemann, B. (1990). *Gesammelte Mathematishe Werke*, reprinted with comments by Raghavan Narasimhan. Springer.

Simmons, G.F. (2007). *Calculus Gems*. Mathematical Association of America.

CHAPTER 3

The Darboux integral

While the Riemann integral is relatively straightforward to understand, proving theorems, particularly existence theorems, using its definition is rather awkward. Determining that there is a limiting value for the Riemann sums typically requires the use of special sequences of Riemann sums. Once a potential value for the integral is determined, we need a way to connect elements of the special sequence of Riemann sums to any generic Riemann sum with small mesh. The connection is almost always established by means of refining partitions. The process is relatively straightforward for increasing functions, but can be difficult for more general functions.

In 1870, sixteen years after Riemann published his definition, Gaston Darboux defined an integral that makes many of these issues take care of themselves. The cost (and you should expect there to be a cost) is the need to work with supremums and infimums. For mathematicians who have worked with these concepts for a while, this cost seems almost trivial. For students encountering the notion of supremum and infimum for the first time, the cost appears rather more substantial. If you are not relatively comfortable working with these ideas, you would be well advised to spend some time reviewing Appendix A.2 (page 311 and following) which provides a quick overview.

An important reason supremums are useful for our purposes is that the supremum, $\sup S$, of a non-empty, bounded set S of real numbers always exists. Among other benefits, the supremum's existence eliminates the need to use special sequences to establish a value for the integral. Supremums and infimums also make it possible to prove results like lemmas 1 and 2 (page 30) for non-monotonic functions. The key to proving the two lemmas is the fact that, when f is an increasing function, $f(x_{i-1}) \leq f(t) \leq f(x_i)$ for any $t \in [x_{i-1}, x_i]$. This suggests that we might try selecting tags m_i and M_i so that $f(m_i)$ and $f(M_i)$ are the minimum and maximum values of f

over the interval. While the idea is appealing, it turns out to be problematic. In addition to introducing an additional layer of notation, the sums corresponding to tags $\{m_i\}$ and $\{M_i\}$ will not telescope like those in lemma 3 (page 31). Maybe we could work around these obstacles were it not for an even more fundamental difficulty. The function f may not have a maximum or minimum value over $[x_{i-1}, x_i]$ (see exercise 1). Supremums and infimums solve this problem as well.

Definition 5 (Darboux Integral). *Let f be a bounded function defined on $[a, b]$ and let $\mathcal{P} = \{[x_{k-1}, x_k]\}_{k=1}^{n} = \{I_k\}$ be a partition of $[a, b]$. The **lower and upper Darboux sums** of f over $\mathcal{P}$ are, respectively,*

$$S_{\underline{D}}(f, \mathcal{P}) = \sum_{i=1}^{n} \inf_{[x_{i-1}, x_i]} f \cdot (x_i - x_{i-1}) = \sum_{\mathcal{P}} \inf_{I_k} f \; \Delta x_k$$

and

$$S_{\overline{D}}(f, \mathcal{P}) = \sum_{i=1}^{n} \sup_{[x_{i-1}, x_i]} f \cdot (x_i - x_{i-1}) = \sum_{\mathcal{P}} \sup_{I_k} f \; \Delta x_k.^{1}$$

*The **lower and upper Darboux integrals** of f over $[a, b]$ are*

$$\underline{D}\!\int_{a}^{b} f = \sup_{\mathcal{P}} \; S_{\underline{D}}(f, \mathcal{P}) = \sup \; \{ S_{\underline{D}}(f, \mathcal{P}) : \mathcal{P} \text{ is a partition of } [a, b] \}$$

and

$$\overline{D}\!\int_{a}^{b} f = \inf_{\mathcal{P}} \; S_{\overline{D}}(f, \mathcal{P}).$$

*If $\underline{D}\!\int_{a}^{b} f = \overline{D}\!\int_{a}^{b} f$, then f is **Darboux integrable** over $[a, b]$ and the **Darboux integral** of f over $[a, b]$ is*

$$D\!\int_{a}^{b} f = \underline{D}\!\int_{a}^{b} f.$$

In the exercises, you are asked to verify that all the relevant supremums and infimums are defined, that $S_{\underline{D}}(f, \mathcal{P}) \leq S_{\overline{D}}(f, \mathcal{P})$, and that $\underline{D}\!\int_{a}^{b} f \leq \overline{D}\!\int_{a}^{b} f$. (See exercises 2 and 9.)

Given any partition $\mathcal{P}$ of $[a, b]$, Riemann's approach to the integral must consider an infinite number of Riemann sums: one for each choice of tags.

---

^{1} $\inf_{I_k} f = \inf \{ f(x) \mid x \in I_k \}$ and $\sup_{I_k} f = \sup \{ f(x) \mid x \in I_k \}$.

In contrast, the Darboux approach only considers two sums, the upper and lower. When a tagged partition is used in a Darboux sum, the tags are ignored. The drawback is that the two Darboux sums, because they involve infimums and supremums, are conceptually more complicated than are Riemann sums.

On the other hand, the Riemann definition of the integral requires an unspecified value A and, as we have seen, it can take some effort to verify its existence. The Darboux integral provides two potential values for the integral, $\overline{D}\!\int_a^b f$ and $\underline{D}\!\int_a^b f$, and f is Darboux integrable if these two values match. Consequently, there is no mention of mesh, ε, or δ in the Darboux definition. So, if we are willing to overlook the complication of working with supremums and infimums, the Darboux definition is much simpler than the Riemann definition.

3.1 Darboux integrability

How does this translate to computations and proofs? To address this question, we will consider the same examples we used to introduce the Riemann integral. Keep your finger or a bookmark in Chapter 2 and compare the work.

Example 11 (Constant functions). Let $f(x) = c$ on $[a, b]$. Let $\mathcal{P} = \{I_k\}$ be any partition of $[a, b]$. Since f is constant, both the supremum and infimum over any subinterval are c. Thus

$$S_{\overline{D}}(f, \mathcal{P}) = \sum_{\mathcal{P}} \sup_{I_k} f \; \Delta x_k = \sum_{\mathcal{P}} c \; \Delta x_k = c \sum_{\mathcal{P}} \Delta x_k = c \, (b - a).$$

Similarly, $S_{\underline{D}}(f, \mathcal{P}) = c \, (b - a)$. Since $c \, (b - a)$ is the only possible value of a Darboux sum,

$$\underline{D}\!\int_a^b f = \sup_{\mathcal{P}} \; S_{\underline{D}}(f, \mathcal{P}) = c \, (b - a) = \overline{D}\!\int_a^b f.$$

This means that the constant function $f(x) = c$ is Darboux integrable over $[a, b]$ with $D\!\int_a^b f = c \, (b - a)$.

Example 12 (Dirichlet). The Dirichlet function

$$d(x) = \begin{cases} 1, & x \in \mathbb{Q} \\ 0, & x \notin \mathbb{Q}, \end{cases}$$

where $\mathbb{Q}$ is the set of rational numbers, is not Darboux integrable over any non-degenerate[2] interval $[a, b]$. To see why, suppose that $\mathcal{P} = \{I_k\}$ is a partition of $[a, b]$. Over any non-degenerate subinterval the supremum of d is 1 and the infimum is 0. Thus

$$S_{\overline{D}}(d, \mathcal{P}) = \sum_{\mathcal{P}} \sup_{I_k} d \; \Delta x_k = \sum_{\mathcal{P}} \Delta x_k = b - a$$

and

$$S_{\underline{D}}(d, \mathcal{P}) = \sum_{\mathcal{P}} \inf_{I_k} d \; \Delta x_k = 0.$$

Again, the upper and lower Darboux sums have only one possible value each. Thus

$$\overline{D}\int_a^b d = \inf_{\mathcal{P}} S_{\overline{D}}(d, \mathcal{P}) = (b - a)$$

and

$$\underline{D}\int_a^b d = \sup_{\mathcal{P}} S_{\underline{D}}(d, \mathcal{P}) = 0.$$

The Dirichlet function, $d(x)$, is not Darboux integrable.

Example 13 (Identity function). Let f be the identity function $f(x) = x$ on $[0, 2]$ and take $\mathcal{P}_n = \{I_k\} = \left\{ \left[\frac{2k-2}{n}, \frac{2k}{n} \right] \right\}_{k=1}^n$. Since f is increasing,

$$\inf_{I_k} f = f\left(\frac{2k-2}{n} \right) = \frac{2k-2}{n}$$

and

$$\sup_{I_k} f = f\left(\frac{2k}{n} \right) = \frac{2k}{n}.$$

Thus

$$S_{\underline{D}}(f, \mathcal{P}_n) = \sum_{k=1}^n \frac{2k-2}{n} \frac{2}{n} = 2 - \frac{2}{n}$$

and

$$S_{\overline{D}}(f, \mathcal{P}_n) = \sum_{k=1}^n \frac{2k}{n} \frac{2}{n} = 2 + \frac{2}{n}.$$

---

[2] In this context, non-degenerate means $a < b$ so that the interval consists of more than a single point.

Now, for all n,

$$2 - \frac{2}{n} = S_{\underline{D}}(f, \mathcal{P}_n) \le \sup_{\mathcal{P}} S_{\underline{D}}(f, \mathcal{P}) = {}^{\underline{D}}\!\!\int_0^2 f$$

$$\le {}^{\overline{D}}\!\!\int_0^2 f = \inf_{\mathcal{P}} S_{\overline{D}}(f, \mathcal{P}) \le S_{\overline{D}}(f, \mathcal{P}_n) = 2 + \frac{2}{n}.$$

Hence ${}^{\underline{D}}\!\!\int_0^2 f = {}^{\overline{D}}\!\!\int_0^2 f = 2$ so that f is Darboux integrable over $[0, 2]$ with ${}^{D}\!\!\int_0^2 f = 2$.

Compare what we just did for the Darboux integral to the amount of effort required to prove the same results for the Riemann integral. While the concepts we are working with are a bit more abstract, the proof itself is far more compact.

In example 13, we used a specific partition to show that the upper and lower integrals are the same. We will be well served by an integrability criterion that helps us do this more generally. Cauchy provided a relevant criterion for Riemann integrals that we adapt here for Darboux integrals. Note its connection to a key idea in the previous example.

Theorem 12 (Cauchy criterion for integrability). *A bounded function f on the interval $[a, b]$ is Darboux integrable if and only if for any $\varepsilon > 0$ it is possible to find a partition $\mathcal{P} = \{I_k\}$ for which $S_{\overline{D}}(f, \mathcal{P}) - S_{\underline{D}}(f, \mathcal{P}) < \varepsilon$ or, equivalently, for which $\sum_{\mathcal{P}} \left(\sup_{I_k} f - \inf_{I_k} f \right) \Delta x_k < \varepsilon$.*

Proof. Suppose that a bounded function satisfies the criterion. Let $\varepsilon > 0$ and suppose that $\mathcal{P}_\varepsilon$ is a partition guaranteed by the criterion. Since

$$^{\underline{D}}\!\!\int_a^b f = \sup_{\mathcal{P}} S_{\underline{D}}(f, \mathcal{P}) \ge S_{\underline{D}}(f, \mathcal{P}_\varepsilon)$$

and

$$^{\overline{D}}\!\!\int_a^b f = \inf_{\mathcal{P}} S_{\overline{D}}(f, \mathcal{P}) \le S_{\overline{D}}(f, \mathcal{P}_\varepsilon),$$

we see that

$$0 \le {}^{\overline{D}}\!\!\int_a^b f - {}^{\underline{D}}\!\!\int_a^b f \le S_{\overline{D}}(f, \mathcal{P}_\varepsilon) - S_{\underline{D}}(f, \mathcal{P}_\varepsilon) < \varepsilon.$$

Because the inequality holds for any $\varepsilon > 0$, we can conclude that ${}^{\overline{D}}\!\!\int_a^b f = {}^{\underline{D}}\!\!\int_a^b f$ and that f is Darboux integrable over $[a, b]$.

Conversely, suppose that f is Darboux integrable over $[a, b]$ so that ${}^{\underline{D}}\!\!\int_a^b f = {}^{\overline{D}}\!\!\int_a^b f$. Let $\varepsilon > 0$. Since ${}^{\underline{D}}\!\!\int_a^b f$ is the *least* upper bound of

$\{S_{\underline{D}}(f, P) : P$ a partition of $[a, b]\}$, we know that $\underline{D}\!\int_a^b f - \varepsilon/2$ is not an upper bound. Thus there must be a partition P_1 so that $\underline{D}\!\int_a^b f - \varepsilon/2 < S_{\underline{D}}(f, P_1)$. Similarly, there is a partition P_2 satisfying $\overline{D}\!\int_a^b f + \varepsilon/2 > S_{\overline{D}}(f, P_2)$. Let $Q = P_1 \cup P_2$ and use lemma 13 (below) to observe that

$$S_{\overline{D}}(f, Q) - S_{\underline{D}}(f, Q) \leq S_{\overline{D}}(f, P_2) - S_{\underline{D}}(f, P_1)$$

$$< \left(\overline{D}\!\int_a^b f + \varepsilon/2 \right) - \left(\underline{D}\!\int_a^b f - \varepsilon/2 \right) = \varepsilon$$

as required.   □

Lemma 13 (below) is an analog of lemma 3 (page 31). While the proof that $f(x) = x$ is Darboux integrable did not need the five sets of inequalities used in the Riemann-integrability proof, analogous results still have an important role to play. Lemma 13 is more broadly applicable since the assumption that f is increasing is dropped.

Lemma 13 (Refinements). *Suppose that f is a bounded function on the interval $[a, b]$, that P is a partition of $[a, b]$, and that Q is a refinement of P. Then $S_{\underline{D}}(f, P) \leq S_{\underline{D}}(f, Q) \leq S_{\overline{D}}(f, Q) \leq S_{\overline{D}}(f, P)$.*

Proof. The inner inequality is a consequence of exercise 2.

For the first inequality, begin by assuming that Q is created from P by the addition of a single division point y that falls in the interval $\left[x_{j-1}, x_j \right]$ of P. Since all the other subintervals of P and Q are equal, their corresponding terms will cancel when the lower Darboux sums are subtracted. The remaining terms come from the intervals $\left[x_{j-1}, y \right]$ and $\left[y, x_j \right]$ for Q and $\left[x_{j-1}, x_j \right]$ for P. Since $\inf A \geq \inf B$ whenever $A \subseteq B$ (why?),

$$S_{\underline{D}}(f, Q) - S_{\underline{D}}(f, P) = \left(\inf_{[x_{j-1}, y]} f \cdot \left(y - x_{j-1} \right) + \inf_{[y, x_j]} f \cdot \left(x_j - y \right) \right)$$
$$- \inf_{[x_{j-1}, x_j]} f \cdot \left(x_j - x_{j-1} \right)$$
$$\geq \left(\inf_{[x_{j-1}, x_j]} f \cdot (y - x_{j-1}) + \inf_{[x_{j-1}, x_j]} f \cdot (x_j - y) \right)$$
$$- \inf_{[x_{j-1}, x_j]} f \cdot \left(x_j - x_{j-1} \right)$$
$$= 0.$$

Hence $S_{\underline{D}}(f, P) \leq S_{\underline{D}}(f, Q)$.

The general result for the first inequality now follows by induction. The proof of the right inequality is similar and is left as an exercise. □

We close out this section by addressing the Darboux integrability of increasing and continuous functions.

Theorem 14 (Increasing functions). *If f is an increasing function defined on $[a, b]$ then f is Darboux integrable over $[a, b]$.*

Proof. Let $\varepsilon > 0$ and let $\mathcal{P} = \{[x_{k-1}, x_k]\}_{k=1}^n$ be any partition of $[a, b]$ with $\|\mathcal{P}\| < \frac{\varepsilon}{f(b)-f(a)}$. Since f is increasing, $\inf_{[x_{k-1}, x_k]} f = f(x_{k-1})$ and $\sup_{[x_{k-1}, x_k]} f = f(x_k)$. Hence

$$S_{\overline{D}}(f, \mathcal{P}) - S_{\underline{D}}(f, \mathcal{P}) = \sum_{k=1}^n f(x_k)\, \Delta x_k - \sum_{k=1}^n f(x_{k-1})\, \Delta x_k$$

$$\leq \sum_{k=1}^n (f(x_k) - f(x_{k-1}))\, \|\mathcal{P}\|$$

$$= (f(b) - f(a))\, \|\mathcal{P}\| < \varepsilon.$$

Darboux integrability follows from the Cauchy criterion. □

Theorem 15 (Continuous functions). *Suppose that f is continuous on the interval $[a, b]$. Then f is Darboux integrable over $[a, b]$.*

Proof. Let $\varepsilon > 0$. Since $[a, b]$ is a compact set, f is uniformly continuous on $[a, b]$ and we can find a $\delta > 0$ so that $|f(x) - f(y)| < \frac{\varepsilon}{b-a}$ for all $x, y \in [a, b]$ satisfying $|x - y| < \delta$. Now choose $\mathcal{P} = \{I_k\}$ to be a partition of $[a, b]$ with $\|\mathcal{P}\| < \delta$. Then for all k between 1 and n,

$$\sup_{I_k} f - \inf_{I_k} f \leq \frac{\varepsilon}{b-a}$$

(see exercise 11). Hence

$$S_{\overline{D}}(f, \mathcal{P}) - S_{\underline{D}}(f, \mathcal{P}) = \sum_{\mathcal{P}} \sup_{I_k} f \cdot \Delta x_k - \sum_{\mathcal{P}} \inf_{I_k} f \cdot \Delta x_k$$

$$= \sum_{\mathcal{P}} \left(\sup_{I_k} f - \inf_{I_k} f \right) \cdot \Delta x_k$$

$$\leq \frac{\varepsilon}{b-a} \sum_{\mathcal{P}} \Delta x_k = \varepsilon.$$

We conclude that f is Darboux integrable on $[a, b]$ since f satisfies the Cauchy criterion. □

If you have not already done so, you should take some time to compare the proofs for the Riemann and Darboux integrals. In every case, you will find that by working with the slightly more abstract concepts of infimum and supremum, the proofs are shorter and often easier to understand.

3.2 Comparing Riemann and Darboux integration

Thus far, we have essentially followed the same path for the Riemann and Darboux integrals. The proofs are different, but the results are the same. Here is a tantalizing result for the Darboux integral that has no counterpart for the Riemann integral.

Theorem 16. *Let F be differentiable on $[a, b]$ with a bounded derivative. Then*

$$\underline{D}\int_a^b F' \leq F(b) - F(a) \leq \overline{D}\int_a^b F'.$$

Proof. Exercise 15.   □

This theorem generalizes the fundamental theorem of calculus for the Riemann integral. When F' is Darboux integrable, we conclude that $\underline{D}\int_a^b F' = F(b) - F(a)$. When F' is not Darboux integrable, we know that the upper and lower Darboux integrals straddle $F(b) - F(a)$. There is no analogous result for the Riemann integral since, in the Riemann context, nothing plays a role similar to the upper and lower Darboux integrals.

So, are there any functions that are Darboux integrable but not Riemann integrable? It turns out that the answer is no. But before we establish this fact, we prove a result in the spirit of lemma 2 (page 30).

Lemma 17 (Refinement bounds). *Suppose that f is defined on $[a, b]$ with $|f| < B$. Let $\mathcal{P}$ be a partition of $[a, b]$. If $\mathcal{Q}$ is a refinement of $\mathcal{P}$ created by introducing N additional division points, then*

1. $0 \leq S_{\overline{D}}(f, \mathcal{P}) - S_{\overline{D}}(f, \mathcal{Q}) \leq 2BN \|\mathcal{P}\|$ and

2. $0 \leq S_{\underline{D}}(f, \mathcal{Q}) - S_{\underline{D}}(f, \mathcal{P}) \leq 2BN \|\mathcal{P}\|$.

Proof. Lemma 13 shows that both differences are at least zero.

To verify the right side of the second inequality, consider for the moment a single subinterval $[x_{i-1}, x_i]$ of $\mathcal{P}$. Since $\mathcal{Q} = \{[y_{j-1}, y_j]\}_{j=1}^m$ is a refinement of $\mathcal{P}$, there are values j and k so that $x_{i-1} = y_{j-1} < \cdots < y_k = x_i$.

In other words, $[x_{i-1}, x_i]$ is made up of one or more subintervals of Q. If $j = k$, then no additional division points were added in $[x_{i-1}, x_i]$ and

$$\inf_{[x_{i-1},x_i]} f \cdot (x_i - x_{i-1}) - \inf_{[y_{k-1},y_k]} f \cdot (y_k - y_{k-1}) = 0.$$

Alternatively, if $j < k$, then

$$\left[\inf_{[y_{j-1},y_j]} f \cdot (y_j - y_{j-1}) + \cdots + \inf_{[y_{k-1},y_k]} f \cdot (y_k - y_{k-1}) \right]$$
$$- \inf_{[x_{i-1},x_i]} f \cdot (x_i - x_{i-1})$$
$$= \left(\inf_{[y_{j-1},y_j]} f - \inf_{[x_{i-1},x_i]} f \right) \cdot (y_j - y_{j-1}) + \cdots$$
$$+ \left(\inf_{[y_{k-1},y_k]} f - \inf_{[x_{i-1},x_i]} f \right) \cdot (y_k - y_{k-1})$$
$$\leq 2B \cdot (y_j - y_{j-1}) + \cdots + 2B \cdot (y_k - y_{k-1}) = 2B(x_i - x_{i-1}) \leq 2B \|\mathcal{P}\|.$$

Since N division points were added to make Q, there are at most N intervals from $\mathcal{P}$ for which the difference is nonzero. Thus

$$S_{\underline{D}}(f, Q) - S_{\underline{D}}(f, \mathcal{P}) \leq 2BN \|\mathcal{P}\|.$$

A similar analysis (exercise 16) proves the right-hand side of the first inequality. □

Note the similarities between the preceding proof and the proof of theorem 5 (page 32). In both cases we bound the function differences and telescope the intervals.

Theorem 18 (Riemann-Darboux equivalence). *Let f be defined on $[a, b]$. Then f is Riemann integrable if and only if f is Darboux integrable. Moreover, $^R\!\!\int_a^b f = {}^D\!\!\int_a^b f$. In other words, the Riemann and Darboux integrals are the same.*

Proof. First, suppose that f is Riemann integrable. To prove that f is Darboux integrable, we need to control the upper and lower Darboux sums using tagged partitions.

Let $\varepsilon > 0$ be given and choose $\delta > 0$ so that $\left| {}^R\!\!\int_a^b f - S_R(f, \mathcal{P}) \right| < \varepsilon$ for any tagged partition $\mathcal{P}$ of $[a, b]$ with mesh $\|\mathcal{P}\| < \delta$. Fix a particular partition $\mathcal{P}_0 = \{I_k\}$ with $\|\mathcal{P}_0\| < \delta$ and select tags $\{s_k\}$ and $\{t_k\}$ so that

$$f(s_k) - \varepsilon < \inf_{I_k} f \leq \sup_{I_k} f < f(t_k) + \varepsilon.$$

If we denote the corresponding tagged partitions by $\mathcal{P}_s$ and $\mathcal{P}_t$, then

$$S_{\overline{D}}(f, \mathcal{P}_0) - S_{\underline{D}}(f, \mathcal{P}_0) = \sum_{\mathcal{P}} \sup_{I_k} f \cdot \Delta x_k - \sum_{\mathcal{P}} \inf_{I_k} f \cdot \Delta x_k$$

$$< \sum_{\mathcal{P}} (f(t_k) + \varepsilon) \cdot \Delta x_k - \sum_{\mathcal{P}} (f(s_k) - \varepsilon) \cdot \Delta x_k$$

$$= S_R(f, \mathcal{P}_t) - S_R(f, \mathcal{P}_s) + 2\varepsilon(b - a).$$

Now $S_R(f, \mathcal{P}_t)$ and $S_R(f, \mathcal{P}_s)$ are both within ε of $^R\!\!\int_a^b f$, so they are within 2ε of each other. Thus

$$S_{\overline{D}}(f, \mathcal{P}_0) - S_{\underline{D}}(f, \mathcal{P}_0) < 2\varepsilon + 2\varepsilon(b - a).$$

Since there is a partition satisfying the Cauchy criterion (see exercise 17b), we conclude that f is Darboux integrable.

To prove the converse, suppose that f is Darboux integrable and let $\varepsilon > 0$ be given. Since f is Darboux integrable, f is bounded by some value B and we can find a partition $\mathcal{P}_0$ so that $S_{\overline{D}}(f, \mathcal{P}_0) - S_{\underline{D}}(f, \mathcal{P}_0) < \varepsilon/2$. For future reference, let N be the number of division points in $\mathcal{P}_0$. Suppose that $\mathcal{P}$ is any tagged partition of $[a, b]$ with mesh $\|\mathcal{P}\| < \delta$, where δ will be determined later, and let $\mathcal{Q} = \mathcal{P}_0 \cup \mathcal{P}$. By lemma 13 (page 56),

$$S_{\overline{D}}(f, \mathcal{Q}) - S_{\underline{D}}(f, \mathcal{Q}) \leq S_{\overline{D}}(f, \mathcal{P}_0) - S_{\underline{D}}(f, \mathcal{P}_0) < \varepsilon/2$$

and, since

$$S_{\underline{D}}(f, \mathcal{Q}) \leq {}^D\!\!\int_a^b f \leq S_{\overline{D}}(f, \mathcal{Q}),$$

we can conclude that $\left| {}^D\!\!\int_a^b f - S_{\underline{D}}(f, \mathcal{Q}) \right| < \varepsilon/2$ as well. Then by lemma 17 (page 58),

$$0 \leq {}^D\!\!\int_a^b f - S_{\underline{D}}(f, \mathcal{P})$$

$$\leq \left| {}^D\!\!\int_a^b f - S_{\underline{D}}(f, \mathcal{Q}) \right| + \left| S_{\underline{D}}(f, \mathcal{Q}) - S_{\underline{D}}(f, \mathcal{P}) \right|$$

$$< \varepsilon/2 + 2BN \|\mathcal{P}\|.$$

Taking $\|\mathcal{P}\| < \delta = \frac{\varepsilon}{4BN}$, the last expression is less than ε and we can conclude that

$$^D\!\!\int_a^b f - \varepsilon < S_{\underline{D}}(f, \mathcal{P}).$$

By a similar argument,

$$S_{\overline{D}}(f, \mathcal{P}) < {}^{D}\!\!\int_{a}^{b} f + \varepsilon.$$

Then

$$^{D}\!\!\int_{a}^{b} f - \varepsilon < S_{\underline{D}}(f, \mathcal{P}) \leq S_{R}(f, \mathcal{P}) \leq S_{\overline{D}}(f, \mathcal{P}) < {}^{D}\!\!\int_{a}^{b} f + \varepsilon$$

implies that

$$\left| S_{R}(f, \mathcal{P}) - {}^{D}\!\!\int_{a}^{b} f \right| < \varepsilon.$$

Hence f is Riemann integrable with ${}^{R}\!\!\int_{a}^{b} f = {}^{D}\!\!\int_{a}^{b} f$. □

Since the Riemann and Darboux integrals are the same, what is the advantage of having the two integrals? In fact, since the two integrals are not different, a better question is: What is the advantage to having two approaches to the Riemann-Darboux integral? This question has a fairly straightforward answer. The Riemann approach is conceptually more natural and therefore easier to understand. The Darboux integral has more efficient tools for verifying the properties of the integral.

Lest you fear that the integrals to be introduced later in the text are also different approaches to the same integral, be assured that this is not the case. Later integrals will extend the Riemann-Darboux integral. In addition to employing different approaches, these other integrals will produce a larger set of integrable functions and more robust convergence theorems.

3.3 Additional integrability results

We will complete our investigation of the Riemann-Darboux integral by developing a pair of theorems that provide a precise set of integrability criteria for the Riemann integral. It is strongly recommended that you complete exercise 14 before proceeding. Doing this exercise will prepare you to understand the issues in the proof of theorem 19 (below).

By the Cauchy criterion, if f is integrable then $\sum_{\mathcal{P}}(\sup_{I_k} f - \inf_{I_k} f) \Delta x_k$ must be small for a suitably chosen partition $\mathcal{P}$. The idea in theorem 19 is that for this to occur, $\sup_{I_k} f - \inf_{I_k} f$ can only be "large" on a set of subintervals with very small total length. The proof essentially combines the bound-and-telescope method used for continuous functions with a modification of the technique used for increasing functions. The variables h and l below are meant to suggest respectively the height of the difference $\sup_{I_k} f - \inf_{I_k} f$ and the total length of the subintervals where this difference is "large."

Theorem 19 (Height-width bounds). *Let f be a bounded function on $[a, b]$. Then f is Riemann-Darboux integrable if and only if for any choice of positive values h and l, we can find a partition $\mathcal{P}$ of $[a, b]$ for which the sum of the lengths of the subintervals where $\sup_{I_k} f - \inf_{I_k} f$ exceeds h is less than l.*

Proof. Begin by assuming that f is Darboux integrable and suppose that $h, l > 0$ are given. By the Cauchy criterion, we can find a partition $\mathcal{P} = \{I_k\}$ for which $S_{\overline{D}}(f, \mathcal{P}) - S_{\underline{D}}(f, \mathcal{P}) < hl$. Split $\mathcal{P}$ into two disjoint sets $\mathcal{P}_>$, the intervals $I_k \in \mathcal{P}$ for which $\sup_{I_k} f - \inf_{I_k} f$ exceeds h, and $\mathcal{P}_\le$, the intervals for which $\sup_{I_k} f - \inf_{I_k} f$ is bounded by h. Then

$$
\begin{aligned}
hl &> S_{\overline{D}}(f, \mathcal{P}) - S_{\underline{D}}(f, \mathcal{P}) \\
&= \sum_{\mathcal{P}} \left(\sup_{I_k} f - \inf_{I_k} f \right) \Delta x_k \\
&= \sum_{\mathcal{P}_\le} \left(\sup_{I_k} f - \inf_{I_k} f \right) \Delta x_k + \sum_{\mathcal{P}_>} \left(\sup_{I_k} f - \inf_{I_k} f \right) \Delta x_k \\
&\ge \sum_{\mathcal{P}_>} \left(\sup_{I_k} f - \inf_{I_k} f \right) \Delta x_k \\
&\ge h \sum_{\mathcal{P}_>} \Delta x_k.
\end{aligned}
$$

We conclude that the sum of the lengths of the subintervals where $\sup_{I_k} f - \inf_{I_k} f$ exceeds h is less than l.

For the converse, suppose that f is bounded by B and that, for any choice of positive values h and l, we can find a partition $\mathcal{P} = \{I_k\}$ of $[a, b]$ for which the sum of the lengths of the subintervals for which $\sup_{I_k} f - \inf_{I_k} f$ exceeds h is less than l. Let $\varepsilon > 0$ and take $h = \frac{\varepsilon}{2(b-a)}$ and $l = \frac{\varepsilon}{4B}$. Using the same notation as above,

$$
\begin{aligned}
S_{\overline{D}}(f, \mathcal{P}) - S_{\underline{D}}(f, \mathcal{P}) &= \sum_{\mathcal{P}_\le} \left(\sup_{I_k} f - \inf_{I_k} f \right) \Delta x_k + \sum_{\mathcal{P}_>} \left(\sup_{I_k} f - \inf_{I_k} f \right) \Delta x_k \\
&\le \sum_{\mathcal{P}_\le} h \Delta x_k + \sum_{\mathcal{P}_>} 2B \Delta x_k \\
&\le h(b - a) + 2Bl = \varepsilon.
\end{aligned}
$$

Hence f is Darboux integrable by the Cauchy criterion.   □

What could cause $\sup_{I_k} f - \inf_{I_k} f$ to exceed h? One possibility is that the interval is sufficiently long that f can change gradually from one value to another. This case is not of particular interest since we can easily eliminate this behavior by choosing smaller intervals for the partition. The other possibility is that I_k contains a discontinuity of f. Theorem 19 tells us that in some sense the set of discontinuities cannot be very large. The next definition and theorem make this intuition precise.

Definition 6 (Measure zero). *A set S has **measure zero** if for any $\varepsilon > 0$ we can find a countable set of open intervals $\{I_k\}$ so that*

1. $S \subset \cup_k I_k$ and

2. $\sum_k l\,(I_k) < \varepsilon$

where $l\,(I_k)$ is the length of I_k. In other words, there is a countable cover of S by open intervals of total length less than ε.

Example 14. Any finite set has measure zero.

Suppose that $S = \{a_1, a_2, \ldots, a_n\}$. Let $\varepsilon > 0$ be given and take $I_k = \left(a_k - \frac{\varepsilon}{3n}, a_k + \frac{\varepsilon}{3n}\right)$. Then $S \subset \cup_{k=1}^n I_k$ and $\sum_{k=1}^n l\,(I_k) = \sum_{k=1}^n \frac{2\varepsilon}{3n} = \frac{2}{3}\varepsilon$.

Example 15. The set $S = \left\{\frac{1}{n}\right\}_{n=1}^\infty$ has measure zero.

For any $\varepsilon > 0$, take $I_n = \left(\frac{1}{n} - \frac{\varepsilon}{4\cdot 2^n}, \frac{1}{n} + \frac{\varepsilon}{4\cdot 2^n}\right)$. Then $S \subset \cup_{n=1}^\infty I_n$ and $\sum_{n=1}^\infty l\,(I_n) = \sum_{n=1}^\infty \frac{\varepsilon}{2\cdot 2^n} = \frac{1}{2}\varepsilon$.

Over 30 years after Darboux's integral was introduced to the mathematical world, Henri Lebesgue used the concept of sets of measure zero to provide a precise description of the functions that are Riemann-Darboux integrable.

Theorem 20 (Lebesgue, 1902). *Let f be a bounded function on $[a, b]$. Then f is Darboux integrable over $[a, b]$ if and only if the set of discontinuities of f in $[a, b]$ has measure zero. (Alternatively, we say that f is continuous almost everywhere on $[a, b]$. This is often abbreviated to f is continuous a.e. on $[a, b]$.)*

Proof. Let $\mathcal{D}$ be the set of discontinuities of f in $[a, b]$ and suppose that $x \in \mathcal{D}$. If I is an interval containing x as an interior point of I, then $\sup_I f - \inf_I f > 0$. Thus $\sup_I f - \inf_I f > \frac{1}{n}$ for sufficiently large values of n.

Alternatively, if x is a boundary point between two subintervals, then by considering the interval I formed by taking the union we can find an integer n such that $\sup_I f - \inf_I f > \frac{1}{n}$ for at least one of the subintervals. (See exercise 19.)

Suppose that f is Darboux integrable. Given any $\varepsilon > 0$ and an integer n, theorem 19 tells us that we can find a partition $\mathcal{P}_n$ so that $\sum_{I \in \mathcal{P}_{n,>}} l(I) < \frac{\varepsilon}{2 \cdot 2^n}$ where $\mathcal{P}_{n,>}$ is the set of subintervals I from $\mathcal{P}_n$ for which $\sup_I f - \inf_I f > \frac{1}{n}$. Since the subintervals of $\mathcal{P}_{n,>}$ are closed rather than open, modify them to become open intervals by extending them on each end by $\frac{\varepsilon}{4m \cdot 2^n}$ where m is the number of intervals in $\mathcal{P}_{n,>}$.

Denote the set of expanded open intervals by $\mathcal{P}_{n,>}^*$. Then the union of the closed subintervals in $\mathcal{P}_{n,>}$ is contained in the union of the open intervals in $\mathcal{P}_{n,>}^*$. Moreover, $\sum_{I \in \mathcal{P}_{n,>}^*} l(I) < \frac{\varepsilon}{2^n}$. Together, all the subintervals in $\cup_{n=1}^\infty \mathcal{P}_{n,>}^*$ form a countable set of open intervals whose union contains $\mathcal{D}$. Computing the sum of the lengths of all the open subintervals, we find that

$$\sum_n \sum_{I \in \mathcal{P}_{n,>}^*} l(I) < \sum_{n=1}^\infty \frac{\varepsilon}{2^n} = \varepsilon.$$

Thus $\mathcal{D}$, the set of discontinuities of f in $[a,b]$, has measure zero.

For the converse, suppose that f is bounded by B and that the set $\mathcal{D}$ of discontinuities of f in $[a,b]$ has measure zero. Let $\varepsilon > 0$ and choose a countable set of open intervals $\{I_k\}$ whose union contains $\mathcal{D} \cup \{a,b\}$ and for which $\sum_k l(I_k) < \varepsilon/4B$. Let C be the set of points in (a,b) where f is continuous. Then for any $x \in C$ we can find an open interval J_x containing x and for which $\sup_{J_x} f - \inf_{J_x} f < \frac{\varepsilon}{4(b-a)}$. Since $\{I_k\} \cup \{J_x : x \in C\}$ is an open cover of the compact set $[a,b]$, we can find a finite subcover. Express the subcover as $\{I_k : k \in F_D\} \cup \{J_x : x \in F_C\}$ where F_D and F_C are finite sets of indices. If needed, remove excess intervals until no point of $[a,b]$ falls into more than two intervals of the cover. (See exercise 21.)

Ignore any endpoints of intervals in the finite subcover that fall outside of (a,b) and use the endpoints that fall inside $[a,b]$ as division points of a partition $\mathcal{P}$ of $[a,b]$. We will show that $\mathcal{P}$ fulfills the Cauchy criterion.

Now every subinterval of $\mathcal{P}$ will be contained in the closure of at least one of the intervals in the finite subcover and no point in $[a,b]$ will be in more than two intervals. (See exercise 22.) Hence

$$S_{\overline{D}}(f, \mathcal{P}) - S_{\underline{D}}(f, \mathcal{P})$$

$$= \sum_{I \in \mathcal{P}} \left(\sup_I f - \inf_I f \right) l(I)$$

$$\leq \sum_{k \in F_D} \left(\sup_{I_k} f - \inf_{I_k} f \right) l(I_k) + \sum_{x \in F_C} \left(\sup_{J_x} f - \inf_{J_x} f \right) l(J_x)$$

$$\leq 2B \sum_k l(I_k) + \frac{\varepsilon}{4(b-a)} \sum_{x \in F_C} l(J_x)$$

$$< 2B \frac{\varepsilon}{4B} + \frac{\varepsilon}{4(b-a)} 2(b-a) = \varepsilon.$$

We conclude that f is Darboux integrable by the Cauchy criterion.   □

The power of this theorem is illustrated in the following examples.

Example 16. Any bounded function f with a finite number of discontinuities in $[a, b]$ is Riemann-Darboux integrable over $[a, b]$ since any finite set has measure zero.

Example 17. The function

$$f(x) = \begin{cases} \frac{1}{n}, & x \in \left(\frac{1}{n+1}, \frac{1}{n}\right], n \in \mathbb{N} \\ 0, & \text{otherwise} \end{cases}$$

is Riemann-Darboux integrable over $[0, 1]$ since the set of discontinuities of f is $\{1, \frac{1}{2}, \frac{1}{3}, \frac{1}{4}, \ldots\}$ which is a set of measure zero.

Example 18. The Dirichlet function is not Riemann integrable since it is discontinuous everywhere.

3.4 Exercises

3.0 Darboux integral: filling the gaps

1. Give an example of a bounded function on $[0, 1]$ that has no maximum or minimum value. What are the supremum and infimum of your function over $[0, 1]$? (Your function cannot be continuous.)

2. Assume that f is defined and bounded on $[a, b]$.

   (a) Explain why $\sup_I f$ and $\inf_I f$ exist for any interval $I \subset [a, b]$. (Why are the relevant sets nonempty and bounded?)

(b) Explain why $S_{\underline{D}}(f,\mathcal{P})$ and $S_{\overline{D}}(f,\mathcal{P})$ exist for any partition of $[a,b]$.

(c) Explain why $S_{\underline{D}}(f,\mathcal{P}) \le S_{\overline{D}}(f,\mathcal{P})$ for any partition of $[a,b]$.

(d) Explain why $\inf_{\mathcal{P}} S_{\overline{D}}(f,\mathcal{P})$ and $\sup_{\mathcal{P}} S_{\underline{D}}(f,\mathcal{P})$ are defined in the definition of the Darboux integral.

3. Prove that if $\mathcal{P}$ is any tagged partition of $[a,b]$ then $S_{\underline{D}}(f,\mathcal{P}) \le S_R(f,\mathcal{P}) \le S_{\overline{D}}(f,\mathcal{P})$.

4. Use exercise 3 to prove that if f is both Riemann and Darboux integrable over $[a,b]$, then $^R\!\int_a^b f = {}^D\!\int_a^b f$.

5. Verify the standard integral properties for the Darboux integral.

   (a) **Uniqueness.** The value of the Darboux integral is unique (if it exists).

   (b) **Linearity.** If f and g are Darboux integrable over the interval $[a,b]$ and $c \in \mathbb{R}$, then so are $f+g$ and cf. Moreover, $^D\!\int_a^b (f+g) = {}^D\!\int_a^b f + {}^D\!\int_a^b g$ and $^D\!\int_a^b cf = c\,{}^D\!\int_a^b f$.

   (c) **Monotonicity.** If f and g are Darboux integrable over the interval $[a,b]$ with $f(x) \le g(x)$ for $x \in [a,b]$, then $^D\!\int_a^b f \le {}^D\!\int_a^b g$.

   (d) **Triangle inequality.** If f and $|f|$ are Darboux integrable over $[a,b]$, then $\left| {}^D\!\int_a^b f \right| \le {}^D\!\int_a^b |f|$.

3.1 Darboux integrability: filling the gaps

6. In example 3 (page 28)

   (a) Verify that $\sum_{k=1}^{n} \frac{2k-2}{n} \frac{2}{n} = 2 - \frac{2}{n}$ and $\sum_{k=1}^{n} \frac{2k}{n} \frac{2}{n} = 2 + \frac{2}{n}$.

   (b) Explain why $^D\!\int_0^2 f = {}^{\overline{D}}\!\int_0^2 f = 2$.

7. Verify that
$$S_{\overline{D}}(f,\mathcal{P}) - S_{\underline{D}}(f,\mathcal{P}) = \sum_{\mathcal{P}} \left(\sup_{[x_{i-1},x_i]} f - \inf_{[x_{i-1},x_i]} f \right) \Delta x.$$

8. In the proof in lemma 13 (page 56)

   (a) Why is $\inf A \ge \inf B$ whenever $A \subseteq B$?

   (b) Use induction on the number of inserted division points to complete the proof that $S_{\underline{D}}(f,\mathcal{P}) \le S_{\underline{D}}(f,\mathcal{Q})$.

   (c) Prove that $S_{\overline{D}}(f,\mathcal{Q}) \le S_{\overline{D}}(f,\mathcal{P})$.

9. Use lemma 13 (page 56) to explain why $^D\!\int_a^b f \le {}^{\overline{D}}\!\int_a^b f$.

10. Suppose that f is an increasing function on $[a,b]$. Verify that $f(b)$ satisfies the definition of $\sup_{[a,b]} f$.

11. Assume that $|f(x) - f(y)| < B$ for all $x, y \in S$.

    (a) Prove that $\sup_S f - \inf_S f \leq B$. (Use contraposition.)
    (b) Give an example of a function for which $\sup_S f - \inf_S f = B$.

3.1 Darboux integrability: deeper reflections

12. Suppose f is Darboux integrable over $[a, b]$ and $[b, c]$. Prove that f is Darboux integrable over $[a, c]$ and $^D\!\!\int_a^c f = {}^D\!\!\int_a^b f + {}^D\!\!\int_b^c f$. (First show that f is Darboux integrable over $[a, c]$ using the Cauchy criterion. Then verify the value of $^D\!\!\int_a^c f$ by showing that $\left| {}^D\!\!\int_a^b f + {}^D\!\!\int_b^c f - {}^D\!\!\int_a^c f \right| < \varepsilon$ for any $\varepsilon > 0$.)

13. Suppose f is Darboux integrable over $[a, b]$ and $a < c < b$. Prove that f is Darboux integrable over $[a, c]$ and $[c, b]$. (Use the Cauchy criterion.)

14. Using only the Cauchy criterion and theorem 15 (page 57), prove that if f is bounded with a finite number of discontinuities on $[a, b]$, then f is Darboux integrable over $[a, b]$. (Use the boundedness of f to constrain the difference between upper and lower sums near points of discontinuity. Use the integrability of f on intervals of continuity to control the difference between the upper and lower sums elsewhere.)

3.2 Relationship between integrals: filling the gaps

15. Let F be differentiable on $[a, b]$. Prove that $\underline{D}\int_a^b F' \leq F(b) - F(a) \leq \overline{D}\int_a^b F'$. (Use exercise 3 and the ideas in the proof of theorem 9 on page 39.)

16. Complete the proof of lemma 17 by proving that $0 \leq S_{\overline{D}}(f, \mathcal{P}) - S_{\overline{D}}(f, \mathcal{Q}) \leq 2BN \, \|\mathcal{P}\|$.

17. In the proof of theorem 18 (page 59)

    (a) Use the definitions of supremum and infimum to explain why it is possible to select tags $\{s_k\}$ and $\{t_k\}$ so that $f(s_k) - \varepsilon < \inf_{[x_{k-1}, x_k]} f \leq \sup_{[x_{k-1}, x_k]} f < f(t_k) + \varepsilon$.
    (b) The proof shows that for any $\varepsilon > 0$ we can find a partition $\mathcal{P}_0$ for which $S_{\overline{D}}(f, \mathcal{P}_0) - S_{\underline{D}}(f, \mathcal{P}_0) < 2\varepsilon + 2\varepsilon(b-a)$. Explain why this is sufficient. Why do you suppose that a proof was not be adjusted to make the expression come out as $S_{\overline{D}}(f, \mathcal{P}_0) - S_{\underline{D}}(f, \mathcal{P}_0) < \varepsilon$?
    (c) Supply the details to prove that $0 \leq S_{\overline{D}}(f, \mathcal{P}) - {}^D\!\!\int_a^b f < \varepsilon$.
    (d) When proving the converse, we selected an arbitrary partition $\mathcal{P}$ with mesh $\|\mathcal{P}\| < \delta$ where $\delta > 0$ was to be selected later. What are the pros and cons of using this type of exposition?

(e) Supply the details to prove that $S_{\overline{D}}(f, \mathcal{P}) < {}^{D}\!\!\int_{a}^{b} f + \varepsilon$.

(f) We never explicitly proved that ${}^{D}\!\!\int_{a}^{b} f = {}^{R}\!\!\int_{a}^{b} f$ when f is Riemann integrable. Explain why this is not a problem.

3.3 Additional integrability results: filling the gaps

18. In the first part of the proof of theorem 19, under what circumstances could $\sum_{\mathcal{P}_{>}} \left(\sup_{I_k} f - \inf_{I_k} f \right) \Delta x_k$ and $\sum_{\mathcal{P}_{>}} h \Delta x_k$ be equal?

19. Suppose that f is discontinuous at x.

(a) Prove that $\sup_I f - \inf_I f > 0$ for any interval I containing x in its interior. (Use contraposition.)

(b) Prove that, for any nondegenerate intervals of the form $I_1 = [\alpha, x]$ and $I_2 = [x, \beta]$, either $\sup_{I_1} f - \inf_{I_1} f > 0$ or $\sup_{I_2} f - \inf_{I_2} f > 0$ (or both). (Consider $I_1 \cup I_2$.)

20. Suppose that $f : [a, b] \to \mathbb{R}$ is continuous at $x \in (a, b)$. Explain why there is an interval J_x with $x \in J_x \subseteq (a, b)$ such that $\sup_{J_x} f - \inf_{J_x} f < \frac{\varepsilon}{2(b-a)}$.

21. Let $\mathcal{C}$ be a finite cover of $[a, b]$ by open intervals.

(a) Suppose that $(x_1, y_1), (x_2, y_2), (x_3, y_3) \in \mathcal{C}$ and that $z \in (x_1, y_1) \cap (x_2, y_2) \cap (x_3, y_3)$. Prove that one of the three intervals is not needed in the cover of $[a, b]$.

(b) Explain why there must be a subset of intervals from $\mathcal{C}$ that is still a cover of $[a, b]$ but for which no point in $[a, b]$ lies in more than two intervals.

22. The set $\mathcal{C} = \{(-0.1, 0.3), (0.2, 0.9), (0.5, 0.8), (0.75, 1.1)\}$ is an open cover of $[0, 1]$.

(a) What is the partition $\mathcal{P}$ that would be generated from this cover in the proof of theorem 20 (page 63)?

(b) For each subinterval I of $\mathcal{P}$ find an interval J from $\mathcal{C}$ so that I is contained in the closure of J.

23. In the proof of theorem 20 (page 63), why is $\sum_{x \in F_c} l(J_x) \leq 2(b - a)$?

3.3 Additional integrability results: deeper reflections

24. Prove that any countable set has measure zero.

25. Prove that the union of two sets of measure zero has measure zero.

26. Using the two previous exercises, prove that the set of irrational numbers in $[0, 1]$ does not have measure zero.

27. Prove that if A is a set of measure zero and $B \subset A$, then B has measure zero.

28. Give an alternative proof of the fact that any function f that is increasing on $[a, b]$ is Darboux integrable over $[a, b]$. (Use the following ideas.)

    (a) Explain why f can only have jump discontinuities. In other words, $\lim_{x \to d^-} f(x) < \lim_{x \to d^-} f(x)$ at any point d of discontinuity.
    (b) Explain why f can only have a countable number of discontinuities.

29. Let
$$f(x) = \begin{cases} \frac{1}{2^n}, & x = \frac{k}{2^n} \text{ with } k \text{ odd, } n \in \mathbb{N} \\ 0, & \text{otherwise.} \end{cases}$$

    Give four proofs that f is Riemann-Darboux integrable over $[0, 1]$.

    (a) Use the definition of the Riemann integral. Suppose that $\mathcal{P}$ is a tagged partition with mesh $\|\mathcal{P}\| \leq \frac{1}{2^N}$. At most, how many times can a tag produce the value $\frac{1}{2^N}$? At most, what is the value of the term $f(t_k) \Delta x_k$ associated with each subinterval in the partition? Explain why changing the tags or using smaller subintervals can only decrease the Riemann sum. (This is a hard exercise.)
    (b) Use the Cauchy criterion. Choose a particular partition $\mathcal{P}_n$ and bound $S_{\overline{D}}(f, \mathcal{P}_n) - S_{\underline{D}}(f, \mathcal{P}_n)$ above by $\frac{n^2}{2^n} + \frac{1}{n}$ or by some similar bound that goes to zero as $n \to \infty$. Explain why this guarantees integrability.
    (c) Use theorem 20 (page 63). Prove that f is continuous at every irrational number.
    (d) Use theorem 11 (page 42).

30. Prove that
$$r(x) = \begin{cases} \frac{1}{n}, & x = \frac{m}{n} \text{ with } \frac{m}{n} \text{ in lowest terms,} \\ 0, & \text{otherwise,} \end{cases}$$

    is continuous except on a set of measure zero. (Use theorems 11 and 20 on pages 42 and 63.)

31. We did not revisit the fundamental theorems for the Darboux integral since they the same as for the Riemann integral. However, FTC-1 can be extended a bit from the statement given on page 39. Prove the following generalization: Suppose that F is differentiable on $[a, b]$ and that F' is bounded and continuous on $[a, b]$ except on a set of measure zero. Then $D\!\int_a^x F' = F(x) - F(a)$ for all $x \in [a, b]$.

3.4 Related ideas: deeper reflections

32. Use theorem 19 (page 62) to prove the converse of exercise 38 from Chapter 2: If f is Cauchy integrable over $[a, b]$ then f is Riemann-Darboux integrable over $[a, b]$. (Blend the following ideas into a well-written proof.)

   (a) Given $\delta, h > 0$, explain how to modify any partition $\mathcal{P} = \{[x_{i-1}, x_i]\}_{i=1}^{n}$ with $x_i - x_{i-1} = \Delta x = \frac{b-a}{n} < \delta/2$ to make partitions $\mathcal{P}_l = \{[s_{i-1}, s_i]\}_{i=1}^{n}$ and $\mathcal{P}_u = \{[t_{i-1}, t_i]\}_{i=1}^{n}$ with $\|\mathcal{P}_l\|, \|\mathcal{P}_u\| < \delta$ and such that, for at least half of the intervals for which $\sup_{[x_{k-1}, x_k]} f - \inf_{[x_{k-1}, x_k]} f$ exceeds h,

   i. $f(s_k) \approx \inf_{[x_{k-1}, x_k]} f$,
   ii. $f(t_k) \approx \sup_{[x_{k-1}, x_k]} f$,
   iii. $s_{k+1} = t_{k+1} = x_{k+1}$, and
   iv. both $s_{k+1} - s_k = \Delta s_k$ and $t_{k+1} - t_k = \Delta t_k$ are at least Δx.

   (b) Assume that f is not Riemann-Darboux integrable. Use h and l to select ε such that $\|\mathcal{P}_l\|, \|\mathcal{P}_u\| < \delta$ while the Cauchy sums differ by more than ε.

3.5 References

Bressoud, David M. (2006). *A Radical Approach to Real Analysis* (2nd ed.). Mathematical Association of America.

Burk, F.E. (2007). *A Garden of Integrals*. Mathematical Association of America.

Darboux, G. (1875). Mémoire sur les functions discontinues. *Ann. Sci. Ecole Normale Superieure* **4** (2): 57–112.

DePree, J. and C. Swartz. (1988). *Introduction to Real Analysis*. John Wiley & Sons.

Gelbaum, B.R. and J.M.H. Olmsted (2003). *Counter examples in Analysis*. Dover.

Lay, S. (2004). *Analysis with an Introduction to Proof* (4th ed.). Prentice Hall.

CHAPTER 4

A Functional zoo

As has been noted previously, the study of integration since the time of Newton and Leibniz has been driven by a recognition that there are many functions that are not as well behaved as polynomials. This chapter introduces you to a set of such functions. Even in the absence of questions of integration, this collection of functions illustrates the wide variety of somewhat unexpected behaviors that functions can exhibit.

4.1 Dirichlet and friends

We begin with functions related to the Dirichlet function. As a reminder, the Dirichlet function is defined as

$$d\,(x) = \begin{cases} 1, & x \in \mathbb{Q} \\ 0, & x \notin \mathbb{Q}. \end{cases}$$

The Dirichlet function is bounded and discontinuous everywhere. In point of fact, $d\,(x)$ probably should not be called *the* Dirichlet function since Dirichlet actually defined a class of related functions which take on different values for rational and irrational inputs. However, the 0–1 function has become such a standard that it has garnered the title of "The Dirichlet Function."

There are two closely related functions. The first is sometimes called the snowflake function though one can also find instances where it is referred to as the Dirichlet function. The snowflake function is defined by

$$s\,(x) = \begin{cases} \frac{1}{n}, & x = \frac{m}{n} \text{ with } \frac{m}{n} \text{ in lowest terms} \\ 0, & \text{otherwise.} \end{cases}$$

The snowflake function is continuous at every irrational number and discontinuous at every rational number. (See exercise 1.) This provides us with an example of a non-monotone, bounded function with a dense set of discontinuities. Since its set of discontinuities has measure zero, s is Riemann-Darboux integrable.

If we take the reciprocal of every non-zero value of s, we obtain the function

$$r(x) = \begin{cases} n, & x = \frac{m}{n} \text{ with } \frac{m}{n} \text{ in lowest terms} \\ 0, & \text{otherwise.} \end{cases}$$

The function r, like the Dirichlet function, is discontinuous everywhere. Moreover, r is unbounded on every nondegenerate[1] interval.

4.2 Trigonometric series

Since questions about trigonometric series triggered much of the mathematical research on integration, it is fitting that we include two examples of trigonometric series.

4.2.1 Fourier's function

Suppose that you want to model the steady state temperature of the infinite lamina bounded above and below by $x = 0$ and $x = 1$ and on the left by $w = 0$. (Here we follow Fourier's convention of using z for the temperature, w for the horizontal axis, and x for the vertical axis.) The top and bottom boundary are held at a constant temperature of 0 and the temperature on the left boundary is described by the function $f(x)$. Assuming no loss of heat from the face of the lamina, the lamina's temperature is modeled by the initial value problem

$$\frac{\partial^2 z}{\partial w^2} + \frac{\partial^2 z}{\partial x^2} = 0,$$
$$z(w,0) = z(w,1) = 0, \quad w > 0,$$
$$z(0,x) = f(x).$$

Joseph Fourier proposed a method for solving this type of differential equation.[2] Begin by noting that functions of the form $z(w,x) = e^{-cw} \sin cx$

---

[1] By nondegenerate we mean that the interval has non-empty interior. The interval $[1, 1]$ is an example of a degenerate interval.

[2] We have modified Fourier's heat problem slightly to make the computations simpler. Fourier's original problem used upper and lower boundaries of $x = -1$ and $x = 1$. To satisfy the original boundary conditions, Fourier used terms of the form $e^{-(2n-1)\frac{\pi w}{2}} \cos(2n-1)\frac{\pi x}{2}$ from which the series $f(x) = \sum_{n=1}^{\infty} \frac{(-1)^n 4}{(2n-1)\pi} \cos\left((2n-1)\frac{\pi x}{2}\right)$ was derived.

satisfy the differential equation $\frac{\partial^2 z}{\partial w^2} + \frac{\partial^2 z}{\partial x^2} = 0$. If we restrict our attention to the case where c has the form $c = n\pi$, then these functions also satisfy $z(w, 0) = z(w, 1) = 0$, $w > 0$. Fourier's idea is that a solution to a particular initial value problem can be generated by taking infinite linear combinations of such functions so that

$$z(w, x) = \sum_{n=1}^{\infty} a_n e^{-n\pi w} \sin(n\pi x)$$

where the coefficients a_n are chosen so that $\sum_{n=1}^{\infty} a_n \sin(n\pi x) = z(0, x) = f(x)$. Similar ideas had been suggested previously, but the process remained theoretical until Fourier provided a method for computing the sequence $\{a_n\}$ of coefficients. Fourier's approach involved what we would now call inner product techniques.

Recall that for integers n and m,

$$\int_0^1 \sin(n\pi x) \sin(m\pi x) \, dx = \begin{cases} 1/2, & n = m \\ 0, & n \neq m. \end{cases}$$

Suppose that $f(x) = \sum_{n=1}^{\infty} a_n \sin(\pi n x)$. Then

$$\int_0^1 f(x) \sin(m\pi x) \, dx = \int_0^1 \left(\sum_{n=1}^{\infty} a_n \sin(n\pi x) \right) \sin(m\pi x) \, dx$$

$$= \sum_{n=1}^{\infty} a_n \int_0^1 \sin(n\pi x) \sin(m\pi x) \, dx$$

$$= a_1 \cdot 0 + a_2 \cdot 0 + \cdots + a_{m-1} \cdot 0$$

$$+ a_m \cdot \frac{1}{2} + a_{m+1} \cdot 0 + \cdots$$

$$= a_m/2.$$

Thus the equation

$$a_n = 2 \int_0^1 f(x) \sin(n\pi x) \, dx$$

provides a method to compute the coefficients we need to generate our solution

$$z(w, x) = \sum_{n=1}^{\infty} a_n e^{-n\pi w} \sin(n\pi x).$$

When the left boundary is held at a constant temperature of $f(x) = 1$, we find that $a_n = 0$ when n is even and that $a_n = \frac{4}{n\pi}$ when n is odd. Thus

$$f(x) = \sum_{n=1}^{\infty} \frac{4}{(2n-1)\pi} \sin((2n-1)\pi x)$$

and the steady state temperature of the lamina is given by

$$z(w,x) = \sum_{n=1}^{\infty} \frac{4}{(2n-1)\pi} e^{-(2n-1)\pi w} \sin((2n-1)\pi x).$$

The practiced eye will recognize multiple issues with the work we have just done.

1. While our physical experience tells us that heat distribution solutions should exist, how do we know that our mathematical model has a solution? While unproven, the assumption that a mathematical solution exists at least seems reasonable.

2. If the mathematical problem does have a solution, by what right do we expect the solution to take the form $\sum_{n=1}^{\infty} a_n e^{-n\pi w} \sin(n\pi x)$? More generally, why should we be able to express the solution as a sum of functions that factor so that each part depends on only one of the variables? Moreover, how do we know that integer values for c in $e^{-cw} \sin cx$ are sufficient? How do we know that we do not also need to use functions like $e^{-cw} \cos cx$ or some other types of functions? Unlike the assumption of the existence of a mathematical solution, the expectation that f can be expressed as $f(x) = \sum_{n=1}^{\infty} a_n \sin(\pi n x)$ smacks of wishful thinking.

3. What justifies interchanging the infinite summation and the integration? We have seen multiple examples that tell us that interchanging limits and integrals can produce erroneous results.

4. Even if we can legitimately interchange the series and the integral to compute the coefficients, how do we know that, except for some obvious special cases such as $x = 0$, the series even converges? It seems reasonable that when the series converges, it should converge to $f(x)$, but how do we know?

Such are the questions that drove the study of real analysis in general and integration in particular. We will not address these questions here, but some related ideas are investigated in the exercises and in the final chapter. (See for example exercise 9.)

If $f(x) = \sum_{n=1}^{\infty} \frac{4}{(2n-1)\pi} \sin\left((2n-1)\pi x\right)$ evaluates to 1 on the interval $(0, 1)$ and evaluates to 0 at the endpoints, then f must also satisfy

$$f(x) = \begin{cases} 1, & x \in (-2, -1) \cup (0, 1) \\ 0, & x = -2, -1, 0, 1, 2 \\ -1, & x \in (-1, 0) \cup (1, 2). \end{cases}$$

This type of behavior in a function was quite unexpected at the time that Fourier did his work. Functions were expected to have smooth graphs like those of polynomials or rational functions. Lagrange thought that perhaps this "un-function-like" behavior was a consequence of the fact that the series does not converge for all values of x and therefore does not determine a true function. This line of thinking is given credence by noticing that the coefficients are multiples of the odd terms of the diverging harmonic series. In fact, Lagrange's conjecture was wrong: the series converges for all values of x. But that is a story for another time and place.

The issues that arose when studying the heat equations are not confined to that problem. They also appear in the study of mathematical models of other physical phenomena such as the shape of a vibrating string. (See exercise 10.)

4.2.2 Weierstrass's function

As a second illustration of a trigonometric series function with unexpected behavior we consider a function related to an 1872 example of Karl Weierstrass. Let

$$g(x) = \sum_{n=1}^{\infty} \frac{1}{2^n} \cos\left(c^n \pi x\right)$$

where c is an integer. This trigonometric series converges uniformly to a continuous function since the individual terms are bounded by $\frac{1}{2^n}$. When $c > 2$, the series

$$-\sum_{n=1}^{\infty} \left(\frac{c}{2}\right)^n \pi \sin\left(c^n \pi x\right)$$

corresponding to term-by-term differentiation converges for all values of x of the form $x = \frac{k\pi}{c^m}$ and diverges for all x of the form $x = \frac{(k+1/2)\pi}{c^m}$. It seems reasonable to conjecture that g is differentiable where $x = \frac{k\pi}{c^m}$ and not differentiable where $x = \frac{(k+1/2)\pi}{c^m}$. Other values of x seem more problematic.

In fact, when c is an odd integer greater than 7, g is nowhere differentiable. To show that g is not differentiable at x_0 we will prove that given any $\delta > 0$, we can find a value x_1 satisfying $|x_1 - x_0| < \delta$ for which $\left|\frac{g(x_1)-g(x_0)}{x_1-x_0}\right|$ is larger than any specified bound.

To that end, select $m \in \mathbb{N}$ such that $\frac{3}{2c^m} < \delta$. Then we can find an integer N so that

$$1 \le |c^m x_0 - N| \le \tfrac{3}{2}.$$

In this case, $\cos\left(\pi\left(c^m x_0 - N\right)\right) \le 0$. Moreover, if we choose x_1 so that $c^m x_1 = N$, then $|x_1 - x_0| \le \frac{3}{2c^m}$ and

$$\cos\left(c^m \pi x_1\right) - \cos\left(c^m \pi x_0\right) = \cos \pi N - \cos\left(\pi N + \pi\left(c^m x_0 - N\right)\right)$$
$$= (-1)^N \left[1 - \cos\left(\pi\left(c^m x_0 - N\right)\right)\right]$$

with $1 - \cos\left(\pi\left(c^m x_0 - N\right)\right) \ge 1$.

Now suppose that n is an integer greater than m. Because we are assuming c is an odd integer,

$$\cos\left(c^n \pi x_1\right) - \cos\left(c^n \pi x_0\right)$$
$$= \cos c^{n-m} \pi N - \cos\left(c^{n-m} \pi N + c^{n-m} \pi\left(c^m x_0 - N\right)\right)$$
$$= (-1)^N \left[1 - \cos\left(c^{n-m} \pi\left(c^m x_0 - N\right)\right)\right]$$

with $1 - \cos\left(c^{n-m} \pi\left(c^m x_0 - N\right)\right) \ge 0$. Thus all the terms in

$$\sum_{n=m}^{\infty} \frac{\frac{1}{2^n}\cos\left(c^n \pi x_1\right) - \frac{1}{2^n}\cos\left(c^n \pi x_0\right)}{x_1 - x_0}$$

have the same sign. Since $\left|\cos\left(c^m \pi x_1\right) - \cos\left(c^m \pi x_0\right)\right| \ge 1$ and $|x_1 - x_0| \le \frac{3}{2c^m}$, we see that

$$\left|\sum_{n=m}^{\infty} \frac{1}{2^n}\frac{\cos\left(c^n \pi x_1\right) - \cos\left(c^n \pi x_0\right)}{x_1 - x_0}\right| \ge \frac{2}{2^m}\frac{1}{|x_1 - x_0|} \ge \frac{4}{3}\left(\frac{c}{2}\right)^m.$$

For n less than m, the mean value theorem guarantees an x_2 between x_1 and x_0 for which

$$\left|\frac{\cos\left(c^n \pi x_1\right) - \pi \cos\left(c^n \pi x_0\right)}{x_1 - x_0}\right| = \left|-\pi c^n \sin \pi c^n x_2\right| \le \pi c^n.$$

Thus

$$\left|\sum_{n=0}^{m-1} \frac{1}{2^n}\frac{\cos\left(c^n \pi x_1\right) - \cos\left(c^n \pi x_0\right)}{x_1 - x_0}\right|$$
$$\le \sum_{n=0}^{m-1} \pi\left(\frac{c}{2}\right)^n = \pi\frac{\left(\frac{c}{2}\right)^m - 1}{\frac{c}{2} - 1} < \pi\frac{\left(\frac{c}{2}\right)^m}{\frac{c}{2} - 1}.$$

Putting the two halves together,

$$\left| \sum_{n=0}^{\infty} \frac{\frac{1}{2^n} \cos\left(c^n \pi x_1\right) - \frac{1}{2^n} \cos\left(c^n \pi x_0\right)}{x_1 - x_0} \right| \geq \frac{4}{3}\left(\frac{c}{2}\right)^m - \pi \frac{\left(\frac{c}{2}\right)^m}{\frac{c}{2} - 1}$$

$$= \left(\frac{c}{2}\right)^m \left[\frac{4}{3} - \frac{2\pi}{c - 2}\right].$$

As long as $c > 7$, we note that $\frac{4}{3} - \frac{2\pi}{c-2} > 0$ and $\left(\frac{c}{2}\right)^m$ can be made arbitrarily large. Thus $\left|\frac{g(x_1) - g(x_0)}{x_1 - x_0}\right|$ is unbounded on any neighborhood of x_0. Consequently, g cannot be differentiable at x_0.

4.3 Friends of Cantor

Most real analysis texts include Georg Cantor's proof that the real numbers are uncountable. In keeping with one of the recurring themes of this text, Cantor's work had its origins in the study of convergence sets of trigonometric functions. Our purpose here is to introduce some additional results of this investigation: the Cantor set, its relatives, and some associated functions.

4.3.1 The basics

The Cantor set is generated by starting with the unit interval $C_0 = I_0 = [0, 1]$ and successively removing the middle thirds of all of the intervals in the previous stage.

$C_0 = I_{0,1} = [0, 1]$
$C_1 = I_{1,1} \cup I_{1,2} = \left[\frac{0}{3}, \frac{1}{3}\right] \cup \left[\frac{2}{3}, \frac{3}{3}\right]$
$C_2 = I_{2,1} \cup I_{2,2} \cup I_{2,3} \cup I_{2,4} = \left[\frac{0}{9}, \frac{1}{9}\right] \cup \left[\frac{2}{9}, \frac{3}{9}\right] \cup \left[\frac{6}{9}, \frac{7}{9}\right] \cup \left[\frac{8}{9}, \frac{9}{9}\right]$

$\cdots$

$C_n = I_{n,1} \cup I_{n,2} \cup \cdots \cup I_{n,2^n} = \left[\frac{0}{3^n}, \frac{1}{3^n}\right] \cup \left[\frac{2}{3^n}, \frac{3}{3^n}\right] \cup \cdots \cup \left[\frac{3^n - 1}{3^n}, \frac{3^n}{3^n}\right]$

$\cdots$

The intermediate sets C_n each consist of 2^n closed intervals of length $\frac{1}{3^n}$. (See Figure 4.1.) The Cantor set is the intersection of all these sets: $C = \cap_n C_n$.

The Cantor set has many surprising properties.

1. **The Cantor set is closed** as the intersection of closed sets. (OK—not so surprising.)

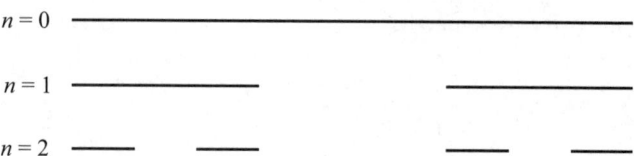

Figure 4.1. The first three steps in constructing the Cantor set

2. **The Cantor set is nowhere dense.** In other words the closure of C (in this case, C itself) contains no intervals. To see why, suppose that $x, y \in C$. Then there is an n large enough that $\frac{1}{3^n} < |x - y|$. Since x and y cannot both belong to the same interval in C_n, there must be an open interval I in the complement of C_n located between x and y. As the complement of C contains the complement of C_n, I is also in the complement of C. Thus C cannot contain the interval (x, y).

3. **Every point $x \in C$ is a limit of a sequence of points in the complement of C and of a sequence of points from $C \setminus \{x\}$.** If $x \in C$, then $x \in C_n$ for every natural number n. The subinterval of C_n to which x belongs will have its middle third removed at the next stage. The midpoint of the removed interval is in the complement of C and both endpoints of the removed interval belong to C. Thus for every n, we can chose $a_n \notin C$ and $b_n \in C \setminus \{x\}$ so that both $|a_n - x|$ and $|b_n - x|$ are less than $\frac{1}{3^n}$.

4. **The Cantor set is uncountable.** (See exercise 16.)

5. **The Cantor set has measure zero.** Since C_n consists of 2^n (closed) intervals each of length $\frac{1}{3^n}$, we can cover C_n with open intervals of total length at most $2 \cdot \frac{2^n}{3^n}$. As C is contained in C_n for every n, we can cover C with a finite set of open intervals of arbitrarily small total length.

The Cantor set can also be defined as the set of numbers in $[0, 1]$ that can be expressed in ternary notation using only 0 and 2. (See exercise 15.)

4.3.2 Variations

Suppose that instead of removing intervals of length $\frac{1}{3^n}$ at the nth stage, intervals of length α^n are removed. Then in the limit a total length of $\sum_{n=0}^{\infty} 2^n \alpha^{n+1} = \frac{\alpha}{1-2\alpha}$ will be removed from the unit interval. The remaining Cantor-like set will have all of the same properties as the basic

Cantor set but, rather than being a set of measure zero, will have measure[3] $1 - \frac{\alpha}{1-2\alpha} = \frac{1-3\alpha}{1-2\alpha}$. By suitably choosing $0 < \alpha \le \frac{1}{3}$, we can arrange for the Cantor-like set to have any measure between zero and 1.

4.3.3 The Cantor function

As with the Cantor set, the Cantor function is built up iteratively. Begin by defining f_0 by setting $f_0(0) = 0$, $f_0(1) = 1$ and using linear interpolation in between. In this base case, we have taken a somewhat circuitous route to defining $f_0(x) = x$ on $[0, 1] = C_0$.

At each stage in the construction of the Cantor set, C_n consists of 2^n intervals from which the middle third will be removed to make the next set C_{n+1}. To move from f_n to f_{n+1}, we set $f_{n+1} = f_n$ on the complement of C_n and modify the function on each of the subintervals of C_n. Consider one of the 2^n subintervals, say $I_{n,k} = \left[\frac{k}{3^n}, \frac{k+1}{3^n}\right]$, from C_n. The function f_{n+1} is defined to agree with f_n at the endpoints of $I_{n,k}$. On the interval to be removed, f_{n+1} is assigned the average of the values at the two endpoints. Linear interpolation is used on the remaining subintervals. We illustrate the process by constructing f_1 and f_2.

Following the construction process, we begin with the endpoints: $f_1(0) = f_0(0) = 0$ and $f_1(1) = f_0(1) = 1$. The average of the values 0 and 1 is $\frac{1}{2}$ so $f_1(x) = \frac{1}{2}$ for $x \in \left[\frac{1}{3}, \frac{2}{3}\right]$. On $\left[\frac{0}{3}, \frac{1}{3}\right]$, f_1 interpolates between 0 and $\frac{1}{2}$ and on $\left[\frac{2}{3}, \frac{3}{3}\right]$, f_1 interpolates between $\frac{1}{2}$ and 1.

To create f_2, we set $f_2 = f_1$ on $\left[\frac{1}{3}, \frac{2}{3}\right]$, the complement of C_1, and modify f_1 on the subintervals $\left[\frac{0}{3}, \frac{1}{3}\right]$ and $\left[\frac{2}{3}, \frac{3}{3}\right]$. Make f_2 constant ($\frac{1}{4}$ and $\frac{3}{4}$) on the removed subintervals $\left[\frac{1}{9}, \frac{2}{9}\right]$ and $\left[\frac{7}{9}, \frac{8}{9}\right]$ and interpolate on the remaining intervals of C_1: $\left[\frac{0}{9}, \frac{1}{9}\right]$, $\left[\frac{2}{9}, \frac{3}{9}\right]$, $\left[\frac{6}{9}, \frac{7}{9}\right]$, and $\left[\frac{8}{9}, \frac{9}{9}\right]$.

Another way to think about the process is that f_{n+1} agrees with f_n on the intervals where f_n is constant and f_{n+1} is made constant on the middle third of each interval where f_n is strictly increasing. The first three iterations of the process are displayed in Figure 4.2.

When x is in the complement of C, the sequence $\{f_n(x)\}$ is constant from some point on so $\{f_n(x)\}$ converges. The vertical distance between successive plateaus of f_n is $\frac{1}{2^n}$. These facts can be used to show that $\{f_n\}$ converges uniformly on all of $[0, 1]$. (See exercise 17.) The function to which $\{f_n\}$ converges is called the Cantor function which we denote by c.

---

[3] The concept of measure will be more fully developed in the next chapter.

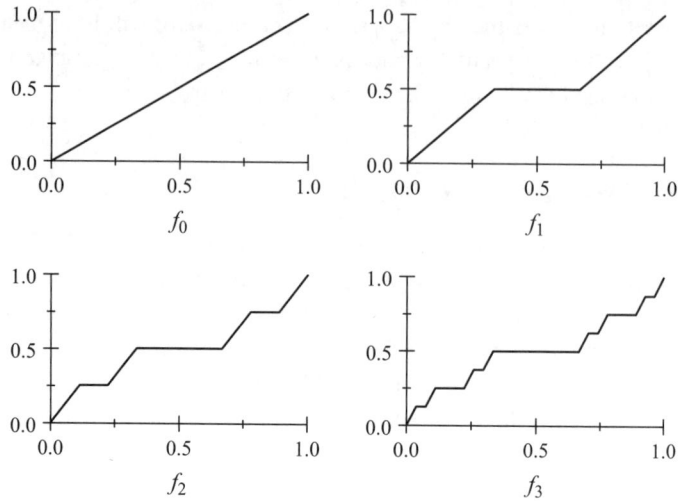

Figure 4.2. Building Cantor's function

Back in Chapter 2 we saw that

$$F(x) = \begin{cases} 0, & 0 \le x < 1/2 \\ 1, & 1/2 \le x \le 1 \end{cases}$$

is differentiable on $[0, 1]$ except at $x = 1/2$. Since (re)defining a function at a single point will not change its Riemann integral, we can make sense of ${}^{R}\!\!\int_0^x F'$ but ${}^{R}\!\!\int_0^x F' \ne F(x) - F(0)$ for $x \in \left(\frac{1}{2}, 1\right)$. This failure is not a complete surprise since, by FTC-2, ${}^{R}\!\!\int_a^x f$ must be a continuous function of x when f is Riemann integrable.

Now consider

$$G(x) = \begin{cases} 0, & 0 \le x < 1/2 \\ x - \frac{1}{2}, & 1/2 \le x \le 1. \end{cases}$$

G is also differentiable on $[0, 1]$ except at $x = 1/2$. In this case, G is continuous and ${}^{R}\!\!\int_0^x G' = G(x) - G(0)$ for all $x \in [0, 1]$. (See exercise 18.) The function G illustrates a more general phenomenon. Suppose that F is continuous on the interval $[a, b]$ and is differentiable except on a finite set. As long as F' is Riemann integrable when we assign values at the missing points, ${}^{R}\!\!\int_a^x F' = F(x) - F(a)$ for all $x \in [a, b]$. (See exercise 19.)

Can we extend this process to allow an infinite number of points where the function is not differentiable? More specifically, suppose that we can assign values to F' at those points where F is not differentiable in a manner that (1) is consistent with the values of F' at points of differentiability and (2) guarantees that F' is Riemann integrable. Is the continuity of F sufficient to guarantee that $\mathcal{R}\!\!\int_a^x F' = F(x) - F(a)$ for all $x \in [a, b]$?

No. The Cantor function provides a counterexample. Since the Cantor function is constant on the complement of C, the Cantor function, c, is differentiable with $c'(x) = 0$ on the complement of the Cantor set. Since C has measure zero, it is natural to extend c' to be zero there as well. When this is done, $\mathcal{R}\!\!\int_0^1 c' = 0$ while $c(1) - c(0) = 1$. In fact, as long as the values chosen for c' on the Cantor set are bounded, c' will be Riemann integrable with $\mathcal{R}\!\!\int_0^x c' = 0$ for $x \in [0, 1]$. (See exercise 20.)

4.4 Volterra's example

In 1881, Vito Volterra provided an example of a function f that is differentiable on all of $[0, 1]$ and for which f' is bounded but not Riemann integrable on $[a, b]$. According to Lebesgue's criterion (theorem 20, page 63), such a function will have a derivative that is defined and bounded everywhere while being discontinuous on a set of positive measure.

Volterra's function is built on a Cantor-like set K constructed by removing intervals of length α^n at stage n with $0 < \alpha < \frac{1}{3}$. Critically, the set K will have a measure of $\frac{1-3\alpha}{1-2\alpha} > 0$. Most of our effort will be spent in defining the function V on the complement of K.

The basic building block of the construction is the function

$$g(x) = \begin{cases} x^2 \sin \frac{1}{x}, & x > 0 \\ 0, & x = 0. \end{cases}$$

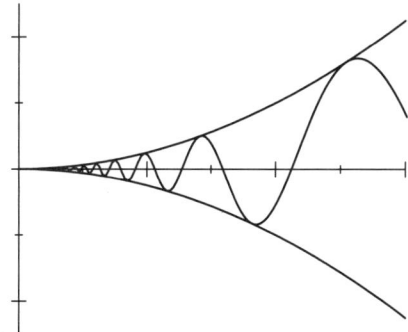

Figure 4.3. The function $g(x)$ bounded by $-x^2$ and x^2

The key properties of g are

1. The function g is differentiable for all $x \geq 0$.

2. The derivative is bounded by $|g'(x)| \leq 3$.

3. Given any $\alpha > 0$, there are $x, z \in (0, \alpha)$ for which $g'(x) = 1$ and $g'(z) = 0$.

Now suppose that we are given an interval (a, b). Then there is a largest value $0 < c < \frac{b-a}{2}$ so that $g'(c) = 0$. Define $h_{a,b}(x)$ on (a, b) by

$$h_{a,b}(x) = \begin{cases} g(x-a), & x \in (a, a+c) \\ g(c), & x \in [a+c, b-c] \\ g(b-x), & x \in (b-c, b). \end{cases}$$

The left portion of $h_{a,b}$ consists of a copy of g that has been shifted to begin at $x = a$. The right portion is a mirror image ending at b. We have spliced a constant function in the middle in such a way that the left and right derivatives at $a + c$ and $b - c$ are all zero. (See Figure 4.4.) Thus $|h_{a,b}(x)| \leq \min\left\{(x-a)^2, (x-b)^2\right\}$ and $h_{a,b}$ is differentiable on (a, b) with $|h'_{a,b}(x)| \leq 3$.

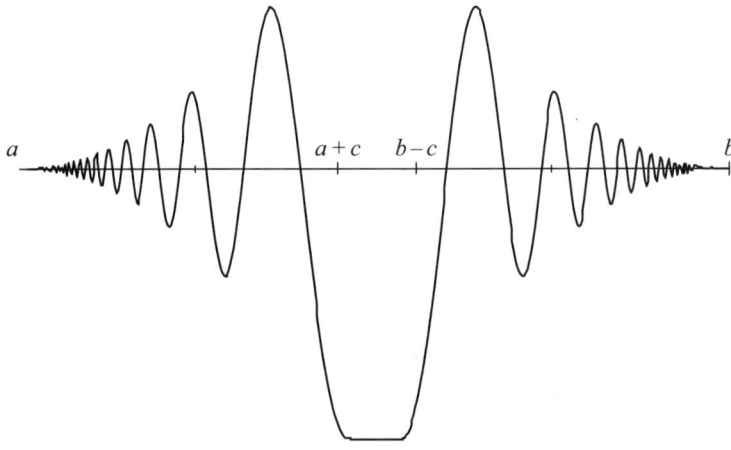

Figure 4.4. $h_{a,b}$

The Volterra function V is defined by taking $V(x) = 0$ for all $x \in K$ and $V(x) = h_{a,b}(x)$ for x in any interval (a, b) removed to form K.

By its construction, V is continuously differentiable on any interval in the complement of K. Suppose then that $x_0 \in K$. If $x \in K$, then

$$\frac{V(x) - V(x_0)}{x - x_0} = 0.$$

If $x \notin K$, then x belongs to some interval (a, b) in the complement of K. Since $x \in (a, b)$ and $x_0 \notin (a, b)$, one of $(x - a)^2$ or $(x - b)^2$ is smaller than $(x - x_0)^2$. Hence,

$$|V(x) - V(x_0)| = |h_{a,b}(x)| \le \min\left\{ (x-a)^2, (x-b)^2 \right\} < (x - x_0)^2.$$

Since

$$\left| \frac{V(x) - V(x_0)}{x - x_0} \right| \le |x - x_0|,$$

we conclude that $V'(x_0) = 0$ for all $x_0 \in K$.

To see that V' is not continuous at any $x_0 \in K$, note that x_0 is the limit of endpoints of removed subintervals. Moreover, arbitrarily close to each endpoint of a removed subinterval there is a point where V' is 1. Thus, while $V'(x_0) = 0$, there is a sequence of points $\{x_n\}$ approaching x_0 for which $V'(x_n) = 1$. Since V' is discontinuous on K and K has positive measure, V' is not Riemann integrable.

We will soon see definitions of the integral for which V' is integrable with $\int_0^x V' = V(x)$.

4.5 Exercises

4.1 Dirichlet and friends: filling the gaps

1. Prove that Dirichlet's snowflake function is discontinuous at every rational number and continuous at every irrational number. (In any finite interval, there are only finitely many rational numbers with denominator less than $1/\varepsilon$.)

2. Prove that the function

$$r(x) = \begin{cases} n, & x = \frac{m}{n} \text{ with } \frac{m}{n} \text{ in lowest terms} \\ 0, & \text{otherwise} \end{cases}$$

is unbounded on every nondegenerate interval. (Explain why, given any B, a nondegenerate interval must contain an x for which $f(x) > B$.)

4.2 Trigonometric series: filling the gaps

3. Show that if $f(x) = 1$ for $x \in (0, 1)$ then, according to the computational technique of Fourier,

   (a) the even coefficients in $f(x) = \sum_{n=1}^{\infty} a_n \sin(n\pi x)$ are zero
   (b) $a_n = \frac{4}{n\pi}$ when n is odd.

4. Suppose that $f(x) = \sum_{n=1}^{\infty} \frac{4}{(2n-1)\pi} \sin((2n-1)\pi x)$ evaluates to 1 on the interval $(0, 1)$. Prove that f must also satisfy

$$
f(x) = \begin{cases} 1, & x \in (-2, -1) \cup (0, 1) \\ 0, & x = -2, -1, 0, 1, 2 \\ -1, & x \in (-1, 0) \cup (1, 2). \end{cases}
$$

5. Prove that $g(x) = \sum_{n=1}^{\infty} \frac{1}{2^n} \sin(c^n x)$ is well defined and continuous by showing that the series converges uniformly.

6. Prove that $\sum_{n=1}^{\infty} \left(\frac{c}{2}\right)^n \pi \sin(c^n \pi x)$ converges for $x = \frac{k\pi}{c^m}$ and diverges for $x = \frac{(k+1/2)\pi}{c^m}$.

7. Suppose that $c, m, N \in \mathbb{N}$, $1 \le |c^m x_0 - N| \le \frac{3}{2}$, and $c^m x_1 = N$.

   (a) Fill in the details to prove that

   $$
   \cos(c^m \pi x_1) - \cos(c^m \pi x_0) = (-1)^N \left[1 - \cos(\pi(c^m x_0 - N))\right].
   $$

   (b) If n is an integer greater than m, provide the details to show that

   $$
   \cos(c^m \pi x_1) - \cos(c^m \pi x_0) = (-1)^N \left[1 - \cos(c^{n-m} \pi(c^m x_0 - N))\right].
   $$

8. Prove that if f is differentiable at x_0, then $\frac{f(x) - f(x_0)}{x - x_0}$ is bounded on some neighborhood of x_0.

4.2 Trigonometric series: deeper reflections

9. Functions of the form $e^{-cw} \cos cx$ also satisfy $\frac{\partial^2 z}{\partial w^2} + \frac{\partial^2 z}{\partial x^2} = 0$. Show that if we use the techniques of Fourier to express $f(x) = 1$, $x \in (0, 1)$, as $f(x) = \sum_{n=1}^{\infty} a_n \cos(n\pi x)$, then all of the coefficients are zero. Comment.

10. A vibrating string can be modeled by the initial value problem

$$
\frac{\partial^2 y}{\partial x^2} = \frac{1}{c^2} \frac{\partial^2 y}{\partial t^2},
$$

$$
y(x, 0) = f(x)
$$

where $y(x, t)$ is the y-coordinate of the string x units from the origin at time t. The parameter c depends on the weight and tension of the string and $f(x)$ represents the starting position from which the string is released at time $t = 0$.

(a) Show that functions of the form $\cos ac\pi t \sin a\pi x$ satisfy this differential equation.[4]

(b) Ignoring potential issues related to convergence, use the ideas of Fourier to find a trigonometric series that describes the shape of a string of length 1 at time t if the string begins in the shape of a triangle

$$f(x) = \begin{cases} 2x, & x \in [0, 1/2] \\ 2 - 2x, & x \in [1/2, 1]. \end{cases}$$

(c) Use a computerized graphics system to animate an approximation to your solution.

(d) Show that functions of the form $\sin ac\pi t \cos a\pi x$ also satisfy the differential equation. What happens when we try to use functions of this form to construct a solution?

11. Prove that if f is bounded on $[a, b]$ and differentiable at x_0, then $\frac{f(x) - f(x_0)}{x - x_0}$ is bounded on $[a, b]$.

12. Trigonometric series are not the only series that exhibit bad behavior. Let

$$g_n(x) = \frac{x^3}{(1 + nx^2)(1 + (n-1)x^2)}$$

and define

$$G(x) = \sum_{n=1}^{\infty} g_n(x).$$

(a) Use partial fractions and telescoping series to verify that $G(x) = x$.

(b) Compute $g_n'(x)$.

(c) Show that $\sum_{n=1}^{\infty} g_n'(0)$ converges but is not equal to $G'(0)$.

This behavior is more problematic than a series for which $\sum_{n=1}^{\infty} f_n'(c)$ does not converge as convergence can give one a false sense that all is well.

4.3 Friends of cantor: filling the gaps

13. In property (5) of the Cantor set, why is the total length of the covering open intervals $2 \cdot \frac{2^n}{3^n}$ rather than $\frac{2^n}{3^n}$?

---

[4] Note that c is a parameter in this initial value problem so that a plays the role analogous to that of c in Fourier's original problem.

14. Prove that the Cantor function is monotone increasing.

15. While we commonly express numbers in the decimal system, it is possible to use other bases. The Cantor set can be understood using the ternary (base-three) system. With $a_i \in \{0, 1, 2\}$, we define

$$(0.a_1 a_2 a_3 \ldots)_3 = \sum_{n=1}^{\infty} \frac{a_n}{3^n}.$$

   (a) Show that $(0.1111\ldots)_3 = \frac{1}{2}$.
   (b) Show that both $(0.1000\ldots)_3$ and $(0.02222\ldots)_3$ represent $\frac{1}{3}$.
   (c) Find two ternary representations for $\frac{2}{3}$.
   (d) Use the ternary representation to describe the sets that are removed to form C_1 and C_2, the first and second stages in constructing the Cantor set.
   (e) Give a ternary description (with proof of equivalence) of the Cantor set.
   (f) Use this description to find an element of C that is not the endpoint of a removed interval.

16. Use the description of C in exercise 15 to prove that C is uncountable. (Use a diagonalization argument similar to the one that proves $[0, 1]$ is uncountable.)

17. Prove that the sequence of functions that define the Cantor function converges uniformly. (How far apart are $f_n\left(\frac{k}{3^n}\right)$ and $f_n\left(\frac{k+1}{3^n}\right)$? Use this fact to bound $\left| f_m(x) - f_p(x) \right|$ when $x \in \left[\frac{k}{3^n}, \frac{k+1}{3^n}\right]$.)

18. Let
$$G(x) = \begin{cases} 0, & 0 \le x < 1/2 \\ x - \frac{1}{2}, & 1/2 \le x \le 1. \end{cases}$$

   (a) Compute $G'(x)$ for $x \in [0, 1]$. Explain any special considerations.
   (b) Compute $\text{R}\!\int_0^x G'$ for $x \in \left[0, \frac{1}{2}\right]$.
   (c) Compute $\text{R}\!\int_{\frac{1}{2}}^x G'$ for $x \in \left[\frac{1}{2}, 1\right]$.
   (d) Compute $\text{R}\!\int_0^x G'$ for $x \in \left[\frac{1}{2}, 1\right]$.

19. Suppose that F is continuous on $[a, b]$ and differentiable except at a finite number of points. Suppose further that, by suitably defining F' at the points where F is not differentiable, F' is Riemann integrable on $[a, b]$. Prove that $\text{R}\!\int_a^x F' = F(x) - F(a)$ for all $x \in [a, b]$. (Use induction.)

20. Let $c(x)$ be the Cantor function. Define c' on the Cantor set in such a way that c' remains bounded. Prove that c' is Darboux integrable with $D\!\int_0^x c' = 0$ for all $x \in [0, 1]$. (Given $\varepsilon > 0$, use C_n to show that it is possible to select a partition $\mathcal{P}$ of $[0, x]$ in such a way that $S_{\overline{D}}(f, \mathcal{P}) < \varepsilon$.)

4.3 Friends of Cantor: deeper reflections

21. Prove that the length of the curve $y = c(x)$ is 2. (Use the fact that $\sqrt{\Delta x^2 + \Delta y^2} \le \Delta x + \Delta y$ to conclude that the length of the Cantor function is bounded above by 2. Explain why the length of the Cantor function is greater than the length of any f_n used in its construction. Find a lower bound for the length of f_n by replacing the diagonal segments with vertical segments.)

As with the Cantor set, the Cantor function can be defined using ternary expansions of the numbers in $[0, 1]$. (See exercise 15.)
Given $x \in [0, 1]$

(a) Express x in ternary notation.
(b) If the ternary expansion of x contains a 1, replace every digit after the first 1 with 0.
(c) Replace every 2 with a 1.
(d) Interpret the result as a binary number.

This process defines a function c^*. The following two exercises investigate c^* and its relationship to c.

22. Use the preceding definition of c^* for the following computations.

(a) Verify that $(0.111111\ldots)_3 = \frac{1}{2}$ and use this information to compute $c^*\left(\frac{1}{2}\right)$.
(b) Verify that $(0.020202\ldots)_3 = \frac{1}{4}$ and use this information to compute $c^*\left(\frac{1}{4}\right)$.
(c) The value $\frac{1}{3}$ can be expressed both as $(0.100000\ldots)_3$ and as $(0.022222\ldots)_3$. Verify that both expressions produce the same value for $c^*\left(\frac{1}{3}\right)$.
(d) Verify that c^* is well defined. In other words, show that in those cases where x has two possible ternary expansions, both expansions produce the same value for $c^*(x)$.

23. Show that $c^* = c$ on $[0, 1]$.

(a) Verify that c^* is increasing (but not strictly increasing).

(b) Verify that c^* is constant on every interval in the complement of the Cantor set.

(c) Verify that $c^*(x) = c(x)$ for x in the complement of the Cantor set. (If x is in the complement of C, then x is in the complement of C_n for some n. Verify that $f_n(x) = c^*(x)$ where $\{f_n\}$ is the sequence of functions used to define the Cantor function in Section 3.)

(d) Use parts (a) and (c) to verify that $c^*(x) = c(x)$ on $[0, 1]$.

24. There is a third way of defining the Cantor function on $[0, 1]$.

(a) Let $g_0(x) = x$.

(b) For any integer $n \geq 1$, define

$$
g_{n+1}(x) = \begin{cases}
\frac{1}{2}g_n(3x), & x \in \left[0, \frac{1}{3}\right] \\
\frac{1}{2}, & x \in \left[\frac{1}{3}, \frac{2}{3}\right] \\
\frac{1}{2}g_n(3x) + \frac{1}{2}, & x \in \left[\frac{2}{3}, 1\right].
\end{cases}
$$

Prove that $g_n(x) = f_n(x)$ where $\{f_n\}$ is the sequence of functions used to define the Cantor function in Section 3. Conclude that $\lim_{n \to \infty} g_n(x) = c(x)$, the Cantor function. (Use induction.)

4.4 Volterra's example: filling the gaps

25. Define

$$
g(x) = \begin{cases}
x^2 \sin \frac{1}{x}, & x > 0 \\
0, & x = 0.
\end{cases}
$$

(a) Use the definition of the derivative to prove that g is differentiable at $x = 0$.

(b) Compute $g'(x)$ for $x > 0$ and verify that $|g'(x)| \leq 3$ for $x \in [0, 1]$.

(c) Explain why any interval of the form $(0, \alpha)$ contains points x and z for which $g'(x) = 1$ and $g'(z) = 0$.

4.4 Volterra's example: deeper reflections

26. Suppose that x belongs to a subinterval (a, b) of length α^n or more that is removed in the construction of K. Define $V_n(x) = h_{a,b}(x)$ for such x and define V_n to be 0 elsewhere.

(a) Explain why V_n is differentiable on all of $[0, 1]$.

(b) Prove that $\int_0^x V_n' = V_n(x)$ for all $x \in [0, 1]$.

(c) Prove that V_n converges uniformly to V on $[0, 1]$. (Hint: How large can $|V(x)|$ be if x comes from a removed interval of length less than α^n?)

(d) Prove that V_n' converges to V' on $[0, 1]$.

(e) Discuss $\lim_{n \to \infty} {}^R\!\!\int_0^x V_n'$ and ${}^R\!\!\int_0^x \lim_{n \to \infty} V_n'$.

4.6 References

Allen, E.S. (1941). The scientific work of Vito Volterra. *Amer. Math. Monthly* **48**: 516–519. JSTOR 2303385.

Boyce, W.E. & DiPrima, R.C. (2005). *Elementary Differential Equations and Boundary Value Problems* (8th ed.). John Wiley & Sons.

Burk, F.E. (2007). *A Garden of Integrals*. Mathematical Association of America.

Dauben, J.W. (1979). *Georg Cantor: His Mathematics and Philosophy of the Infinite*. Harvard University Press.

Dirichlet, G.L. (1969). *Werke*, reprint. Chelsea.

Fourier, J. (1822). *The Analytical Theory of Heat*. translated by Alexander Freeman (1878, re-released 2003). Dover Publications.

Gelbaum, B.R. and J.M.H. Olmsted (2003). *Counter examples in Analysis*. Dover.

Gonzalez-Velasco, E.A. (1992). Connections in mathematical analysis: the case of Fourier series. *American Mathematical Monthly* **99** (5): 427–441. JSTOR 2325087.

Goodstein, J. R. (2007). *The Volterra Chronicles: The Life and Times of an Extraordinary Mathematician 1860–1940*. History of Mathematics 31, American Mathematical Society.

Zygmund, A. (2002). *Trigonometric Series* (3rd ed.). Cambridge University Press.

CHAPTER 5

Another Approach: Measure Theory

> *I have to pay a certain sum, which I have collected in my pocket. I take the bills and coins out of my pocket and give them to the creditor in the order I find them until I have reached the total sum. This is the Riemann integral. But I can proceed differently. After I have taken all the money out of my pocket I order the bills and coins according to identical values and then I pay the several heaps one after the other to the creditor. This is my integral.*
>
> — Henri Lebesgue in a letter to Paul Montel

In his 1902 Ph.D. thesis, Henri Lebesgue introduced an approach to integration that resolves many of the convergence issues that we have noted with the Riemann-Darboux integral. Lebesgue partitioned the y-axis rather than the x-axis.[1] Figure 5.1 shows an interval from a partition of the range of f. The set of points that f sends into this subinterval is marked on the x-axis. The effect of this approach is to gather together those points for which f has approximately the same value.

Preimages play a critical role in this development. Given any set A, the **preimage** of A under the function f is the set $f^{-1}(A) = \{x \in \mathbb{R} : f(x) \in A\}$. For the time being, we will only consider preimages of intervals.

Now suppose that we are given a function f that is defined on $[a, b]$ and for which $\alpha < f < \beta$. Take any partition $\mathcal{P} = \{[y_{k-1}, y_k]\}_{k=1}^{n}$ of $[\alpha, \beta]$ and set $E_k = f^{-1}((y_{k-1}, y_k])$, $k = 1, 2, \ldots, n$. Note that $\{E_k\}_{k=1}^{n}$ consists of disjoint sets whose union is all of $[a, b]$. The lower and upper Lebesgue

---

[1] What follows is in the spirit of Lebesgue's work but, for the sake of comparison, notation from previous chapters is used.

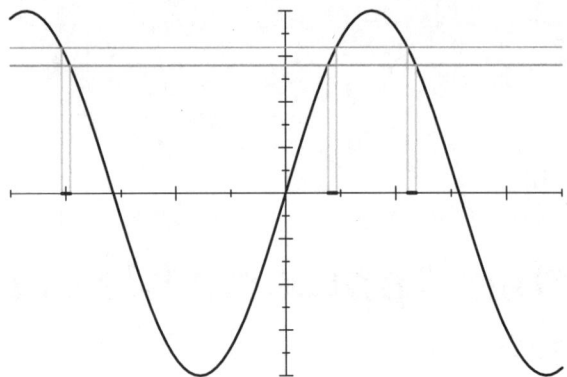

Figure 5.1. The preimage of an interval

sums are

$$S_{\underline{L}}(f,\mathcal{P}) = \sum_{k=1}^{n} y_{k-1} \cdot m(E_k) \text{ and } S_{\overline{L}}(f,\mathcal{P}) = \sum_{k=1}^{n} y_k \cdot m(E_k)$$

where $m(A)$ is the measure of the set A.[2]

Example 19 (Constant function). Suppose that $f(x) = c$ is a constant function on $[a,b]$. Noting that $c - 1 < f < c + 1$, let $\mathcal{P} = \{[y_{k-1}, y_k]\}_{k=1}^{n}$ be a partition of $[c - 1, c + 1]$. Taking k_c to be the unique integer for which $y_{k_c-1} < c \leq y_{k_c}$, the preimages associated with this partition are $E_{k_c} = [a,b]$ and $E_k = \emptyset$ for $k \neq k_c$. Thus $S_{\underline{L}}(f,\mathcal{P}) = y_{k_c-1} \cdot (b-a)$ and $S_{\overline{L}}(f,\mathcal{P}) = y_{k_c}\cdot(b-a)$. As the mesh $\|\mathcal{P}\|$ of the partition becomes smaller, both the upper and lower sum will approach $c \cdot (b-a)$. Even though we have not yet defined the Lebesgue integral, we conclude that $L\!\int_a^b f = c \cdot (b-a)$.

Example 20 (Dirichlet function). Let $d(x)$ be the Dirichlet function

$$d(x) = \begin{cases} 1, & x \in \mathbb{Q} \\ 0, & x \notin \mathbb{Q} \end{cases}$$

on $[0,1]$. Note that $-1 < d < 2$ and take a partition $\mathcal{P} = \{[y_{k-1}, y_k]\}_{k=1}^{n}$ of $[-1,2]$. Now there are unique integers k_0 and k_1 for which $y_{k_0-1} < 0 \leq y_{k_0}$ and $y_{k_1-1} < 1 \leq y_{k_1}$. As long as there is a division point in $(0,1)$, the associated preimages are $E_{k_0} = [0,1]\setminus\mathbb{Q}$ and $E_{k_1} = [0,1]\cap\mathbb{Q}$. For all other values of k, $E_k = \emptyset$. As we have seen previously,

---

[2] For now, think of measure as a generalization of length.

$m\left(E_{k_1}\right) = m\left([0,1] \cap \mathbb{Q}\right) = 0$ so that $m\left(E_{k_0}\right) = m\left([0,1] \setminus \mathbb{Q}\right) = 1.$[3] Thus $S_{\underline{L}}\left(f, \mathcal{P}\right) = y_{k_0-1} \cdot 1 + y_{k_1-1} \cdot 0 = y_{k_0-1}$ and $S_{\overline{L}}\left(f, \mathcal{P}\right) = y_{k_0} \cdot 1 + y_{k_1} \cdot 0 = y_{k_0}$. Both of these values are within $\|\mathcal{P}\|$ of 0. Hence d is Lebesgue integrable with $^L\!\int_a^b d = 0$.

Given a function $f : [a,b] \to (\alpha, \beta)$, it is fairly straightforward to verify that for any pair of partitions $\mathcal{P}_1$ and $\mathcal{P}_2$ of $[\alpha, \beta]$, we have $S_{\underline{L}}\left(f, \mathcal{P}_1\right) \leq S_{\overline{L}}\left(f, \mathcal{P}_2\right)$. (See exercises 2 and 3.) Also, for any partition $\mathcal{P}$ of $[\alpha, \beta]$,

$$S_{\overline{L}}\left(f, \mathcal{P}\right) - S_{\underline{L}}\left(f, \mathcal{P}\right) = \sum_{k=1}^{n} y_k \cdot m\left(E_k\right) - \sum_{k=1}^{n} y_{k-1} \cdot m\left(E_k\right)$$

$$= \sum_{k=1}^{n} (y_k - y_{k-1}) \cdot m\left(E_k\right)$$

$$\leq \|\mathcal{P}\| \sum_{k=1}^{n} m\left(E_k\right) = \|\mathcal{P}\| \left(b - a\right).$$

Since we can arrange for $\|\mathcal{P}\|$ to be arbitrarily small, we conclude that the upper and lower Lebesgue sums will converge and that every bounded function has an integral in the sense of Lebesgue. This seems too good to be true.

It is.

5.1 Measurable sets I

While the process of computing the measure of an arbitrary set may be complicated, the idea that every set should have a measure seems rather straightforward. Moreover, the measure should satisfy some basic properties.

1. **Intervals.** For any interval $[a,b]$, $m\left([a,b]\right) = b - a$. In particular, the measure of a single point is zero.

2. **Monotonicity.** For any sets $A \subset B$, $m\left(A\right) \leq m\left(B\right)$.

3. **Translation invariance.** If $B = A + t = \{a + t : a \in A\}$ for some $t \in \mathbb{R}$ (in other words, B is a translation by t of the set A), then $m\left(B\right) = m\left(A\right)$.

4. **Countable additivity.** If a set E is the disjoint union of a countable collection of sets $\{E_k\}$ then $m\left(E\right) = \sum_k m\left(E_k\right)$.

---

[3] We have not yet justified the statement that $m\left([0,1] \setminus \mathbb{Q}\right) = 1$, but we will.

The restriction to countable disjoint unions is necessary. In addition to the problem of making sense of an uncountable sum, the combination of properties (1) and (4) for unrestricted sums would imply that all sets have zero measure.

While all appears to be in good order, there are significant issues lurking below the surface.

Example 21 (Vitali, 1905). Define two real numbers x and y to be **rationally equivalent** if $x - y \in \mathbb{Q}$. For example, π and $\pi + \frac{1}{2}$ are rationally equivalent while π and $\sqrt{2}$ are not. This relation is an equivalence relation that divides $\mathbb{R}$ into disjoint subsets of rationally equivalent real numbers called equivalence classes. (See exercise 6.) Given an element x of an equivalence class, adding (or subtracting) any rational number will result in another member of the same equivalence class. Thus every equivalence class will have infinitely many members in $[0, 1]$. For each equivalence class, select a single representative element from the interval $[0, 1]$. For example, we might represent the class containing π by $\pi - 3$ and the class containing $\sqrt{2}$ by $\sqrt{2} - \frac{5}{4}$. Denote the set of representatives by V.[4]

We make the following observations.

1. **If s and t are distinct rational numbers, then $V + s$ and $V + t$ are disjoint.** Suppose that s and t are rational numbers with $x \in (V + s) \cap (V + t)$. Then $x - s$ and $x - t$ are elements of V and, since s and t are rational numbers, both $x - s$ and $x - t$ represent the equivalence class that contains x. As the equivalence class of x has only one representative in V, we conclude that $x - s = x - t$ so that $s = t$.

2. **For any $x \in [0, 1]$ there is a rational number $s \in [-1, 1]$ such that $x \in V + s$.** Given $x \in [0, 1]$, denote the representative of x by r_x. Since both x and r_x belong to $[0, 1]$, we know that $|x - r_x| \leq 1$ or, equivalently, $s = x - r_x \in [-1, 1]$. Now x and r_x differ by a rational number so that s is rational with $x = r_x + s \in V + s$.

3. **If $x \in [0, 1]$ and $s \in [-1, 1]$, then $x + s \in [-1, 2]$.**

Now let $\{s_n\}$ be an enumeration of all of the rational numbers in $[-1, 1]$. Since $V + s_n$ and $V + s_m$ are disjoint when $n \neq m$, countable additivity and

---

[4] If you are familiar with the concept, you may have noticed that we have just used the Axiom of Choice.

the translation invariance of measure imply that

$$m\left(\cup_n (V + s_n)\right) = \sum_n m\left(V + s_n\right) = \sum_n m\left(V\right).$$

At this point there are two options. Either $m(V) = 0$ (in which case $m\left(\cup_n (V + s_n)\right) = 0$) or $m(V) > 0$ (which implies that $m\left(\cup_n (V + s_n)\right)$ is infinite).

From our previous discussion we know

$$[0, 1] \subset \cup_n (V + s_n) \subset [-1, 2]$$

so that monotonicity and the interval property imply that

$$1 \le m\left(\cup_n (V + s_n)\right) \le 3.$$

We have arrived at a contradiction.

What are we to make of this situation? We see that it is impossible for all four of the properties to hold for a measure defined on all sets. Since we want our measure to satisfy the four properties, we have to accept the fact that some sets of real numbers, like V, will not be measurable. We will investigate which sets are measurable later in this chapter. And as we shall see, the set of measurable sets has an impact on the set of functions that are Lebesgue integrable.

5.2   Outer measure

Since the seemingly obvious fact that all sets are measurable is demonstrably false, we turn our attention to the question of computing the measure of a set. We will come back to the problem of identifying which sets are measurable later. We base the measure of an arbitrary set on the measure of intervals.

Definition 7 (Outer measure). *Let $A \subset \mathbb{R}$. The **Lebesgue outer measure** of A is*

$$\mu^*(A)$$

$$= \inf\left\{\sum_k l\left(I_k\right) : \{I_k\} \text{ is a countable cover of } A \text{ by finite open intervals}\right\}$$

where $l(I)$ is the length of the interval I.

Example 22 (Outer measure of $[a, b]$). Consider the interval $[a, b]$. We claim that $\mu^* ([a, b]) = b - a$. For any $\varepsilon > 0$ we can cover $[a, b]$ by a single open interval $(a - \varepsilon, b + \varepsilon)$ whose length is $b - a + 2\varepsilon$. Thus, we see that $\mu^* ([a, b]) \leq b - a$.

To show that $\mu^* ([a, b]) \geq b - a$, suppose that $\{I_k\}$ is a cover of $[a, b]$ by open intervals. Since $[a, b]$ is compact,[5] we can select a finite subcover $\{(a_j, b_j)\}_{j=1}^{n}$. For definiteness, we will also assume that the intervals have been ordered by their left endpoints and that (by removing unneeded intervals) no left endpoint is repeated and no interval is contained in another. In order to cover $[a, b]$, the intervals $\{(a_j, b_j)\}_{j=1}^{n}$ must overlap. In other words,

$$a_1 < a, \ a_2 < b_1, \ a_3 < b_2, \ldots, a_n < b_{n-1}, \text{ and } b < b_n.$$

These inequalities imply that

$$\sum_k l\,(I_k) \geq \sum_{j=1}^{n} l\left((a_j, b_j)\right)$$
$$= (b_1 - a_1) + (b_2 - a_2) + \cdots + (b_n - a_n)$$
$$> (b_1 - a) + (b_2 - a_2) + \cdots + (b - a_n)$$
$$> (b_1 - a) + (b_2 - b_1) + \cdots + (b - b_{n-1})$$
$$= b - a.$$

Since $b - a$ is a lower bound on the sums of the lengths of covering intervals, $\mu^* ([a, b]) \geq b - a$. Combining the two inequalities, $\mu^* ([a, b]) = b - a$.

In fact, the Lebesgue outer measure of any interval I is its length. This fact can be verified by modifying the preceding argument, but it is far easier to prove using the monotonicity of μ^*.

We have verified one of the four properties that a measure should satisfy. The verifications of monotonicity and translation invariance are relatively straightforward and are left as exercises. This leaves the property of countable additivity. From Vitali's example, we know that this property fails for certain combinations of sets. Can we rescue anything in this regard? Yes. We can prove subadditivity.

---

[5] See Appendix A.1 for a review of the definition and basic properties of a compact set.

Theorem 21 (Subadditivity). *Suppose that $\{A_k\}_{k=1}^{\infty}$ is a countable collection of sets. Then*

$$\mu^* \left(\cup_k A_k\right) \leq \sum_k \mu^* \left(A_k\right).$$

Proof. If $\mu^* \left(A_k\right)$ is infinite for any k, the inequality automatically holds. So suppose that $\mu^* \left(A_k\right)$ is finite for all k and let $\varepsilon > 0$. Since $\mu^* \left(A_k\right)$ is defined by an infimum, we can choose a countable cover $\left\{I_{k,j}\right\}_j$ of A_k consisting of open intervals such that

$$\mu^* \left(A_k\right) \leq \sum_j l \left(I_{k,j}\right) < \mu^* \left(A_k\right) + \frac{\varepsilon}{2^k}.$$

Then $\left\{I_{k,j}\right\}_{k,j}$ is a countable cover of $\cup_k A_k$ by open intervals. As $\mu^* \left(\cup_k A_k\right)$ also is defined by an infimum,

$$\mu^* \left(\cup_k A_k\right) \leq \sum_{k,j} l \left(I_{k,j}\right) < \sum_k \left(\mu^* \left(A_k\right) + \frac{\varepsilon}{2^k}\right) = \sum_k \mu^* \left(A_k\right) + \varepsilon.$$

But $\varepsilon > 0$ is arbitrary, so that

$$\mu^* \left(\cup_k A_k\right) \leq \sum_k \mu^* \left(A_k\right) \qquad \qquad \square$$

Note that we have strict inequality in the case of Vitali's example. For the set V constructed in that example, $\mu^* \left(V\right) > 0$ else

$$1 = \mu^* \left([0,1]\right) \leq \mu^* \left(\cup_n \left(V + s_n\right)\right) = \sum_n \mu^* \left(V\right) = 0.$$

Since $\mu^* \left(V\right) > 0$, we conclude that

$$\mu^* \left(\cup_n \left(V + s_n\right)\right) \leq \mu^* \left([-1,2]\right) = 3$$
$$< \infty = \sum_n \mu^* \left(V\right) = \sum_n \mu^* \left(V + s_n\right).$$

5.3 Measurable sets II

We now see that the crux of the problem lies in identifying the sets for which we have countable additivity. Constantin Carathéodory (1873–1950) provided the key criterion.[6] If the measure of the set E is always the sum of

---

[6] Lebesgue took an alternative approach which is outlined in Exercises 71 and following.

the measures of its countable disjoint parts, then all the more will $\mu^*(E) = \mu^*(A) + \mu^*(B)$ when A and B are disjoint sets with $E = A \cup B$.[7] Equivalently, $\mu^*(E) = \mu^*(E \cap C) + \mu^*(E \cap C^c)$ for any set C.[8] Carathéodory's insight was to apply the Golden Rule. Instead of asking if every set would "play nicely" with E, we ask if E will "play nicely" with every set.

Definition 8 (Carathéodory's measurability condition). *A set $E \subseteq \mathbb{R}$ is (Lebesgue) **measurable** if*

$$\mu^*(A) = \mu^*(A \cap E) + \mu^*(A \cap E^c)$$

for every subset $A \subseteq \mathbb{R}$. When E is measurable, we write $\lambda(E) = \mu^*(E)$.

Note that, by subadditivity, we only need to check that $\mu^*(A) \geq \mu^*(A \cap E) + \mu^*(A \cap E^c)$.

Example 23 (Every finite set is measurable). Suppose that $E = \{x_1, x_2, \ldots, x_n\}$. Let A be a subset of $\mathbb{R}$. Now $\left\{ \left(x_i - \frac{\varepsilon}{4n}, x_i + \frac{\varepsilon}{4n} \right) \right\}_{i=1}^{n}$ is a cover of E and thus of $E \cap A$ with total length $\frac{\varepsilon}{2}$. Hence $\mu^*(A \cap E) < \frac{\varepsilon}{2}$. Choose a countable cover $\{I_k\}$ of A by open intervals such that

$$\sum_k l(I_k) \leq \mu^*(A) + \frac{\varepsilon}{2}.\,^9$$

Then $\{I_k\}$, being a cover of A, is also a cover of $A \cap E^c$ so that $\mu^*(A \cap E^c) \leq \mu^*(A) + \frac{\varepsilon}{2}$. Hence

$$\mu^*(A \cap E) + \mu^*(A \cap E^c) \leq \mu^*(A) + \varepsilon.$$

Since the inequality holds for all $\varepsilon > 0$, we conclude that $\mu^*(A) \geq \mu^*(A \cap E) + \mu^*(A \cap E^c)$. The reverse inequality is a consequence of subadditivity so that the finite set E is measurable.[10]

Before spending too much energy identifying various measurable sets, we should check to see that Carathéodory's definition gets us what we are looking for: countable additivity. We don't even know that additivity works for a pair of disjoint measurable sets. Let's start there.

---

[7] We will see this move repeatedly in what follows. Wanting to know a result for countable unions, we start with the union of a pair of sets.

[8] A^c denotes the complement of the set A. In other words, $A^c = \{x \in \mathbb{R} : x \notin A\}$.

[9] If $\mu^*(A) = \infty$, we cannot have strict inequality.

[10] There are easier ways to verify that finite sets are measurable. (See Exercise 18.) This proof was chosen to illustrate the definition of measurability.

If E and F are disjoint, measurable sets, we can use the measurability of F to conclude that

$$\mu^* (E \cup F) = \mu^* ((E \cup F) \cap F) + \mu^* ((E \cup F) \cap F^c)$$
$$= \mu^* (F) + \mu^* (E).$$

Additivity for any finite set of disjoint, measurable sets follows by induction. Hence, if $\{E_k\}_{k=1}^\infty$ is a sequence of disjoint, measurable sets,

$$\sum_{k=1}^{n} \mu^* (E_k) = \mu^* \left(\cup_{k=1}^{n} E_k \right) \leq \mu^* \left(\cup_{k=1}^{\infty} E_k \right) \leq \sum_{k=1}^{\infty} \mu^* (E_k).$$

Since this inequality is true for all values of n,

$$\sum_{k=1}^{\infty} \mu^* (E_k) \leq \mu^* \left(\cup_{k=1}^{\infty} E_k \right) \leq \sum_{k=1}^{\infty} \mu^* (E_k).$$

Countable additivity holds for disjoint measurable sets!

Unfortunately, we don't know very many measurable sets nor do we even know that the countable union of disjoint measurable sets is itself a measurable set. In fact, we don't even know that the union of a pair of disjoint measurable sets is again measurable. We will deal with that question in the next section.

Meanwhile, let's increase our collection of known measurable sets by showing that $(-\infty, a)$ is measurable.

Example 24 ($(-\infty, a)$ is measurable). Let A be a subset of $\mathbb{R}$. Given $\varepsilon > 0$, choose a countable cover $\{I_k\}$ of A by open intervals such that

$$\sum_k l (I_k) \leq \mu^* (A) + \varepsilon.$$

Since $A \subseteq \cup_k I_k$, monotonicity and subadditivity imply that

$$\mu^* (A \cap (-\infty, a)) + \mu^* (A \cap [a, +\infty))$$
$$\leq \mu^* ((\cup_k I_k) \cap (-\infty, a)) + \mu^* ((\cup_k I_k) \cap [a, +\infty))$$
$$\leq \sum_k \mu^* (I_k \cap (-\infty, a)) + \sum_k \mu^* (I_k \cap [a, +\infty)).$$

Now the intersection of two intervals is again an interval and the measure of any interval is its length. Moreover, all the terms in the summations are nonnegative so the terms can be reordered. Therefore we can continue the chain of inequalities with

$$= \sum_k l\left(I_k \cap (-\infty, a)\right) + \sum_k l\left(I_k \cap [a, +\infty)\right)$$

$$= \sum_k \left[l\left(I_k \cap (-\infty, a)\right) + l\left(I_k \cap [a, +\infty)\right)\right]$$

$$= \sum_k l\left(I_k\right) \leq \mu^*\left(A\right) + \varepsilon.$$

Since $\varepsilon > 0$ was arbitrary, we conclude that

$$\mu^*\left(A \cap (-\infty, a)\right) + \mu^*\left(A \cap [a, +\infty)\right) \leq \mu^*\left(A\right).$$

The reverse inequality is a consequence of subadditivity. Therefore, the interval $(-\infty, a)$ is measurable.

It is apparent that using Carathéodory's definition to verify that sets are measurable tends to be quite tedious. Think about the increased complication of using the definition to prove that (a, b) is a measurable set. We need some additional tools.

5.4   Sigma algebras

The concept of a sigma algebra is a powerful tool for verifying measurability. While the definition of a sigma algebra is stated generally, our primary concern is with sigma algebras for which the X in the definition is either $\mathbb{R}$ or a subset of $\mathbb{R}$.

Definition 9 (Sigma algebra). *Let X be a set. A collection $\mathcal{A}$ of subsets of X is a **sigma algebra** if*

1. The empty set belongs to $\mathcal{A}$.

2. If $A \in \mathcal{A}$, then $A^c \in \mathcal{A}$.

3. Given a countable collection of sets $\{A_k\}$ from $\mathcal{A}$, we have $\cup_k A_k \in \mathcal{A}$.

By using complements, the third property of a sigma algebra also implies that $\cap_k A_k \in \mathcal{A}$ for any countable collection of sets from $\mathcal{A}$. (See exercise 28.)

Before proving that the set of measurable sets is a sigma algebra, we pause to demonstrate the power of the concept. For purposes of illustration, assume for now that the set of Lebesgue measurable sets is a sigma algebra.

Example 25. The following sets are measurable.

1. $[a, +\infty) = (-\infty, a)^c$.

2. $[a, b) = (-\infty, b) \cap [a, +\infty)$.

3. $(a, b) = \cup_n \left[a + \frac{1}{n}, b\right)$.

4. $\{a\} = \cap_n \left(a - \frac{1}{n}, a + \frac{1}{n}\right)$.

5. Any countable set.

With a bit more effort, one can prove that all open and all closed sets are measurable. (See exercise 37.)

Compare the work just expended to the toil that would have been required to prove the same facts directly from the definition. The difference in effort justifies the non-trivial mental energy we will expend to prove that the set of Lebesgue measurable sets is a sigma algebra.

The first two conditions of a sigma algebra are fairly straightforward exercises (25 and 26). As one might expect, significantly more effort is required to verify that the set of Lebesgue measurable sets is closed under countable unions. We begin by verifying that the union of two measurable sets is again measurable.

To that end, assume that E and F are measurable sets. Applying the measurability of E to an arbitrary set A,

$$\mu^* (A) = \mu^* (A \cap E) + \mu^* (A \cap E^c).$$

Then using the measurability of F with the set $A \cap E^c$,

$$\mu^* (A \cap E^c) = \mu^* ((A \cap E^c) \cap F) + \mu^* ((A \cap E^c) \cap F^c).$$

Combining these results and using subadditivity twice, we find that

$$
\begin{aligned}
\mu^* (A) &= \mu^* (A \cap E) + \mu^* ((A \cap E^c) \cap F) + \mu^* ((A \cap E^c) \cap F^c) \\
&\geq \mu^* ((A \cap E) \cup (A \cap E^c \cap F)) + \mu^* (A \cap E^c \cap F^c) \\
&= \mu^* (A \cap (E \cup F)) + \mu^* \left(A \cap (E \cup F)^c\right) \\
&\geq \mu^* \left((A \cap (E \cup F)) \cup \left(A \cap (E \cup F)^c\right)\right) \\
&= \mu^* (A).
\end{aligned}
$$

Since equality must hold throughout the last chain, we conclude that the union of two measurable sets is measurable. The measurability of finite unions of measurable sets follows by induction.

Now suppose that we have a sequence $\{F_k\}$ of disjoint, measurable sets. We will show that $\cup_k F_k$ is measurable. The proof relies on an intermediate result.

Let A be an arbitrary subset of the real numbers. We claim that

$$\mu^* \left(A \cap \left(\cup_{k=1}^n F_k\right)\right) = \sum_{k=1}^n \mu^* \left(A \cap F_k\right).$$

The claim is trivially true when $n = 1$. Assuming the result for $n - 1$, the measurability of F_n and the fact that the elements of $\{F_k\}$ are disjoint imply that

$$\begin{aligned}
\mu^* \left(A \cap \left(\cup_{k=1}^n F_k\right)\right) &= \mu^* \left(A \cap \left(\cup_{k=1}^n F_k\right) \cap F_n\right) \\
&\quad + \mu^* \left(A \cap \left(\cup_{k=1}^n F_k\right) \cap F_n^c\right) \\
&= \mu^* \left(A \cap F_n\right) + \mu^* \left(A \cap \left(\cup_{k=1}^{n-1} F_k\right)\right) \\
&= \mu^* \left(A \cap F_n\right) + \sum_{k=1}^{n-1} \mu^* \left(A \cap F_k\right) \\
&= \sum_{k=1}^n \mu^* \left(A \cap F_k\right).
\end{aligned}$$

Having previously verified that finite unions of measurable sets are measurable, we see that

$$\begin{aligned}
\mu^* (A) &= \mu^* \left(A \cap \left(\cup_{k=1}^n F_k\right)\right) + \mu^* \left(A \cap \left(\cup_{k=1}^n F_k\right)^c\right) \\
&\geq \mu^* \left(A \cap \left(\cup_{k=1}^n F_k\right)\right) + \mu^* \left(A \cap \left(\cup_{k=1}^\infty F_k\right)^c\right) \\
&= \sum_{k=1}^n \mu^* \left(A \cap F_k\right) + \mu^* \left(A \cap \left(\cup_{k=1}^\infty F_k\right)^c\right).
\end{aligned}$$

Because this inequality holds for all n, we can use subadditivity to conclude that

$$\begin{aligned}
\mu^* (A) &\geq \sum_{k=1}^\infty \mu^* \left(A \cap F_k\right) + \mu^* \left(A \cap \left(\cup_{k=1}^\infty F_k\right)^c\right) \\
&\geq \mu^* \left(\cup_{k=1}^\infty \left(A \cap F_k\right)\right) + \mu^* \left(A \cap \left(\cup_{k=1}^\infty F_k\right)^c\right) \\
&= \mu^* \left(A \cap \left(\cup_{k=1}^\infty F_k\right)\right) + \mu^* \left(A \cap \left(\cup_{k=1}^\infty F_k\right)^c\right).
\end{aligned}$$

The inequality in the other direction is a consequence of subadditivity. We conclude that the union of countably many disjoint, measurable sets is measurable.

What happens when the sets are not disjoint? Suppose that we have a sequence $\{E_k\}$ of measurable sets. Set $F_n = E_n \cap \left(\cup_{k=1}^{n-1} E_k\right)^c$. Then $\{F_k\}$ is a sequence of disjoint, measurable sets with $\cup_k E_k = \cup_k F_k$. Since $\cup_k F_k$ is measurable, so is $\cup_k E_k$. Any countable union of measurable sets is measurable.

The collection of measurable sets is a sigma algebra!

5.5   Measurable sets III

At this point, we know that $(-\infty, a)$ is a measurable set and that the collection of measurable sets is a sigma algebra. Given any collection S of subsets of $\mathbb{R}$, there is a smallest[11] sigma algebra containing S. This sigma algebra is called the sigma algebra **generated by** S. (See exercise 45.) Since we know that all intervals of the form $(-\infty, a)$ are measurable, we know that the sigma algebra $\mathcal{M}$ of measurable sets must contain the sigma algebra generated by $\{(-\infty, a) : a \in \mathbb{R}\}$. This sigma algebra is important enough to merit its own name.

Definition 10 (Borel sets). *The **Borel sigma algebra**, $\mathcal{B}$, is the sigma algebra generated by $\{(-\infty, a) : a \in \mathbb{R}\}$. The elements of $\mathcal{B}$ are called **Borel sets**.*

There are many other options for a collection of sets that will generate $\mathcal{B}$: finite open intervals, half-open intervals, open sets, compact sets, and more. (See exercise 46.)

We know that $\mathcal{B} \subseteq \mathcal{M}$. Is $\mathcal{B} = \mathcal{M}$? No, but as we shall see, the two sigma algebras are almost equal. The key to this investigation is an alternative description of measurable sets. Loosely speaking, theorem 22 (below) tells us that measurable sets are those sets that can be "closely approximated" by open or closed sets.[12]

---

[11] Smallest in this context means contained in any other sigma algebra satisfying the condition in question.

[12] This theorem also connects Carathéodory's and Lebesgue's senses of measurability. See Exercises 71 and following.

Theorem 22 (Measurability). *Let $E \subseteq \mathbb{R}$. The following are equivalent.*

1. *E is Lebesgue measurable in the sense of Carathéodory.*

2. *Given $\varepsilon > 0$, there is an open set G such that $E \subseteq G$ and $\mu^* (G \backslash E) < \varepsilon$.*[13]

3. *Given $\varepsilon > 0$, there is an closed set F such that $F \subseteq E$ and $\mu^* (E \backslash F) < \varepsilon$.*

Proof. Let E be a measurable set. First suppose that $\lambda(E) < \infty$ and let $\varepsilon > 0$. Then there is a countable cover of E by open intervals $\{I_k\}$ such that

$$\lambda(E) \le \lambda(\cup_k I_k) \le \sum_k l(I_k) < \lambda(E) + \varepsilon.$$

If we set $G = \cup_k I_k$, then $E \subseteq G$ and $\lambda(G) < \lambda(E) + \varepsilon$. The measurability of E now implies that

$$\lambda(E) + \varepsilon > \lambda(G) = \lambda(G \cap E) + \lambda(G \backslash E) = \lambda(E) + \lambda(G \backslash E).$$

Statement (2) follows by subtraction.

If $\lambda(E) = \infty$, then apply the same argument to $E_k = E \cap [-k, k]$ to find open sets G_k containing E_k such that $\mu^* (G_k \backslash E_k) < \frac{\varepsilon}{2^k}$. Then $G = \cup_k G_k$ is an open set containing E with $\mu^* (G \backslash E) < \varepsilon$.

To prove that (2) implies (1), suppose that E is a subset of $\mathbb{R}$ and that for any $\varepsilon > 0$ there is an open set G containing E with $\mu^* (G \backslash E) < \varepsilon$. Note that $G^c \subseteq E^c$ and that G, being an open set, is measurable (exercise 37). We will use the measurability of G to establish the measurability of E.

Let A be an arbitrary subset of $\mathbb{R}$. We need to show that $\mu^* (A) \ge \mu^* (A \cap E) + \mu^* (A \cap E^c)$. Since G is measurable,

$$
\begin{aligned}
\mu^* (A \cap E^c) &= \mu^* ((A \cap E^c) \cap G) + \mu^* ((A \cap E^c) \cap G^c) \\
&= \mu^* (A \cap (G \backslash E)) + \mu^* (A \cap G^c) \\
&\le \mu^* (G \backslash E) + \mu^* (A \cap G^c) \\
&< \varepsilon + \mu^* (A \cap G^c).
\end{aligned}
$$

Again appealing to the measurability of G,

$$
\begin{aligned}
\mu^* (A) &= \mu^* (A \cap G) + \mu^* (A \cap G^c) \\
&> \mu^* (A \cap E) + \mu^* (A \cap E^c) - \varepsilon.
\end{aligned}
$$

---

[13] Recall that for sets A and B, $A \backslash B = A \cap B^c$.

Since $\varepsilon > 0$ is arbitrary, we conclude that $\mu^*(A) \geq \mu^*(A \cap E) + \mu^*(A \cap E^c)$ so that E is measurable.

Having established that statements (1) and (2) are equivalent, the equivalence of (1) and (3) follows by taking complements.   □

By definition, the outer measure of any set E is approximated by $\sum_k l(I_k)$ where $\{I_k\}$ is a countable cover of E by open intervals. In this case, the difference between the outer measures of $\cup_k I_k$ and E is arbitrarily small. If you look at the proof of theorem 22, you will see that E is measurable exactly when the outer measure of the difference between the sets $\cup_k I_k$ and E is also small. (Specifically, $\mu^*(\cup_k I_k) - \mu^*(E)$ being small should imply that $\mu^*(\cup_k I_k \setminus E)$ is also small.) Thinking about this another way, the critical sets on which Carathéodory's condition must hold are unions of open intervals (open sets) that barely contain E. A similar comment applies to closed sets barely contained in E. If we take this idea to the limit, we arrive at the following theorem that tells us that Lebesgue measurable sets are almost Borel sets.

Theorem 23. *A subset E of $\mathbb{R}$ is Lebesgue measurable if and only if E can be written as $E = B \cup Z$ where B is a Borel set and Z is a set of measure zero.*

Proof. Exercise 44.   □

5.6 Measurable functions

Think back to what triggered this line of investigation. Given a function $f : [a, b] \rightarrow (\alpha, \beta)$ and a partition $\mathcal{P} = \{[y_{k-1}, y_k]\}_{k=1}^n$ of $[\alpha, \beta]$, we set $E_k = f^{-1}((y_{k-1}, y_k])$, $k = 1, 2, \ldots, n$, and compute the upper and lower Lebesgue sums $S_L(f, \mathcal{P}) = \sum_{k=1}^n y_{k-1} \cdot m(E_k)$ and $S_{\overline{L}}(f, \mathcal{P}) = \sum_{k=1}^n y_k \cdot m(E_k)$ where $m(A)$ is the measure of the set A. At first, this process always seems to produce a well-defined value for the integral. A closer look reveals problems with the notion of the measure of an arbitrary set E.

We have essentially solved the measurement problem by using Lebesgue (outer) measure on measurable sets. This leaves us with the task of applying the ideas of measurement in the context of a function. The critical condition required to make our computations work is that $E_k = f^{-1}((y_{k-1}, y_k])$ be measurable. This leads us to the definition of a measurable function.

Definition 11 (Measurable function). *Let X be a measurable subset[14] of real numbers and let* $f : X \to \mathbb{R}$. *Then* f *is* **measurable** *if the preimage under* f *of every interval is measurable.*

While this definition is straightforward enough, the variety of interval forms makes it slightly awkward to work with. Even if we were to restrict the definition to consider only preimages of the form $f^{-1}((a, b])$, the computations would be needlessly complicated. Sigma algebras prove useful in simplifying this situation. Since the collection of measurable sets is a sigma algebra and preimages interact well with complements and unions, it suffices to verify that preimages of the form $f^{-1}((-\infty, a))$ are measurable sets. When it is more convenient, we can verify instead that all preimages of the form $f^{-1}((-\infty, a])$ or $f^{-1}((a, +\infty))$ are measurable sets. (See exercises 49 and 59.)

Example 26. Let $f(x) = |x|$. Then

$$f^{-1}((-\infty, a)) = \begin{cases} \varnothing, & a \leq 0 \\ (-a, a), & 0 < a. \end{cases}$$

Since these preimages are always measurable sets, f is a measurable function.

Notice how much more awkward our verification would be if we were required to explicitly verify that the preimage of every possible type of interval is measurable. Also note the usefulness of knowing that the collection of measurable sets is a sigma algebra in the next two examples.

Example 27 (Increasing functions are measurable). Suppose X is a measurable set and that $f : X \to \mathbb{R}$ is an increasing function. Let $\alpha = \sup\{x \in X : f(x) < a\}$. (It may be that $\alpha = -\infty$ in which case we understand $(-\infty, -\infty)$ to be the empty set.) Then $f^{-1}((-\infty, a)) = (-\infty, \alpha) \cap X$ or $f^{-1}((-\infty, a)) = (-\infty, \alpha] \cap X$. In either case, the preimage of $(-\infty, a)$ is the intersection of two measurable sets and so is measurable.

Example 28 (Continuous functions are measurable). Suppose X is a measurable set and that $f : X \to \mathbb{R}$ is continuous. Since $(-\infty, \alpha)$ is open, its preimage under the continuous function f is relatively open in X. In other words, $f^{-1}((-\infty, a))$ is the intersection of X with an open set. Since

---

[13] It is not strictly necessary to specify that X is measurable since $X = f^{-1}((-\infty, +\infty))$. We include the restriction to remind ourselves that a measurable function will no longer be measurable if we restrict its domain to a non-measurable subset.

open sets are measurable, $f^{-1}\left((-\infty,a)\right)$ is a measurable set. Thus f is a measurable function.

In a manner similar to the way that measurable sets are closed under complements and unions (and so form a sigma algebra), measurable functions are closed under addition and scalar multiplication (and so form an vector space).

Theorem 24 (Measurable functions form vector space). *Suppose that f and g are measurable functions defined on $[a,b]$ and that c is a real number. Then cf, $f+g$, $\max\{f,g\}$, and $\min\{f,g\}$ are also measurable.*

Proof. The proofs that cf, $\max\{f,g\}$, and $\min\{f,g\}$ are measurable are left as exercises.

To show that $f+g$ is measurable, suppose that $(f+g)(x) < c$. Let $k = c - (f+g)(x)$ so that $g(x) = c - k - f(x)$. Since $k > 0$, we can choose a rational number r so that $f(x) < r < f(x) + k$. Then $g(x) = c - k - f(x) < c - r$.

We have just shown that if $(f+g)(x) < c$, there is a rational number r such that $f(x) < r$ and $g(x) < c - r$. Hence

$$(f+g)^{-1}\left((-\infty,c)\right) \subseteq \cup_{r\in\mathbb{Q}}\left(f^{-1}\left((-\infty,r)\right) \cap g^{-1}\left((-\infty,c-r)\right)\right).$$

The reverse containment is straightforward so, in fact, the two sets are equal. Since f and g are measurable and the collection of measurable sets is a sigma algebra, $(f+g)^{-1}(-\infty,c)$, a countable union of measurable sets, is measurable. Thus $f+g$ is a measurable function. □

The class of measurable functions has an important property that sets it apart from other classes of functions such as continuous and Riemann integrable functions: the set of measurable functions is closed under taking limits.

Theorem 25 (Limits of measurable functions are measurable). *Let $\{f_k\}$ be a sequence of measurable functions that converges pointwise to the function f. Then f is measurable.*

Proof. Let $g_n(x) = \sup_{k\geq n} f_k(x)$. Then

$$g_n^{-1}\left((-\infty,a]\right) = \cap_{k\geq n} f_k^{-1}\left((-\infty,a]\right)$$

which is a measurable set. Thus all of the $\{g_n\}$ are measurable. Now set $h(x) = \inf_n g_n(x)$. Then

$$h^{-1}\left((-\infty,a)\right) = \cup_n g_n^{-1}\left((-\infty,a)\right)$$

so h is measurable. Since $\{f_k\}$ converges to f,

$$h(x) = \inf_{n} \sup_{k \geq n} f_k(x) = \overline{\lim_{k}} f_k(x) = \lim_{k} f_k(x) = f(x),$$

establishing the measurability of f.   □

Note that in the process of proving the previous theorem, we have also verified that for any sequence $\{f_k\}$ of measurable functions, $\sup_k f_k$, $\inf_k f_k$, and $\overline{\lim}_k f_k$ are measurable when they are finite. Similarly, when finite, $\underline{\lim}_k f_k$ is measurable.

In theorem 20 (page 63), we proved that a bounded function is Riemann integrable if and only if it is continuous except on a set of measure zero. Because this type of condition is so ubiquitous in the context of the Lebesgue integral, we introduce a definition to capture the idea.

Definition 12 (Almost everywhere). *A property is said to hold **almost everywhere (a.e.)** if the property holds except on a set of measure zero.*

Theorem 26. *If f is a measurable function and $g = f$ a.e., then g is measurable.*

Proof. Let $c \in \mathbb{R}$, $D = \{x : f(x) \neq g(x)\}$, and $E = \{x : f(x) = g(x)\}$. By hypothesis, $\mu^*(D) = 0$ so both D and E are measurable. Now

$$f^{-1}((-\infty, c)) \cap E = g^{-1}((-\infty, c)) \cap E$$
$$\subseteq g^{-1}((-\infty, c))$$
$$\subseteq \left(g^{-1}((-\infty, c)) \cap E \right) \cup D$$
$$= \left(f^{-1}((-\infty, c)) \cap E \right) \cup D.$$

Thus $g^{-1}((-\infty, c)) = \left(f^{-1}((-\infty, c)) \cap E \right) \cup Y$ where $Y \subseteq D$. As a set with zero outer measure, Y is measurable and since $f^{-1}((-\infty, c))$, E, and Y are all measurable, so is $g^{-1}((-\infty, c))$. We conclude that g is a measurable function.   □

Corollary 27 (of theorem 25). *Let $\{f_k\}$ be a sequence of measurable functions that converges almost everywhere to the function f. Then f is measurable.*

Proof. Exercise 58.   □

This type of situation is typical. Behavior on a set of measure zero usually can be ignored.

5.7 Exercises

5.0 Another approach: filling the gaps

For the following three exercises assume that f is a function defined on $[a, b]$ and that $\alpha < f < \beta$. Also assume that the measures of all relevant sets are well defined.

1. Let $\mathcal{P} = \{[y_{k-1}, y_k]\}_{k=1}^{n}$ be a partition of $[\alpha, \beta]$ and set $E_k = f^{-1}((y_{k-1}, y_k])$.

   (a) Prove that the E_k are disjoint sets.
   (b) Prove that $\cup_{k=1}^{n} E_k = [a, b]$.

2. Let $\mathcal{P}_1$ and $\mathcal{P}_2$ be partitions of $[\alpha, \beta]$ with $\mathcal{P}_2$ a refinement of $\mathcal{P}_1$. Prove that $S_{\underline{L}}(f, \mathcal{P}_1) \leq S_{\underline{L}}(f, \mathcal{P}_2) \leq S_{\overline{L}}(f, \mathcal{P}_2) \leq S_{\overline{L}}(f, \mathcal{P}_1)$. (First assume that $\mathcal{P}_2$ adds a single new division point to $\mathcal{P}_1$ and compute $S_{\underline{L}}(f, \mathcal{P}_2) - S_{\underline{L}}(f, \mathcal{P}_1)$.)

3. Let $\mathcal{P}_1$ and $\mathcal{P}_2$ be partitions of $[\alpha, \beta]$. Use exercise 2 to prove that $S_{\underline{L}}(f, \mathcal{P}_1) \leq S_{\overline{L}}(f, \mathcal{P}_2)$.

5.0 Another approach: deeper reflections

4. Half-open subintervals are used in the definition of the upper and lower Lebesgue sums instead of the closed intervals that are used for the Riemann integral.

   (a) Why?
   (b) Could we have bounded f on $[a, b]$ by $\alpha \leq f \leq \beta$? Explain.

5. If f is a function defined on $[a, b]$ with $\alpha < f < \beta$, then there are infinitely many different choices for α and β. The set of partitions of $[\alpha, \beta]$ used to define the Lebesgue integral will be different if we change the values of α or β. Explain why the choice of α and β does not matter.

5.1 Measurable sets I: filling the gaps

6. Prove that the rationally-equivalent relation (two numbers are rationally equivalent if their difference is rational) is an equivalence relation. In other words, verify that the relation is

   (a) **Reflexive.** Any real number x is rationally equivalent to itself.
   (b) **Symmetric.** If x is rationally equivalent to y, then y is rationally equivalent to x.

(c) **Transitive.** If x is rationally equivalent to y and y is rationally equivalent to z, then x is rationally equivalent to z.

7. Given an equivalence relation on $\mathbb{R}$, an equivalence class is the set of all elements that are related to some fixed element x. Prove that the set of equivalence classes forms a partition of $\mathbb{R}$. In other words, verify that

   (a) If A and B are equivalence classes with $A \neq B$, then $A \cap B = \emptyset$. (Use contraposition.)
   (b) Every element of $\mathbb{R}$ is in some equivalence class.

5.2 Outer measure: filling the gaps

8. Prove that μ^* is monotone. In other words, verify that if A and B are sets with $A \subset B$ then $\mu^*(A) \leq \mu^*(B)$.

9. Prove that the outer measure of any interval is its length. (Use monotonicity. Don't forget infinite intervals.)

10. Prove that μ^* is translation invariant. In other words, show that if $B = A + t = \{a + t : a \in A\}$ then $\mu^*(B) = \mu^*(A)$.

11. If $\mu^*(A_k)$ is infinite for some k, why is $\mu^*(\cup_k A_k) \leq \sum_k \mu^*(A_k)$?

12. Explain why we can conclude that $\mu^*(\cup_k A_k) \leq \sum_k \mu^*(A_k)$ from the knowledge that $\mu^*(\cup_k A_k) \leq \sum_k \mu^*(A_k) + \varepsilon$ for any $\varepsilon > 0$.

5.2 Outer measure: deeper reflections

13. Compute the outer measures of the following sets.

    (a) $\left\{ \frac{1}{n} : n \in \mathbb{N} \right\}$.
    (b) The rational numbers in $[0, 1]$.
    (c) The irrational numbers in $[0, 1]$. (Use your work in (b) and the measure of $[0, 1]$ to find a lower bound on the measure of the irrational numbers in $[0, 1]$.)

5.3 Measurable sets II: filling the gaps

14. Prove that the following statements are equivalent for any subset E of $\mathbb{R}$.

    (a) If A and B are disjoint sets with $E = A \cup B$, then $\mu^*(E) = \mu^*(A) + \mu^*(B)$.
    (b) For any subset C of $\mathbb{R}$, $\mu^*(E) = \mu^*(E \cap C) + \mu^*(E \cap C^c)$.

15. Suppose that I is a finite interval. Explain why
$$l(I) = l(I \cap (-\infty, a)) + l(I \cap [a, +\infty)).$$

16. Given a sequence $\{E_k\}$ of sets, why is $\mu^* \left(\cup_{k=1}^n E_k \right) \leq \mu^* \left(\cup_{k=1}^\infty E_k \right)$?

5.3 Measurable sets II: deeper reflections

17. Prove directly from the definition of measurability that every countable set is Lebesgue measurable.

18. Prove that any set with outer measure zero is Lebesgue measurable.

19. Prove that $[0, 1] \setminus \mathbb{Q}$ is measurable and find its measure.

20. Explain why the translation of any measurable set is again measurable. (Use exercise 10.)

21. Prove that $\mu^* ([a, b] \cup [c, d]) = \mu^* ([a, b]) + \mu^* ([c, d])$ when $[a, b]$ and $[c, d]$ are disjoint.

22. Suppose that A and B are sets satisfying $A \subseteq (-\infty, a)$ and $B \subseteq (a, \infty)$ for some real number a. Prove that $\mu^* (A \cup B) = \mu^* (A) + \mu^* (B)$.

23. The set V constructed in Vitali's example is not measurable according to Carathéodory's definition. Explain how we know this.

24. An alternative approach to the problem identified by Vitali's example would be to give up countable additivity and require only finite additivity. The hope would be that all subsets of $\mathbb{R}$ might be measurable in this less demanding sense. Explain why this hope is doomed.

5.4 Sigma algebras: filling the gaps

25. Prove that $\mathbb{R}$ and $\varnothing$ are measurable.

26. Prove that if E is a measurable set, so is E^c.

27. Assuming that $\mu^* (E \cup F) = \mu^* (E) + \mu^* (F)$ for disjoint, measurable sets E and F, prove that $\mu^* \left(\cup_{k=1}^n E_k \right) = \sum_{k=1}^n \mu^* (E_k)$ for any finite set $\{E_k\}_{k=1}^n$ of disjoint, measurable sets.

28. Use complements to prove that sigma algebras are closed under countable intersections. (Use De Morgan's Law.)

29. Verify that $(A \cap E) \cup ((A \cap E^c) \cap F) = A \cap (E \cup F)$.

30. Assuming that the union of two measurable sets is measurable, use induction to prove that the finite union of measurable sets is measurable.

31. Explain why $A \cap \left(\cup_{k=1}^n F_k \right) \cap F_n = A \cap F_n$ and $A \cap \left(\cup_{k=1}^n F_k \right) \cap F_n^c = A \cap \left(\cup_{k=1}^{n-1} F_k \right)$ when the sets $\{F_k\}$ are disjoint.

32. Suppose that $\{E_k\}$ is a countable collection of measurable sets and define $F_n = E_n \cap \left(\cup_{k=1}^{n-1} E_k\right)^c$.

    (a) Verify that $\{F_k\}$ is a collection of disjoint sets.
    (b) Why is F_k measurable?
    (c) Prove that $\cup_k E_k = \cup_k F_k$.

33. Why is $\mu^* \left(A \cap \left(\cup_{k=1}^n F_k\right)^c\right) \geq \mu^* \left(A \cap \left(\cup_{k=1}^\infty F_k\right)^c\right)$?

34. Show that $\cup_{k=1}^\infty (A \cap F_k) = A \cap \left(\cup_{k=1}^\infty F_k\right)$.

5.4 Sigma algebras: deeper reflections

35. Give two proofs that $(-\infty, a]$ is measurable.

    (a) Use Carathéodory's definition of measurability.
    (b) Use sigma algebras.

36. Give two proofs that $[a, b]$ is measurable.

    (a) Use Carathéodory's definition of measurability.
    (b) Use sigma algebras.

37. Prove that every open set G and every closed set F are measurable. (Show that G is the countable union of open intervals.)

38. Let $X = \{1, 2, 3\}$.

    (a) What is the smallest possible sigma algebra based on X?
    (b) What is the largest possible sigma algebra based on X?

39. Let $X = \{1, 2, 3, 4\}$ and let $S = \{\{1\}, \{1, 2\}\}$. What other sets must be added to S to form a sigma algebra on X?

40. Suppose that $\{\mathcal{A}_\alpha\}$ is a collection of sigma algebras on a set X. (We do not assume that the index set is countable.) Prove that $\cap_\alpha \mathcal{A}_\alpha$ is a sigma algebra.

41. Let f be an arbitrary real-valued function defined on a measurable set X. Prove that the collection of subsets of $\mathbb{R}$ whose inverse images under f are measurable forms a sigma algebra.

5.5 Measurable sets III: filling the gaps

42. Fill in the details of the first part of the proof of theorem 22 (page 104) when $\lambda(E) = \infty$. Specifically, prove that if two sequences of sets $\{A_k\}_{k=1}^\infty$ and $\{B_k\}_{k=1}^\infty$ satisfy $A_k \subseteq B_k$ and $\mu^*(B_k \backslash A_k) < \frac{\varepsilon}{2^k}$, then $\cup_k A_k \subseteq \cup_k B_k$ and $\mu^*((\cup_k B_k) \backslash (\cup_k A_k)) < \varepsilon$.

43. Using the equivalence of statements (1) and (2), complete the proof the proof of theorem 22 (page 104).

    (a) Prove that if E is Lebesgue measurable in the sense of Carathéodory then given $\varepsilon > 0$ there is a closed set F such that $F \subseteq E$ and $\mu^* (E\backslash F) < \varepsilon$. (If E is measurable, so is E^c.)
    (b) Prove that if for any $\varepsilon > 0$ there is a closed set F such that $F \subseteq E$ and $\mu^* (E\backslash F) < \varepsilon$, then E is Lebesgue measurable in the sense of Carathéodory.

44. Let $E \subset \mathbb{R}$.

    (a) Prove that if $E = B \cup Z$ where $B \in \mathcal{B}$ and Z has measure zero, then $E \in \mathcal{M}$, the set of Lebesgue measurable sets. (Use exercise 18.)
    (b) Prove that if E is Lebesgue measurable, then there are sets B and Z where $B \in \mathcal{B}$ and Z has measure zero such that $E = B \cup Z$. (Start with closed sets F_n satisfying $F_n \subseteq E$ and $\mu^* (E\backslash F_n) < \frac{1}{n}$.)

5.5 Measurable sets III: deeper reflections

45. In the definition of the Borel sigma algebra, we took for granted that there is a smallest sigma algebra containing $\{(-\infty, a) : a \in \mathbb{R}\}$. Verify that this is so. Let S be a collection of subsets of $\mathbb{R}$. Prove that there is a smallest sigma algebra containing S.

    (a) Why is there a sigma algebra containing S?
       Let $\mathcal{S}$ be the intersection of all the sigma algebras containing S.
    (b) Explain why $\mathcal{S}$ is a sigma algebra containing S.
    (c) Explain why $\mathcal{S}$ is contained in any other sigma algebra containing S.

46. Prove that the following collections of subsets of $\mathbb{R}$ generate the same sigma algebra. In other words, show that any sigma algebra containing one type of sets contains the other types of sets.

    (a) The set of all intervals of the form $(-\infty, a)$.
    (b) The set of all intervals of the form (a, b).
    (c) The set of all open sets.
    (d) The set of all compact sets.

47. Prove that if $\{A_k\}$ is a sequence of measurable sets for which $\lambda \left(\cup_{k=1}^{\infty} A_k \right) = m > 0$, then for any $\varepsilon > 0$ there is an N so that $\lambda \left(\cup_{k=1}^{N} A_k \right) > m - \varepsilon$. (Rework the sets to be disjoint and use countable additivity.)

48. Prove that if $\{A_k\}$ is a sequence of measurable sets for which $\lambda\left(\cap_{k=1}^{\infty} A_k\right) = 0$, then for any $\varepsilon > 0$ there is an N so that $\lambda\left(\cap_{k=1}^{N} A_k\right) < \varepsilon$. (One approach is to use the previous exercise and relative complements. Alternatively, find a way to use countable additivity.)

5.6 Measurable functions: filling the gaps

49. Suppose that X is a measurable set, $f : X \to \mathbb{R}$, and $f^{-1}\left((-\infty, a)\right)$ is measurable for all $a \in \mathbb{R}$. Use exercise 59 to prove that the following sets are measurable for all values $a, b \in \mathbb{R}$.

   (a) $f^{-1}\left([a, +\infty)\right)$
   (b) $f^{-1}\left((-\infty, a]\right)$
   (c) $f^{-1}\left([a, b)\right)$

50. Suppose that $f : X \to \mathbb{R}$ is an increasing function and that $\alpha = \sup\{x \in X : f(x) < a\}$. Prove that $(-\infty, \alpha) \cap X \subseteq f^{-1}\left((-\infty, a)\right) \subseteq (-\infty, \alpha] \cap X$.

51. Use the ε–δ definition of continuity to prove that if $f : A \to \mathbb{R}$ is continuous and U is an open set of real numbers, then $f^{-1}(U) = V \cap A$ for some open subset V of $\mathbb{R}$.

52. Suppose that f is a measurable function and that c is a real number. Show that cf is measurable.

53. Show that
$$\cup_{r \in \mathbb{Q}} \left(f^{-1}\left((-\infty, r)\right) \cap g^{-1}\left((-\infty, c - r)\right)\right) \subseteq (f + g)^{-1}(-\infty, c).$$

54. Suppose that $f, g : [a, b] \to \mathbb{R}$ are measurable functions. Prove that $\max\{f, g\}$ and $\min\{f, g\}$ are also measurable.

55. Let $g_n(x) = \sup_{k \geq n} f_k(x)$. Explain why
$$g_n^{-1}\left((-\infty, a]\right) = \cap_{k \geq n} f_k^{-1}\left((-\infty, a]\right).$$

56. Let $g_n(x) = \inf_{k \geq n} f_k(x)$. Explain why
$$g_n^{-1}\left((-\infty, a)\right) = \cup_{k \geq n} f_k^{-1}\left((-\infty, a)\right).$$

57. In the proof of theorem 26 (page 108), why is E measurable?

58. Prove Corollary 27 (page 108). (Define new functions g_k and g so that $g_k = f_k$ a.e., $g = f$ a.e., and $\{g_k\}$ converges pointwise to g everywhere.)

5.6 Measurable functions: deeper reflections

59. Let $X \subseteq \mathbb{R}$ and suppose that $f : X \to \mathbb{R}$.

(a) Prove that $f^{-1}(A^c) = X \setminus f^{-1}(A)$ for any $A \subseteq \mathbb{R}$.

(b) Prove that $f^{-1}(\cup_\alpha A_\alpha) = \cup_\alpha f^{-1}(A_\alpha)$ for any (not necessarily countable) collection of subsets $\{A_\alpha\}$ of $\mathbb{R}$.

(c) Prove that $f^{-1}(\cap_\alpha A_\alpha) = \cap_\alpha f^{-1}(A_\alpha)$ for any collection of subsets $\{A_\alpha\}$ of $\mathbb{R}$.

60. Use the definition of measurable function to prove the following functions are measurable.

(a) $f : [0, 1] \to \mathbb{R}$ where $f(x) = 2x + 1$.

(b) $f : \mathbb{R} \to \mathbb{R}$ where $f(x) = x^2$.

(c) The Dirichlet function d.

(d) $f : [0, 2\pi] \to \mathbb{R}$ where $f(x) = \sin x$.

(e)

$$f(x) = \begin{cases} -2 & x \in (-\infty, -1) \\ x^2 - 1 & x \in [-1, 2] \\ 3 & x \in (2, +\infty). \end{cases}$$

61. Give an example of a nonmeasurable function.

62. Give two examples of increasing functions $f, g : \mathbb{R} \to \mathbb{R}$ such that $f^{-1}((-\infty, 1)) = (-\infty, \alpha)$ and $g^{-1}((-\infty, 1)) = (-\infty, \beta]$, where $\alpha = \sup\{x \in X : f(x) < 1\}$ and $\beta = \sup\{x \in X : g(x) < 1\}$.

63. Suppose that f is a measurable function.

(a) Prove that $|f|$ is measurable.

(b) Prove that $f^+ = \frac{1}{2}(|f| + f)$ and $f^- = \frac{1}{2}(|f| - f)$ are both measurable.

(c) Give alternative definitions of f^+ and f^- as piecewise defined functions.

(d) Write f in terms of f^+ and f^-.

64. Suppose that f and g are positive measurable functions with a common domain. Prove that the product fg is measurable. (Mimic the proof for $f + g$.)

65. Suppose that f and g are (not necessarily positive) measurable functions with a common domain. Prove that the product fg is measurable. (Use the previous two exercises.)

66. Prove that any function $f : [a, b] \rightarrow \mathbb{R}$ that is continuous a.e. (the set of discontinuities has measure zero) is a measurable function.

67. Give an example of a function that is continuous a.e. but is not equal a.e. to any continuous function.

68. Give an example of a function that is equal a.e. to a continuous function but is not itself continuous a.e.

69. Suppose the sequence $\{a_k\}$ converges. Explain why $\sup_{k \geq n} a_k$ and $\inf_{k \geq n} a_k$ both exist and are finite.

70. Suppose that $\{f_k\}$ is a sequence of measurable functions which is pointwise bounded. In other words, $\{f_k(x)\}$ is a bounded sequence for any fixed x. Prove that $\inf_k f_k$ and $\underline{\lim}_k f_k$ are measurable.

5.7 Related ideas: an alternative approach to measurability

Lebesgue used an alternate definition of measurability based on both outer and inner measure. The following exercises outline the basic contours of this approach.

The **inner measure** of a subset A of $\mathbb{R}$ is

$$\mu_* (A) = \sup \{\mu^* (F) : F \subseteq A \text{ with } F \text{ compact}\}.$$

71. Prove that $\mu_* (A) \leq \mu^* (A)$ for every subset A of $\mathbb{R}$.

72. Prove that $\mu_* (I) = l(I)$ for any interval I.

A subset A of $\mathbb{R}$ with $\mu^* (A) < \infty$ is **measurable** (in the sense of Lebesgue) if $\mu_* (A) = \mu^* (A)$. In this case we write $\lambda (A) = \mu_* (A) = \mu^* (A)$. If $\mu^* (A) = \infty$, we say that A is **measurable** (in the sense of Lebesgue) if $A \cap [-n, n]$ is measurable for each $n \in \mathbb{N}$.

73. Prove that every set with zero outer measure is measurable in the sense of Lebesgue.

74. Prove that every compact set is measurable in the sense of Lebesgue.

75. Prove that if $\{K_j\}$ is a finite set of disjoint, compact sets, then $\mu^* (\cup_j K_j) = \sum_j \mu^* (K_j)$. (Use the fact that disjoint, compact sets must have a positive distance between them.)

76. Use exercise 75 to prove that if $\{A_i\}$ is a countable set of disjoint measurable sets then $\cup_i A_i$ is measurable and $\lambda (\cup_i A_i) = \sum_i \lambda (A_i)$. (First

assume that $A \subseteq [-n, n]$. Show that $\mu_*\left(\cup_i A_i\right) \geq \sum_j \mu^*\left(K_j\right)$ for an appropriately chosen finite collection of disjoint, compact sets.)

77. Prove that the collection of sets that are measurable in the sense of Lebesgue is a sigma algebra.

78. Use exercises 73 and 77 to prove that the collection of sets that are measurable in the sense of Lebesgue includes all the sets that are measurable in the sense of Carathéodory.

79. Use theorem 22 to prove that the collection of sets that are measurable in the sense of Carathéodory includes all of the sets that are measurable in the sense of Lebesgue.

5.8 References

Bartle, R.G. (1966). *The Elements of Integration*. John Wiley & Sons.

Bressoud, D.M. (2006). *A Radical Approach to Real Analysis* (2nd ed.). Mathematical Association of America.

Burk, F.E. (2007). *A Garden of Integrals*. Mathematical Association of America.

Hewitt, E. and K. Stromberg (1975). *Real and Abstract Analysis*. Springer.

Kharazishvili, A.B. (2004). *Nonmeasurable Sets and Functions*. North-Holland Mathematics Studies, **195**. Elsevier Science B.V.

Taylor, A.E. (2010). *General Theory of Functions and Integration*. Dover.

Vitali, G. (1908). Sui gruppi di punti e sulle funzioni di variabili reali, *Atti dell'Accademia delle Scienze di Torino* (in Italian) **43**: 229–246. archive.org/stream/attidellarealeac43real#page/229.

CHAPTER **6**

The Lebesgue Integral

The theory of measurable sets and functions provides all the supporting structure required to make a formal definition of the Lebesgue integral. With some modest modification, the approach suggested at the beginning of the last chapter (page 92) is ready for development.

Definition 13 (Lebesgue integral). *Let f be a (Lebesgue) measurable function on $[a, b]$ satisfying $\alpha < f \leq \beta < \infty$. Let $\mathcal{P} = \{[y_{k-1}, y_k]\}_{k=1}^{n}$ be a partition of $[\alpha, \beta]$ and set $E_k = f^{-1}((y_{k-1}, y_k])$, $k = 1, 2, \ldots, n$. The (lower)* **Lebesgue sum** *of f with respect to $\mathcal{P}$ is*

$$S_{\underline{L}}(f, \mathcal{P}) = \sum_{k=1}^{n} y_{k-1} \cdot \lambda(E_k)$$

where $\lambda(A)$ is the Lebesgue measure of the set A. The **Lebesgue integral** *of f over $[a, b]$ is*

$$^{L}\!\!\int_{a}^{b} f = \sup_{\mathcal{P}} S_{\underline{L}}(f, \mathcal{P})$$

where the supremum is taken over all partitions of $[\alpha, \beta]$.[1]

This definition may remind you of the definition of the Darboux integral with y_{k-1} playing the role of $\inf_{I_k} f$. This time, however, there is no need for upper sums or upper integrals–the Lebesgue integral always exists for bounded, measurable functions.

6.1 Variations

The resemblance of the definition of the Lebesgue integral to that of the Darboux integral suggests a possible modification of the definition. In fact,

---

[1] While $S_{\overline{L}}(f, \mathcal{P}) = \sum_{k=1}^{n} y_k \cdot \lambda(E_k)$ does not appear in the definition of the Lebesgue integral, it is occasionally used in proofs.

there are several variations on Lebesgue's approach to integration. We will consider two of them before investigating the properties of the Lebesgue integral.

6.1.1 A Darboux-like variation

William H. Young (1863–1942) developed an integral based on the observation that the sets $E_k = f^{-1}((y_{k-1}, y_k])$ partition the domain $[a, b]$ into disjoint measurable sets. We capture this idea in the concept of a measurable partition.

Definition 14 (Measurable partition). *A **measurable partition** of a set S is a finite collection of disjoint, measurable subsets of S whose union is S.*

Since intervals are measurable sets, measurable partitions generalize Riemann-Darboux partitions. But we need to exercise some care here since, unlike in a Riemann partition, the sets in a measurable partition must be disjoint. While the intervals in a Riemann-Darboux partition are not disjoint, they are **non-overlapping** in that a pair of intervals can have at most a single point in common. Moving from non-overlapping to disjoint sets is not a serious obstacle since nothing of significance would have changed had we used partitions of the form $\{[x_0, x_1), [x_1, x_2), [x_2, x_3), \ldots, [x_{n-1}, x_n]\}$ for the Riemann integral. With these observations in mind, we drop the restriction that f is measurable and make the following definition.

Definition 15 (Lebesgue-Young integral). *Let f be a bounded function defined on $[a, b]$. Let $\mathcal{P} = \{E_k\}_{k=1}^n$ be a measurable partition of $[a, b]$. The **lower and upper Lebesgue-Young sums** of f over $\mathcal{P}$ are*

$$S_{\underline{L\text{-}Y}}(f, \mathcal{P}) = \sum_{\mathcal{P}} \inf_{E_k} f \cdot \lambda(E_k)$$

and

$$S_{\overline{L\text{-}Y}}(f, \mathcal{P}) = \sum_{\mathcal{P}} \sup_{E_k} f \cdot \lambda(E_k).$$

*The **lower and upper Lebesgue-Young integrals** of f over $[a, b]$ are*

$$\underline{L\text{-}Y}\int_a^b f = \sup_{\mathcal{P}} S_{\underline{L\text{-}Y}}(f, \mathcal{P})$$

and

$$\overline{L\text{-}Y}\int_a^b f = \inf_{\mathcal{P}} S_{\overline{L\text{-}Y}}(f, \mathcal{P})$$

where the supremum and infimum are taken over the set of all measurable partitions of $[a, b]$.

If $\overset{\text{L-Y}}{\underline{}}\!\int_a^b f = \overline{\text{L-Y}}\!\int_a^b f$, then f is **Lebesgue-Young integrable over** $[a, b]$, and the **Lebesgue-Young integral** of f over $[a, b]$ is

$$\text{L-Y}\!\int_a^b f = \text{L-Y}\!\int_a^b f.$$

In this definition we have replaced the restriction that f be measurable with the requirement that the partitions be measurable.

It is important to understand the difference between the partitions used by the Lebesgue and Lebesgue-Young integrals. The Lebesgue integral partitions the range of f and uses this partition to induce a measurable partition of the domain. The Lebesgue-Young integral considers an arbitrary measurable partition of f's domain. Of course, when f is measurable, the induced partitions formed by the sets $E_k = f^{-1}((y_{k-1}, y_k])$ are measurable so that the Lebesgue partitions of the range generate a subcollection of the possible measurable partitions of the domain. As we saw in the introduction to the previous chapter, when f is bounded and measurable, the gap between the upper and lower sums can be made arbitrarily small using partitions of the form $f^{-1}((y_{k-1}, y_k])$. Consequently, Lebesgue integrable functions are Lebesgue-Young integrable.

6.1.2 A simple variation

A second variation proceeds from the observation that Lebesgue lower sums correspond to functions that take on the value y_{k-1} on the set E_k. A more formal way to express this function is as $\sum_k y_{k-1} \cdot 1_{E_k}$ where 1_A is the function that is 1 for x in A and is zero elsewhere. In other words,

$$1_A(x) = \begin{cases} 1, & x \in A \\ 0, & x \notin A. \end{cases}$$

The function 1_A is known as the **characteristic function** of the set A.[2] The Dirichlet function, d, is an example of a characteristic function where $d = 1_{\mathbb{Q}}$. Functions of the form $\phi = \sum_{k=1}^n \alpha_k \cdot 1_{E_k}$ where $\alpha_k \in \mathbb{R}$ are called **simple functions**. Alternatively, one can define a simple function as one that takes on only a finite set of values $\{\alpha_k\}_{k=1}^n$. In this case, $\phi = \sum_{k=1}^n \alpha_k \cdot 1_{E_k}$ where $E_k = \phi^{-1}(\{\alpha_k\})$.

---

[2] Some texts use the notation χ_A instead of 1_A.

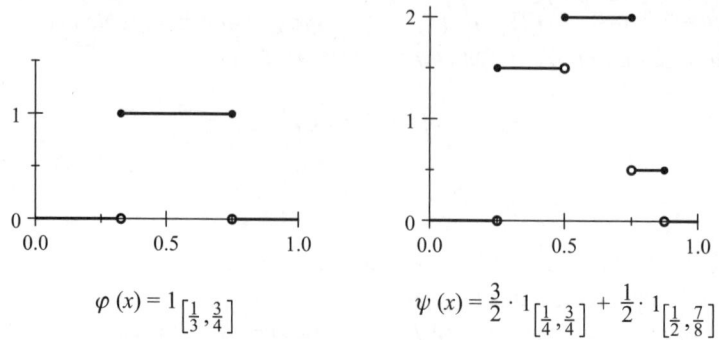

$$\varphi(x) = 1_{\left[\frac{1}{3},\frac{3}{4}\right]} \qquad\qquad \psi(x) = \frac{3}{2} \cdot 1_{\left[\frac{1}{4},\frac{3}{4}\right]} + \frac{1}{2} \cdot 1_{\left[\frac{1}{2},\frac{7}{8}\right]}$$

Figure 6.1. Graphs of simple functions

Example 29. Figure 6.1 displays the graphs of the simple functions $\phi = 1_{\left[\frac{1}{3},\frac{3}{4}\right]}$ and $\psi = \frac{3}{2} \cdot 1_{\left[\frac{1}{4},\frac{3}{4}\right]} + \frac{1}{2} \cdot 1_{\left[\frac{1}{2},\frac{7}{8}\right]}$. By considering its distinct values, the second function can also be expressed as $\psi = \frac{3}{2} \cdot 1_{\left[\frac{1}{4},\frac{1}{2}\right)} + 2 \cdot 1_{\left[\frac{1}{2},\frac{3}{4}\right]} + \frac{1}{2} \cdot 1_{\left(\frac{3}{4},\frac{7}{8}\right]}$.

Now any respectable definition of an integral for real-valued functions ought to satisfy the basic properties

1. $\int_a^b 1_A = \lambda\left([a,b] \cap A\right)$ when A is a measurable set,

2. **Linearity.** $\int_a^b \left(\sum_{k=1}^n \alpha_k f_k\right) = \sum_{k=1}^n \alpha_k \int_a^b f_k$ provided both sides are defined, and

3. **Monotonicity.** If $f \le g$ on $[a,b]$ then $\int_a^b f \le \int_a^b g$.

Combining the first two properties tells us that the integral of any measurable simple function $\phi = \sum_{k=1}^n \alpha_k \cdot 1_{E_k}$ over $[a,b]$ should be

$$\int_a^b \phi = \sum_{k=1}^n \alpha_k \cdot \lambda\left([a,b] \cap E_k\right).$$

If $\mathcal{P} = \{[y_{k-1}, y_k]\}_k$ is a partition of the range of a bounded function f and $E_k = f^{-1}\left((y_{k-1}, y_k]\right)$, then

$$\sum_k y_{k-1} \cdot 1_{E_k} \le f < \sum_k y_k \cdot 1_{E_k}.$$

Monotonicity then implies that

$$\int_a^b \left(\sum_k y_{k-1} \cdot 1_{E_k}\right) \le \int_a^b f < \int_a^b \left(\sum_k y_k \cdot 1_{E_k}\right).$$

These observations are used to extend the definition to more general functions using limits.

Definition 16 (Simple-Lebesgue integral). *The simple-Lebesgue integral is a type of Daniell integral.*[3]

1. *The simple-Lebesgue integral of a measurable simple function $\phi = \sum_{k=1}^{n} \alpha_k \cdot 1_{E_k}$ is*

$$s\text{-}L\int_a^b \phi = \sum_{k=1}^{n} \alpha_k \cdot \lambda\left([a,b] \cap E_k\right).$$

2. *For any nonnegative function f, let $\{\phi_k\}$ be a sequence of positive, measurable simple functions monotonically increasing to f. Then*

$$s\text{-}L\int_a^b f = \lim_k \; s\text{-}L\int_a^b \phi_k.\,{}^4$$

3. *For arbitrary functions, if the integral of at least one of $f^+ = \frac{1}{2}\left(|f| + f\right)$ or $f^- = \frac{1}{2}\left(|f| - f\right)$ is finite, then*

$$s\text{-}L\int_a^b f = s\text{-}L\int_a^b f^+ - s\text{-}L\int_a^b f^-.$$

*The function f is said to be **simple-Lebesgue integrable** if $s\text{-}L\int_a^b |f| < +\infty$ or, equivalently, if the integrals of both f^+ and f^- are finite.*

Using carefully constructed simple functions of the form $\sum_k y_{k-1} \cdot 1_{E_k}$ where $E_k = f^{-1}\left((y_{k-1}, y_k]\right)$, we can show that the Lebesgue and simple-Lebesgue integrals agree for bounded, measurable functions. In other words, Lebesgue integrable functions are simple-Lebesgue integrable and the integrals agree.

Notice that the definition of the simple-Lebesgue integral does not require f to be bounded or measurable. There are, however, two critical issues that must be addressed.

---

[3] The Daniell integral (Percy J. Daniell, 1918) begins with (1) a vector space F of *elementary functions* which is also closed under taking absolute value and (2) a linear function I on F that is monotone ($f \leq g$ implies $I(f) \leq I(g)$). In addition, I must satisfy the condition that if the monotone sequence $\{f_n\}$ decreases to zero, so does $\{I(f_n)\}$. Monotone increasing sequences of functions are then used to extend I to a larger set of functions L^+ and then to the set L of functions f that can be expressed as $f = f_1 - f_2$ with $f_1, f_2 \in L^+$.

[4] Note that $\lim_k \; s\text{-}L\int_a^b \phi_k$ may be infinite.

1. Part (2) of the definition takes for granted that one can find an increasing sequence of measurable simple functions that converges to f. How can we guarantee such a sequence exists?

2. Is the simple-Lebesgue integral well defined? In other words, if $\{\phi_k\}$ and $\{\psi_k\}$ are two increasing sequences of measurable simple functions that converge to f, is $\lim_k {}^{s\text{-}L}\!\int_a^b \phi_k = \lim_k {}^{s\text{-}L}\!\int_a^b \psi_k$?

We will deal with these issues near the end of the next section.

6.2   Reconciling the approaches

In addition to the Riemann-Darboux integral, we currently have three integral definitions in play. How are these integrals related? Specifically, we want to address the following questions.

1. We know that the Dirichlet function is Lebesgue integrable but not Riemann integrable, so not all Lebesgue integrable functions are Riemann integrable. Are there any Riemann integrable functions that are not Lebesgue integrable?

2. We have sketched arguments that Lebesgue integrable functions are Lebesgue-Young and simple-Lebesgue integrable and that all three integrals agree for these functions. Are there functions that are Lebesgue-Young or simple-Lebesgue integrable but not Lebesgue integrable?

3. How are the Lebesgue-Young and simple-Lebesgue integrals related?

A few moments contemplation should lead you to the realization that the first two questions are really questions about the measurability of functions. Measurability is the fundamental requirement for a bounded function to be Lebesgue integrable.

6.2.1   Riemann

Let's begin with the Riemann integral. By theorem 20 (page 63), the set of discontinuities of a Riemann integrable function has measure zero. We will show that all such functions are measurable and therefore Lebesgue integrable.

Theorem 28 (Riemann $\implies$ Lebesgue). *If f is Riemann integrable over $[a, b]$, then f is also Lebesgue integrable over $[a, b]$ with ${}^{L}\!\int_a^b f = {}^{R}\!\int_a^b f$.*

Proof. If f is Riemann integrable, then f is bounded and the set D of the discontinuities of f has measure zero. Let C be the set of points in $[a, b]$ at which f is continuous and let g be the restriction of f to C. Let $c \in \mathbb{R}$. Since g is continuous, $g^{-1}((-\infty, c)) = U \cap C$ for some open set U. Thus $g^{-1}((-\infty, c))$, an intersection of measurable sets, is measurable for all values of c. Now

$$g^{-1}((-\infty, c)) \subseteq f^{-1}((-\infty, c)) \subseteq g^{-1}((-\infty, c)) \cup D,$$

so that $f^{-1}((-\infty, c)) = g^{-1}((-\infty, c)) \cup Z$ for some set $Z \subseteq D$. Since Z has outer measure zero, Z is measurable. Thus $f^{-1}((-\infty, c))$, the union of measurable sets, is measurable. As f is both bounded and measurable, it is Lebesgue integrable.

We will verify that ${}^{L}\!\int_a^b f = {}^{R}\!\int_a^b f$ shortly.   □

6.2.2 Lebesgue-Young

To address the Lebesgue-Young integral, we make use of several properties of upper and lower Lebesgue-Young sums that are analogous to facts about upper and lower Darboux sums. Since measurable partitions are not determined by the endpoints of intervals, we first need to adjust the concept of a refinement.

Definition 17 (Refinement). *Let $\mathcal{P}$ and $\mathcal{Q}$ be measurable partitions of $[a, b]$. Then $\mathcal{Q}$ is a **refinement** of $\mathcal{P}$ if for each $F \in \mathcal{Q}$ there is an $E \in \mathcal{P}$ such that $F \subseteq E$. Equivalently, every $E \in \mathcal{P}$ can be expressed as the disjoint union of sets in $\mathcal{Q}$.*

A standard way to create a common refinement of two measurable partitions $\mathcal{P}$ and $\mathcal{Q}$ is to use $\mathcal{P} \cup \mathcal{Q} = \{E \cap F : E \in \mathcal{P}, F \in \mathcal{Q}, E \cap F \neq \emptyset\}$.[5]

Example 30. $\mathcal{P} = \{[0, 0.5], (0.5, 1]\}$ and $\mathcal{Q} = \{[0, 1] \cap \mathbb{Q}, [0, 1] \cap \mathbb{Q}^c\}$ are two measurable partitions of $[0, 1]$. In this case, $\mathcal{P} \cup \mathcal{Q}$ consists of four sets: the rationals in $[0, 0.5]$, the irrationals in $[0, 0.5]$, the rationals in $(0.5, 1]$, and the irrationals in $(0.5, 1]$.

The proof of the following lemma is similar to that of the corresponding result for Darboux integrals (lemma 13, page 56).

---

[5] Note that there are two possible interpretations of $\mathcal{P} \cup \mathcal{Q}$ depending on the context. If $\mathcal{P}$ and $\mathcal{Q}$ are partitions of the same interval, $\mathcal{P} \cup \mathcal{Q}$ represents the common refinement introduced here. If $\mathcal{P}$ and $\mathcal{Q}$ are partitions of non-overlapping intervals I and J, then $\mathcal{P} \cup \mathcal{Q}$ is the usual union of sets. In this case, $\mathcal{P} \cup \mathcal{Q}$ is a partition of $I \cup J$.

Lemma 29. *Let f be a bounded function on $[a, b]$ and let $\mathcal{P}$ and $\mathcal{Q}$ be measurable partitions of $[a, b]$. If $\mathcal{Q}$ is a refinement of $\mathcal{P}$, then $S_{\underline{L\text{-}Y}}(f, \mathcal{P}) \leq S_{\underline{L\text{-}Y}}(f, \mathcal{Q}) \leq S_{\overline{L\text{-}Y}}(f, \mathcal{Q}) \leq S_{\overline{L\text{-}Y}}(f, \mathcal{P})$.*

Proof. We will prove the first inequality and leave the other two as an exercise.

Suppose that $\mathcal{P} = \{E_i\}$ and $\mathcal{Q} = \{F_k\}$. Since $\mathcal{Q}$ is a refinement of $\mathcal{P}$, we can find an indexing scheme and a set of values $\{n_i\}$ such that each E_i can be expressed as the disjoint union

$$E_i = \cup_{k=n_i+1}^{n_{i+1}} F_k.$$

Then for $k = n_i + 1, n_i + 2, \ldots, n_{i+1}$, $F_k \subseteq E_i$ so that $\inf_{F_k} f \geq \inf_{E_i} f$. Thus

$$S_{\underline{L\text{-}Y}}(f, \mathcal{P}) = \sum_i \inf_{E_i} f \cdot \lambda(E_i)$$

$$= \sum_i \inf_{E_i} f \cdot \sum_{k=n_i+1}^{n_{i+1}} \lambda(F_k)$$

$$\leq \sum_i \sum_{k=n_i+1}^{n_{i+1}} \inf_{F_k} f \cdot \lambda(F_k)$$

$$= S_{\underline{L\text{-}Y}}(f, \mathcal{Q}). \qquad \square$$

Theorem 30 (Lebesgue = Lebesgue-Young). *If f is Lebesgue-Young integrable over $[a, b]$, then f is also Lebesgue integrable over $[a, b]$ and the two integrals agree. (The converse has been established in previous discussion.)*

Proof. Our first task is to verify that f is measurable.

To any measurable partition $\mathcal{P}$ of $[a, b]$ there corresponds a pair of lower and upper simple functions defined by $\phi_{\underline{\mathcal{P}}} = \sum_{E \in \mathcal{P}} \inf_E f \cdot 1_E$ and $\phi_{\overline{\mathcal{P}}} = \sum_{E \in \mathcal{P}} \sup_E f \cdot 1_E$. By construction,

$$\phi_{\underline{\mathcal{P}}} \leq f \leq \phi_{\overline{\mathcal{P}}},$$

$${}^L\!\int_a^b \phi_{\underline{\mathcal{P}}} = \sum_{E \in \mathcal{P}} \inf_E f \cdot \lambda(E) = S_{\underline{L\text{-}Y}}(f, \mathcal{P}),$$

and

$${}^L\!\int_a^b \phi_{\overline{\mathcal{P}}} = \sum_{E \in \mathcal{P}} \sup_E f \cdot \lambda(E) = S_{\overline{L\text{-}Y}}(f, \mathcal{P}).$$

Since f is Lebesgue-Young integrable, we can find two sequences of partitions $\{\mathcal{P}_n\}$ and $\{\mathcal{Q}_n\}$ so that

$$S_{\overline{L\text{-}Y}}(f, \mathcal{Q}_n) - S_{\underline{L\text{-}Y}}(f, \mathcal{P}_n) < \frac{1}{n}.$$

From these sequences, construct a sequence of successive refinements $\{\mathcal{R}_n\}$ by setting $\mathcal{R}_n = \cup_{k \leq n}(\mathcal{P}_k \cup \mathcal{Q}_k)$ and let $\{\phi_n\}$ and $\{\psi_n\}$ be the sequences of upper and lower simple functions defined by $\phi_n = \phi_{\overline{\mathcal{R}_n}}$ and $\psi_n = \phi_{\underline{\mathcal{R}_n}}$. Then

$$^{L}\!\int_a^b (\phi_n - \psi_n) = S_{\overline{L\text{-}Y}}(f, \mathcal{R}_n) - S_{\underline{L\text{-}Y}}(f, \mathcal{R}_n)$$

$$\leq S_{\overline{L\text{-}Y}}(f, \mathcal{Q}_n) - S_{\underline{L\text{-}Y}}(f, \mathcal{P}_n) < \frac{1}{n}.$$

Moreover, because $\mathcal{R}_{n+1}$ is a refinement of $\mathcal{R}_n$, $\{\phi_n\}$ and $\{\psi_n\}$ are respectively decreasing and increasing sequences with $\psi_n \leq f \leq \phi_n$. If we define $\phi = \lim_n \phi_n$ and $\psi = \lim_n \psi_n$, then ϕ and ψ, being limits of measurable functions, are measurable (theorem 25, page 107). We claim that, in addition, $\phi = \psi$ a.e.

Let D be the set of points in $[a, b]$ for which $\phi \neq \psi$. Then

$$D = \{x \in [a,b] : \phi(x) - \psi(x) > 0\}$$

$$= \bigcup_k \left\{ x \in [a,b] : \phi(x) - \psi(x) > \frac{1}{k} \right\}.$$

Set $D_k = \left\{ x \in [a,b] : \phi(x) - \psi(x) > \frac{1}{k} \right\}$ and $D_{k,n} = \{ x \in [a,b] : \phi_n(x) - \psi_n(x) > \frac{1}{k} \}$. Then, for all $n, k \in \mathbb{N}$,

$$\lambda(D_k) \leq \lambda(D_{k,n}) = {}^{L}\!\int_a^b 1_{D_{k,n}} \leq {}^{L}\!\int_a^b k(\phi_n - \psi_n) < \frac{k}{n}.$$

Hence $\lambda(D_k) = 0$ so

$$0 \leq \lambda(D) \leq \sum_k \lambda(D_k) = 0$$

and $\phi = \psi$ a.e.

But $\psi \leq f \leq \phi$, so that $\phi = f$ a.e. as well. Hence, by theorem 26 (page 108), f is measurable and, as a Lebesgue-Young integrable function, f is bounded. Consequently, f is Lebesgue integrable.

To show the Lebesgue and the Lebesgue-Young integrals agree, choose a partition $\mathcal{P} = \{[y_{k-1}, y_k]\}_{k=1}^n$ of an interval $[\alpha, \beta]$ such that $\alpha < f < \beta$. Then $\mathcal{P}$ generates a measurable partition $P^* = \{E_k\} = \{f^{-1}((y_{k-1}, y_k])\}$ of $[a, b]$ for which $y_{k-1} \leq \inf_{E_k} f \leq \sup_{E_k} f \leq y_k$. Hence

$$S_{\underline{L}}(f, \mathcal{P}) \leq S_{\underline{L\text{-}Y}}(f, \mathcal{P}^*) \leq {}^{\text{L-Y}}\!\!\int_a^b f \leq S_{\overline{L\text{-}Y}}(f, \mathcal{P}^*) \leq S_{\overline{L}}(f, \mathcal{P}).$$

Since the Lebesgue integral also falls between $S_{\underline{L}}(f, \mathcal{P})$ and $S_{\overline{L}}(f, \mathcal{P})$ and

$$S_{\overline{L}}(f, \mathcal{P}) - S_{\underline{L}}(f, \mathcal{P}) \leq \|\mathcal{P}\|(b - a),$$

we conclude that

$$^{\text{L-Y}}\!\!\int_a^b f = {}^L\!\!\int_a^b f. \qquad \Box$$

This theorem allows us to take care of a piece of unfinished business from theorem 28 (page 124). By modifying Riemann-Darboux partitions of $[a, b]$ to use half open intervals, we can produce measurable partitions on which the lower/upper Darboux and Lebesgue-Young sums agree. Since the set of partitions by half open intervals is contained in the set of measurable partitions, the properties of supremums and infimums on subsets guarantee that

$$\underline{D}\!\int_a^b f \leq {}^{\underline{L\text{-}Y}}\!\!\int_a^b f \leq {}^{\overline{L\text{-}Y}}\!\!\int_a^b f \leq {}^{\overline{D}}\!\!\int_a^b f.$$

Hence when f is Riemann integrable over $[a, b]$,

$$^R\!\!\int_a^b f = {}^D\!\!\int_a^b f = {}^{\text{L-Y}}\!\!\int_a^b f = {}^L\!\!\int_a^b f.$$

6.2.3 Simple-Lebesgue

Turning to the simple-Lebesgue integral, we first establish that the integral is well defined.

Theorem 31 (The simple-Lebesgue integral is well-defined). *Let f be a positive function on $[a, b]$. Suppose that $\{\phi_k\}$ and $\{\psi_k\}$ are two increasing sequences of positive, measurable simple functions on $[a, b]$ converging to f. Then*

$$\lim_k {}^{\text{S-L}}\!\!\int_a^b \phi_k = \lim_k {}^{\text{S-L}}\!\!\int_a^b \psi_k.$$

Proof. Fix $\varepsilon > 0$.

First suppose that $f < B$. By Egoroff's theorem (theorem 35, page 135 in the next section), we can find measurable sets E and F with $\lambda(E) < \varepsilon/4B$ and $\lambda(F) < \varepsilon/4B$ such that $\{\phi_k\}$ and $\{\psi_k\}$ converge uniformly to f on $[a, b] \setminus E$ and $[a, b] \setminus F$ respectively. Let $G = E \cup F$. Then $\lambda(G) < \varepsilon/2B$ and both $\{\phi_k\}$ and $\{\psi_k\}$ converge uniformly to f on $[a, b] \setminus G$.

Choose N so that both $|\phi_n(x) - f(x)|$ and $|\psi_n(x) - f(x)|$ are less than $\varepsilon/4(b - a)$ for all $x \in [a, b] \setminus G$ and $n > N$. Given $n > N$, $|\phi_n - \psi_n|$ is a simple function satisfying $|\phi_n(x) - \psi_n(x)| < \frac{\varepsilon}{2(b-a)}$ for all $x \in [a, b] \setminus G$ and $|\phi_n(x) - \psi_n(x)| < B$ for all $x \in G$. Hence

$$\left| {}^{s\text{-}L}\!\!\int_a^b \phi_n - {}^{s\text{-}L}\!\!\int_a^b \psi_n \right| \leq {}^{s\text{-}L}\!\!\int_a^b |\phi_n - \psi_n|$$

$$\leq {}^{s\text{-}L}\!\!\int_{[a,b]\setminus G} \frac{\varepsilon}{2(b-a)} + {}^{s\text{-}L}\!\!\int_G B$$

$$\leq \frac{\varepsilon}{2(b-a)}(b - a) + B\frac{\varepsilon}{2B} = \varepsilon.$$

Since $\varepsilon > 0$ and $n > N$ are arbitrary,

$$\lim_n \left| {}^{s\text{-}L}\!\!\int_a^b \phi_n - {}^{s\text{-}L}\!\!\int_a^b \psi_n \right| = 0$$

so

$$\lim_k {}^{s\text{-}L}\!\!\int_a^b \phi_k = \lim_k {}^{s\text{-}L}\!\!\int_a^b \psi_k.$$

Now, remove the restriction that f is bounded and suppose that

$$L < \lim_k {}^{s\text{-}L}\!\!\int_a^b \phi_k.$$

Then we can find an integer n_0 such that $L < {}^{s\text{-}L}\!\!\int_a^b \phi_{n_0}$. Set $B = \max_{[a,b]} \phi_{n_0}$ and define $f^* = \min\{f, B\}$. Define ϕ_k^* and ψ_k^* analogously. Then $\{\phi_k^*\}$ and $\{\psi_k^*\}$ are increasing sequences of positive, measurable simple functions that converge to the bounded function f^*. So by our previous work,

$$\lim_k {}^{s\text{-}L}\!\!\int_a^b \phi_k^* = \lim_k {}^{s\text{-}L}\!\!\int_a^b \psi_k^*.$$

Hence

$$\lim_{k} {}^{s\text{-}L}\!\!\int_{a}^{b} \psi_k \geq \lim_{k} {}^{s\text{-}L}\!\!\int_{a}^{b} \psi_k^*$$

$$= \lim_{k} {}^{s\text{-}L}\!\!\int_{a}^{b} \phi_k^*$$

$$\geq {}^{s\text{-}L}\!\!\int_{a}^{b} \phi_{n_0}^*$$

$$= {}^{s\text{-}L}\!\!\int_{a}^{b} \phi_{n_0} > L.$$

Since $L < \lim_{k} {}^{s\text{-}L}\!\int_{a}^{b} \phi_k$ was arbitrary, we conclude that

$$\lim_{k} {}^{s\text{-}L}\!\!\int_{a}^{b} \psi_k \geq \lim_{k} {}^{s\text{-}L}\!\!\int_{a}^{b} \phi_k.$$

By symmetry,

$$\lim_{k} {}^{s\text{-}L}\!\!\int_{a}^{b} \phi_k = \lim_{k} {}^{s\text{-}L}\!\!\int_{a}^{b} \psi_k$$

as claimed.   □

There are two strategies in the preceding proof that merit comment as they will reappear repeatedly in the remainder of the chapter. The first strategy is the Lebesgue integral's analog to the bound-and-telescope move found in so many Riemann proofs. In the Lebesgue context, the analogous move is to use a general bound on a function on a set with very small measure coupled with a very small bound on the function over a larger set. We will refer to this strategy as *divide and conquer*. This strategy is employed in the previous proof when the integral over $[a, b]$ is split into the integrals over $[a, b] \setminus G$ and G. The connection is best seen in theorem 19 (page 62) where the two types of bound-and-telescope techniques are combined. The analogy is not perfect since the bound-and-telescope technique is typically used to prove that a function is Riemann integrable. We rarely prove integrability results for the Lebesgue integral since bounded, measurable functions are automatically Lebesgue integrable. The technique is more often used to prove the equality of two integrals.

The second strategy is first to prove something under the condition that the function(s) is (are) bounded. The bounded case is then applied to a sequence of truncated functions to obtain the general result. You will encounter these two moves many times in the next two sections.

To round out the analysis of the simple-Lebesgue integral, we note that if an increasing sequence of measurable simple functions converges to a function f, then by theorem 25 (page 107) f must be measurable. This observation provides the missing piece for the following theorem.

Theorem 32 (Lebesgue = simple-Lebesgue). *Let f be a bounded function defined on $[a, b]$. Then f is Lebesgue integrable over $[a, b]$ if and only if f is simple-Lebesgue integrable over $[a, b]$. In this case, the integrals agree.*

Proof. Suppose that f is bounded and simple-Lebesgue integrable. Then f is measurable as the limit of measurable functions. Hence f is Lebesgue integrable.

Conversely, suppose that f is Lebesgue integrable. Given any partition $\mathcal{P} = \{[y_{k-1}, y_k]\}$ of the range of f, the naturally associated simple function $\phi = \sum_k y_{k-1} \cdot 1_{E_k}$, where $E_k = f^{-1}((y_{k-1}, y_k])$, satisfies

$$0 \leq f - \phi < \|\mathcal{P}\|$$

and

$$S_{\underline{L}}(f, \mathcal{P}) = {}^{s\text{-}L}\!\!\int_a^b \phi.$$

Since ${}^{L}\!\!\int_a^b f = \sup_{\mathcal{P}} S_{\underline{L}}(f, \mathcal{P})$, we can use a sequence of partitions $\{\mathcal{P}_n\}$ to generate a sequence $\{\phi_n\}$ of measurable simple functions that converges uniformly to f and so that

$$ {}^{L}\!\!\int_a^b f = \lim_n S_{\underline{L}}(f, \mathcal{P}_n) = \lim_n {}^{s\text{-}L}\!\!\int_a^b \phi_n. $$

We cannot immediately apply the definition of the simple-Lebesgue integral to conclude that ${}^{L}\!\!\int_a^b f = {}^{s\text{-}L}\!\!\int_a^b f$, since $\{\phi_n\}$ may not be an increasing sequence.[6] However, if we take $\widehat{\phi}_n = \max_{k \leq n} \phi_k$, then $\{\widehat{\phi}_n\}$ is an increasing sequence of measurable simple functions converging to f for which

$$ {}^{L}\!\!\int_a^b f = \lim_n {}^{s\text{-}L}\!\!\int_a^b \widehat{\phi}_n = {}^{s\text{-}L}\!\!\int_a^b f. \qquad \square $$

Notice that in the process of proving that the Lebesgue and simple-Lebesgue integrals agree for bounded functions, we have shown that any bounded measurable function is the uniform limit of an increasing sequence

---

[6] Exercises 33 through 36 provide an alternate approach to finding an increasing sequence of measurable simple functions that converges to f.

of measurable simple functions. While the Lebesgue, simple-Lebesgue, and Lebesgue-Young integrals agree on bounded functions, the simple-Lebesgue definition is more flexible than the others as it can be applied to unbounded functions.

Example 31. Define $r : [0, 1] \to \mathbb{R}$ by

$$r(x) = \begin{cases} n, & x = \frac{m}{n} \text{ with } \frac{m}{n} \text{ in lowest terms} \\ 0, & \text{otherwise.} \end{cases}$$

Then r is an unbounded function and so is not Lebesgue integrable. However, if we define ϕ_N on $[0, 1]$ by

$$\phi_N(x) = \begin{cases} n, & x = \frac{m}{n} \text{ with } \frac{m}{n} \text{ in lowest terms, } n \leq N \\ N, & x = \frac{m}{n} \text{ with } \frac{m}{n} \text{ in lowest terms, } n > N \\ 0, & \text{otherwise} \end{cases}$$

$$= \begin{cases} r(x), & r(x) \leq N \\ N, & \text{otherwise,} \end{cases}$$

then $\{\phi_N\}$ is an increasing sequence of measurable simple functions that converges to r. When $n > 0$, $\phi_N^{-1}(\{n\})$ is countable. Thus

$$ {}^{s\text{-}L}\!\!\int_0^1 \phi_N = 0 \cdot \lambda \left(\phi_N^{-1}(\{0\}) \right) + \sum_{n=1}^N n \cdot \lambda \left(\phi_N^{-1}(\{n\}) \right) = 0. $$

Taking the limit, we find that r is simple-Lebesgue integrable with

$$ {}^{s\text{-}L}\!\!\int_0^1 r = \lim_{N \to \infty} 0 = 0. $$

6.2.4   Extended definitions

Since we now know that functions that are Lebesgue-Young or simple-Lebesgue integrable must also be measurable, we can modify the Lebesgue-Young and simple-Lebesgue definitions so that all three integral definitions agree for nonnegative, measurable functions. **We are removing the restriction that f is bounded.**

Definition 18 (Lebesgue integral). *Let* $f : [a, b] \to [0, +\infty)$ *be a (Lebesgue) measurable function. Let* $\mathcal{P}$ *be the partition of* $[0, +\infty)$ *defined by the division points* $\{y_k\}_{k=1}^n$ *and set* $E_k = f^{-1}((y_{k-1}, y_k])$, $k = 1, 2, \ldots, n$, *and* $E_{n+1} = f^{-1}((y_n, +\infty))$. *The (lower) Lebesgue sum* *of f with respect to $\mathcal{P}$ is $S_{\underline{L}}(f,\mathcal{P}) = \sum_{k=1}^{n+1} y_{k-1} \cdot \lambda(E_k)$. The **Lebesgue integral of** f **over** $[a,b]$ is*

$$ {}^{L}\!\int_a^b f = \sup_{\mathcal{P}} S_{\underline{L}}(f,\mathcal{P}) $$

*where the supremum is taken over all partitions $\mathcal{P}$ of $[0, +\infty)$. The function f is said to be **Lebesgue integrable** if ${}^{L}\!\int_a^b f < +\infty$.*

Notice the change made to accommodate unbounded functions. The last interval in the partition of $[0, +\infty)$ is $(y_n, +\infty)$. If f happens to be bounded, then $f^{-1}((y_n, +\infty)) = \emptyset$ when y_n is sufficiently large. Also notice that

$$ S_{\underline{L}}(f,\mathcal{P}) = {}^{L}\!\int_a^b \left(\sum_{k=1}^{n+1} y_{k-1} \cdot 1_{E_k} \right). $$

Definition 19 (Lebesgue-Young integral). *Let f be a nonnegative, (Lebesgue) measurable function defined on $[a,b]$. Let $\mathcal{P} = \{E_k\}_{k=1}^{n}$ be a measurable partition of $[a,b]$. The **lower Lebesgue-Young sum** of f over $\mathcal{P}$ is*

$$ S_{\underline{L\text{-}Y}}(f,\mathcal{P}) = \sum_{\mathcal{P}} \inf_{E_k} f \cdot \lambda(E_k). $$

*The **Lebesgue-Young integral** of f over $[a,b]$ is*

$$ {}^{L\text{-}Y}\!\int_a^b f = \sup_{\mathcal{P}} S_{\underline{L\text{-}Y}}(f,\mathcal{P}) $$

*where the supremum is taken over all measurable partitions $\mathcal{P}$ of $[a,b]$. We say that f is **Lebesgue-Young integrable** if ${}^{L\text{-}Y}\!\int_a^b f < +\infty$.*

Note that, on the basis of our earlier work, we do not need to concern ourselves with the upper sum when f is bounded. If f is not bounded then the upper sum is infinite and therefore not helpful. Also note that for any measurable partition $\mathcal{P} = \{E_k\}_{k=1}^{n}$ of $[a,b]$,

$$ S_{\underline{L\text{-}Y}}(f,\mathcal{P}) = {}^{L}\!\int_a^b \left(\sum_{k=1}^{n} \inf_{E_k} f \cdot 1_{E_k} \right). $$

Both of these definitions are extended to functions that are not nonnegative by taking

$$ {}^{L}\!\int_a^b f = {}^{L}\!\int_a^b f^+ - {}^{L}\!\int_a^b f^- $$

when the integral of at least one of $f^+ = \frac{1}{2}(|f| + f)$ or $f^- = \frac{1}{2}(|f| - f)$ is finite. We say that f is Lebesgue integrable if $^L\!\int_a^b |f| < +\infty$ or, equivalently, if the integrals of both f^+ and f^- are finite. When these changes are made, all three definitions are in complete agreement.

At this point, you may well be thinking "If the three definitions are the same, why did we bother introducing all of them?" This is an excellent question and you probably already have some ideas about how to answer it. Let me provide two reasons.

1. It is always a good mathematical practice to explore alternative ways of approaching a problem. In this case, the alternatives did not produce anything new, but often a modified approach is quite fruitful.

2. Having alternative ways of thinking about the same thing can be very useful. We have seen this already. When we wanted to prove that $^L\!\int_a^b f = {}^R\!\int_a^b f$, we instead proved that $^{L\text{-}Y}\!\int_a^b f = {}^D\!\int_a^b f$. The process would have been far more difficult had we been restricted to using only the original Riemann and Lebesgue definitions. In subsequent sections, we will reap additional benefits from having multiple characterizations of the Lebesgue integral at our disposal.

In texts treating only the Lebesgue integral, it is typically written as $\int_a^b f \, d\lambda$ or $\int_a^b f(x) \, d\lambda(x)$. We will retain the notation $^L\!\int_a^b f$ since we are interested in comparing the properties of the Lebesgue integral with those of various other types of integrals.

6.3   Convergence theorems

You will recall that one of the main historical motivations for a more careful and generalized investigation into integration was the question of convergence. Originally, the focus was on convergence of trigonometric series, but other illustrative examples were soon added to the mix. We now turn our attention to the task of finding conditions under which

$$^L\!\int_a^b \lim_n f_n = \lim_n {}^L\!\int_a^b f_n.$$

We will build our theorems bit by bit, relaxing the conditions as we proceed. We start with a set of three tools. The first two will be used repeatedly and the third (Egoroff's theorem) was used already in the previous section. Notice the appearance of a version of the divide-and-conquer technique in the proof of the next result.

Theorem 33. *If $f = 0$ a.e., then f is Lebesgue integrable with $^L\!\int_a^b f = 0$ for any interval $[a, b]$ in the domain of f.*

Proof. The measurability of f follows from theorem 26 (page 108).

Let $Z = \{x \in [a, b] : f(x) \neq 0\}$. By assumption, $\lambda(Z) = 0$. Now let $\mathcal{P}$ be a measurable partition of $[a, b]$. Divide the partition into those sets that include a point x for which $f^+(x) = 0$ and those that do not. Label the two collections $\mathcal{P}_0$ and $\mathcal{P}_>$. For any $E \in \mathcal{P}_>$, $E \subseteq Z$ so that $\lambda(E) = 0$. Thus

$$S_{\underline{L\text{-}Y}}\left(f^+, \mathcal{P}\right) = \sum_{E \in \mathcal{P}_0} 0 \cdot \lambda(E) + \sum_{E \in \mathcal{P}_>} \inf_E f^+ \cdot \lambda(E) = 0.$$

Since the supremum of $S_{\underline{L\text{-}Y}}\left(f^+, \mathcal{P}\right)$ over all possible measurable partitions is zero,

$$^L\!\int_a^b f^+ = {}^{L\text{-}Y}\!\int_a^b f^+ = 0.$$

The same argument shows that $^L\!\int_a^b f^- = 0$. Hence $^L\!\int_a^b f = 0$. $\square$

Theorem 34. *Suppose that f is a Lebesgue integrable function on $[a, b]$ and that $g = f$ a.e. Then g is also Lebesgue integrable with $^L\!\int_a^b g = {}^L\!\int_a^b f$.*

Proof. First note that g is measurable by theorem 26. Since $g^+ - f^+ = 0$ a.e.,

$$^L\!\int_a^b g^+ = {}^L\!\int_a^b \left(f^+ + (g^+ - f^+)\right)$$

$$= {}^L\!\int_a^b f^+ + {}^L\!\int_a^b \left(g^+ - f^+\right)$$

$$= {}^L\!\int_a^b f^+.$$

The same argument shows that $^L\!\int_a^b g^- = {}^L\!\int_a^b f^-$ so that $^L\!\int_a^b g = {}^L\!\int_a^b f$. $\square$

Theorem 35 (Egoroff, 1911). *Let $\{f_k\}$ be a sequence of real-valued, measurable functions defined on $[a, b]$ and suppose that $\{f_k\}$ converges to f almost everywhere. Then for any $\varepsilon > 0$ there is a measurable subset E of $[a, b]$ with $\lambda(E) < \varepsilon$ such that $\{f_k\}$ converges uniformly to f on the complement of E.*

Proof. For $n, k \in \mathbb{N}$, define

$$A_{n,k} = \left\{ x \in [a, b] : |f_m(x) - f(x)| \geq \frac{1}{n}, \text{ for some } m \geq k \right\}.$$

By construction, $A_{n,k}$ is measurable, $A_{n,k+1} \subseteq A_{n,k}$, and, since $\cap_k A_{n,k}$ is contained in the set of points for which $\{f_k\}$ does not converge to f, it follows that $\lambda\left(\cap_k A_{n,k}\right) = 0$. Consequently, for each n we can select an integer k_n so that $\lambda\left(A_{n,k_n}\right) < \frac{\varepsilon}{2^n}$. (See exercise 43.)

Let $E = \cup_n A_{n,k_n}$. By subadditivity,

$$\lambda(E) \leq \sum_n \lambda\left(A_{n,k_n}\right) < \sum_n \frac{\varepsilon}{2^n} = \varepsilon.$$

Now for any $x \in E^c$, we have $x \notin A_{n,k_n}$ so that $|f_k(x) - f(x)| < \frac{1}{n}$ for $k \geq k_n$. Hence $\{f_k\}$ converges uniformly to f on E^c. $\qquad\square$

With our tools in hand, we are ready to turn to the convergence theorems. The important thing to notice in what follows is that, in the context of the Lebesgue integral, uniform convergence is replaced with the weaker condition of pointwise convergence.

The condition in the next theorem that $\{f_k\}$ be uniformly bounded seems like a new condition that was not required for Riemann integrals. This is not the case as the uniform boundedness condition is implied by Riemann integrability. Riemann-integrable functions are bounded and when a sequence $\{f_k\}$ of bounded functions converges uniformly, $\{f_k\}$ is uniformly bounded. The uniform boundedness condition is not redundant for Lebesgue-integrable functions as they need not even be bounded.

Note the divide-and-conquer move in the following proof.

Theorem 36 (Bounded convergence). *Let $\{f_k\}$ be a uniformly bounded sequence of Lebesgue measurable functions that converges pointwise to f a.e. on $[a, b]$. Then*

$$\lim_k {}^L\!\!\int_a^b f_k = {}^L\!\!\int_a^b \lim_k f_k = {}^L\!\!\int_a^b f.$$

Proof. By assumption, there is a value B such that $|f_k(x)| < B$ for all $k \in \mathbb{N}$ and $x \in [a, b]$. When $\lim_k f_k(x)$ does not exist or is not equal to $f(x)$, redefine $f(x) = 0$. Then $|f(x)| \leq B$ for all $x \in [a, b]$ and, since the set of points on which the value of f has been modified has measure zero, ${}^L\!\!\int_a^b f$ is unchanged.

Now let $\varepsilon > 0$ and use Egoroff's theorem to select a set E with $\lambda(E) < \frac{\varepsilon}{4B}$ such that $\{f_k\}$ converges uniformly to f on $[a,b] \setminus E$. Choose an $N \in \mathbb{N}$ such that $|f_k(x) - f(x)| < \frac{\varepsilon}{2(b-a)}$ for all $k \geq N$ and all $x \in [a,b] \setminus E$. Then for $k \geq N$,

$$\left| {}^L\!\!\int_a^b f_k - {}^L\!\!\int_a^b f \right| \leq {}^L\!\!\int_a^b |f_k - f|$$

$$= {}^L\!\!\int_E |f_k - f| + {}^L\!\!\int_{[a,b] \setminus E} |f_k - f|$$

$$< 2B \cdot \lambda(E) + \frac{\varepsilon}{2(b-a)} \cdot \lambda([a,b]) < \varepsilon.$$

Thus $\lim_k {}^L\!\!\int_a^b f_k = {}^L\!\!\int_a^b f$. $\qquad\qquad\square$

This is a very satisfying theorem as it handles situations like the following two examples which are not explained by convergence theorems for the Riemann integral.

Example 32. Let $f_k(x) = x^k$, $x \in [0,1]$. Since $\{f_k\}$ is bounded by 1 and converges pointwise to $f(x) = 0$ except at $x = 1$, $\lim_k {}^L\!\!\int_0^1 f_k = 0$.

Example 33. Let $\{r_n\}$ be an enumeration of the rational numbers in $[0,1]$. Define d_n on $[0,1]$ by

$$d_n(x) = \begin{cases} 1, & x = r_k \text{ with } n \leq k \leq 2n \\ 0, & \text{otherwise.} \end{cases}$$

Then $\{d_n\}$ is uniformly bounded by 1 and converges pointwise to $f(x) = 0$. Hence $\lim_n {}^L\!\!\int_0^1 d_n = 0$.

Unfortunately, we have no reason to believe that convergent trigonometric series have uniformly bounded partial sums and, as the following example illustrates, we cannot simply drop the condition of a uniform upper bound. If we want to generalize theorem 36, we need to find an alternate, less restrictive condition to take its place.

Example 34. Define the sequence of functions $\{f_k\}$ on $[0,1]$ by

$$f_k(x) = \begin{cases} k, & x \in \left[0, \frac{1}{k}\right] \\ 0, & x \in \left(\frac{1}{k}, 1\right]. \end{cases}$$

Then f_k converges to $f(x) = 0$ almost everywhere. However, ${}^L\!\!\int_0^1 f_k = 1$ for all $k \in \mathbb{N}$ so that $\lim_k {}^L\!\!\int_0^1 f_k \neq {}^L\!\!\int_0^1 f$.

We'll see shortly (theorem 39, page 140) that we can replace the uniform upper bound on the sequence of functions with an integrable bounding function (a measurable function with a finite Lebesgue integral). Before we get to the formal statement of the theorem and its proof, we need two additional stepping stones that are useful in their own right.

Theorem 37 (Monotone convergence, Levi, 1906). *Let $\{f_k\}$ be a monotone increasing sequence of nonnegative, measurable functions that converges a.e. to f on $[a, b]$. Then*

$$^L\!\!\int_a^b f = \lim_k {}^L\!\!\int_a^b f_k$$

where a value of $+\infty$ is permissible.

Proof. By redefining f and each f_k on a set of measure zero, we may assume that $\{f_k\}$ converges to f for all $x \in [a, b]$.

Since each f_k is measurable, there is an increasing sequence of measurable simple functions $\{\phi_{k,j}\}_{j=1}^\infty$ that converges to f_k. (See exercise 36 or the proof of theorem 32.) Now define a new sequence of measurable simple functions $\{\widehat{\phi}_n\}$ by $\widehat{\phi}_n = \max_{1 \le k \le n} \phi_{k,n}$. Then

$$\widehat{\phi}_{n+1} = \max_{1 \le k \le n+1} \phi_{k,n+1} \ge \max_{1 \le k \le n} \phi_{k,n+1} \ge \max_{1 \le k \le n} \phi_{k,n} = \widehat{\phi}_n$$

so that $\{\widehat{\phi}_n\}$ is also an increasing sequence of measurable simple functions. Moreover, for all $k \le n$, we have $\phi_{k,n} \le f_k \le f_n \le f$ so that $\widehat{\phi}_n \le f_n \le f$.

In fact, $\{\widehat{\phi}_n\}$ converges to f for all $x \in [a, b]$. To see why, fix $x \in [a, b]$ and $\varepsilon > 0$. Because $\{f_k(x)\}$ converges to $f(x)$, we can find an N so that $f(x) - \varepsilon/2 < f_N(x) \le f(x)$. Similarly, we can find an M so that $f_N(x) - \varepsilon/2 < \phi_{N,M}(x) \le f_N(x)$. Then for $n \ge \max\{N, M\}$, we have the inequality

$$f(x) - \varepsilon < \phi_{N,M}(x) \le \phi_{N,n}(x) \le \widehat{\phi}_n(x) \le f(x).$$

Since $\varepsilon > 0$ and $x \in [a, b]$ are arbitrary, we can conclude that $\{\widehat{\phi}_n\}$ is an increasing sequence of measurable simple functions that converges to f for all $x \in [a, b]$.

By the definition of the simple-Lebesgue integral, we see that

$$^L\!\!\int_a^b f = {}^{s\text{-}L}\!\!\int_a^b f = \lim_n {}^{s\text{-}L}\!\!\int_a^b \widehat{\phi}_n = \lim_n {}^L\!\!\int_a^b \widehat{\phi}_n.$$

The monotonicity of the Lebesgue integral implies that

$$L\int_a^b \widehat{\phi}_n \le L\int_a^b f_n \le L\int_a^b f$$

so that

$$L\int_a^b f = \lim_n L\int_a^b f_n$$

as claimed. □

Theorem 37 is not directly useful in the context of Fourier series since the sequence of partial sums is certainly not monotone. But even when $\{f_k\}$ is neither increasing nor convergent, we still can say something.

Theorem 38 (Fatou's lemma). *Let $\{f_k\}$ be a sequence of nonnegative, measurable functions. Then*

$$L\int_a^b \lim_k f_k \le \lim_k L\int_a^b f_k.^7$$

Proof. Define $\underline{f}_n = \inf_{k \ge n} f_k$. Then $\left\{\underline{f}_n\right\}$ is an increasing sequence of measurable functions. Since $\underline{f}_n \le f_k$ for all $k \ge n$,

$$L\int_a^b \underline{f}_n \le \inf_{k \ge n} L\int_a^b f_k.$$

Furthermore, $\left\{\inf_{k \ge n} L\int_a^b f_k\right\}_n$ is an increasing sequence of extended real numbers.[8] Taking limits and invoking the monotone convergence theorem,

$$L\int_a^b \lim_k f_k = L\int_a^b \lim_n \underline{f}_n = \lim_n L\int_a^b \underline{f}_n \le \lim_n \inf_{k \ge n} L\int_a^b f_k = \lim_k L\int_a^b f_k.$$

□

This brings us to the main theorem of this section, the dominated convergence theorem. It replaces the constant uniform bound in the bounded convergence theorem with an integrable function.

---

[7] $\lim_k f_k = \lim_k \inf_{j \ge k} f_j$. Since $\{\inf_{j \ge k} f_j\}$ is a monotone increasing sequence, $\lim_k f_k$ will always be defined in the extended real numbers. See Appendix A.3.

[8] The extended real numbers are $\bar{\mathbb{R}} = \mathbb{R} \cup \{-\infty, +\infty\}$.

Theorem 39 (Dominated convergence, Lebesgue, 1910). *Suppose that* $\{f_k\}$ *is a sequence of measurable functions that converges to* f *a.e. on* $[a, b]$. *If there is a measurable function* g *with* $^L\!\int_a^b g < \infty$ *and for which* $|f_k| \leq g$ *a.e. on* $[a, b]$ *for all* k, *then*

$$^L\!\int_a^b f = \lim_k {}^L\!\int_a^b f_k.$$

Proof. We may assume that $\{f_k\}$ converges to f and that $|f_k| \leq g$ on all of $[a, b]$. Define the sequences $\{\underline{f}_k\}$ and $\{\overline{f}_k\}$ by $\underline{f}_k = \inf_{j \geq k} f_j$ and $\overline{f}_k = \sup_{j \geq k} f_j$. Then $\{\underline{f}_k\}$ and $\{\overline{f}_k\}$ are monotone sequences of functions converging to f and satisfying

$$-g \leq \underline{f}_k \leq f_k \leq \overline{f}_k \leq g.$$

Now $\left\{g + \underline{f}_k\right\}$ is a nonnegative, increasing sequence of measurable functions converging to $g + f$, so by the monotone convergence theorem,

$$^L\!\int_a^b g + {}^L\!\int_a^b f = {}^L\!\int_a^b (g + f)$$

$$= \lim_k {}^L\!\int_a^b \left(g + \underline{f}_k\right)$$

$$= \lim_k \left({}^L\!\int_a^b g + {}^L\!\int_a^b \underline{f}_k\right)$$

$$= {}^L\!\int_a^b g + \lim_k {}^L\!\int_a^b \underline{f}_k.$$

Since $^L\!\int_a^b g$ is finite, $\lim_k {}^L\!\int_a^b \underline{f}_k = {}^L\!\int_a^b f$. A similar proof shows that $\lim_k {}^L\!\int_a^b \overline{f}_k = {}^L\!\int_a^b f$. By monotonicity of the Lebesgue integral,

$$^L\!\int_a^b \underline{f}_k \leq {}^L\!\int_a^b f_k \leq {}^L\!\int_a^b \overline{f}_k$$

so that $\lim_k {}^L\!\int_a^b f_k$ exists and is equal to $^L\!\int_a^b f$ as claimed.   □

Since the investigation of convergence properties of the various integrals was historically motivated by questions arising from trigonometric series, we will close the loop and show how the Monotone and dominated convergence theorems apply in the context of series of functions.

Theorem 40 (Interchange of summation and integral). *Suppose* $\{f_k\}$ *is a sequence of measurable functions defined on* $[a, b]$. *Suppose further that the series* $\sum_k {}^L\!\int_a^b |f_k|$ *converges. Then* $\sum_k f_k$ *converges almost everywhere on* $[a, b]$ *and*

$$ {}^L\!\int_a^b \sum_k f_k = \sum_k {}^L\!\int_a^b f_k. $$

Proof. As a positive-term series, $\sum_k |f_k(x)|$ converges in the extended real numbers for every $x \in [a, b]$. Since $\sum_k {}^L\!\int_a^b |f_k| < \infty$, we can conclude that $\sum_k |f_k(x)|$ and so $\sum_k f_k(x)$ must converge to a finite real number for almost every $x \in [a, b]$. Then by the monotone convergence theorem,

$$ {}^L\!\int_a^b \sum_k |f_k| = \sum_k {}^L\!\int_a^b |f_k| < \infty. $$

Moreover,

$$ -\sum_{k=1}^\infty |f_k| \le \sum_{k=1}^n f_k \le \sum_{k=1}^\infty |f_k| $$

for all $n \in \mathbb{N}$. The desired conclusion now follows from the dominated convergence theorem. □

6.4   The fundamental theorems

We begin our investigation of the fundamental theorem of calculus for the Lebesgue integral with a statement of the evaluation form.

Theorem 41 (FTC-1). *If* F *is a differentiable function with a bounded derivative in* $[a, b]$, *then* F' *is Lebesgue integrable on* $[a, b]$ *and*

$$ {}^L\!\int_a^x F' = F(x) - F(a) $$

for all $x \in [a, b]$.

Before turning to the proof, compare this theorem to the corresponding theorem for the Riemann integral. Both theorems require F to be differentiable on all of $[a, b]$. FTC-1 (page 39) for the Riemann integral requires that F' be continuous whereas the Lebesgue version requires only the much weaker assumption that F' be bounded. The bounded version of the theorem is false for the Riemann integral since the Volterra function has a bounded derivative that is not Riemann integrable. We will see an additional weakening of the boundedness hypothesis for the Lebesgue integral later in this section.

Proof. First note that F, being continuous, is measurable. Extend F to $[a, b+1]$ by setting $F(x) = F(b) + F'(b)(x-b)$ for $x \in [b, b+1]$. Then

$$f_n(x) = n\left[F\left(x + \frac{1}{n}\right) - F(x)\right], \quad x \in [a, b]$$

is also measurable with $\{f_n\}$ converging pointwise to F' on $[a, b]$. Hence F' is measurable and, being bounded by assumption, is Lebesgue integrable.

By the mean value theorem, given $t \in [a, b]$ and $n \in \mathbb{N}$, there is a value $c \in \left(t, t + \frac{1}{n}\right)$ such that

$$n\left[F\left(t + \frac{1}{n}\right) - F(t)\right] = F'(c).$$

As we are assuming that F' is bounded, we see that $\{f_n\}$ is uniformly bounded. Thus we can use the bounded convergence theorem, the continuity of F, and FTC-2 for the Riemann integral (page 40) to conclude that

$$\begin{aligned}
{}^L\!\int_a^x F' &= {}^L\!\int_a^x \lim_n n\left[F\left(t + \frac{1}{n}\right) - F(t)\right] dt \\
&= \lim_n n \cdot {}^L\!\int_a^x \left[F\left(t + \frac{1}{n}\right) - F(t)\right] dt \\
&= \lim_n n \cdot \left[{}^L\!\int_a^x F\left(t + \frac{1}{n}\right) dt - {}^L\!\int_a^x F(t)\, dt\right] \\
&= \lim_n n \cdot \left[{}^L\!\int_{a+\frac{1}{n}}^{x+\frac{1}{n}} F(t)\, dt - {}^L\!\int_a^x F(t)\, dt\right] \\
&= \lim_n \left[n \cdot {}^L\!\int_x^{x+\frac{1}{n}} F - n \cdot {}^L\!\int_a^{a+\frac{1}{n}} F\right] \\
&= \lim_n n \cdot {}^R\!\int_x^{x+\frac{1}{n}} F - \lim_n n \cdot {}^R\!\int_a^{a+\frac{1}{n}} F \\
&= F(x) - F(a). \qquad \square
\end{aligned}$$

Make sure that you thoroughly understand the preceding proof and the way it uses $f_n(x) = n\left[F\left(x + \frac{1}{n}\right) - F(x)\right]$. Closely related strategies will be used multiple times in this section with the critical exchange of the limit and the integral justified by the convergence results, including theorem 38, of the previous section.

The proof of the derivative form of the fundamental theorem of calculus for the Lebesgue integral is more complex. We state the theorem now, but we need to do quite a bit of preparatory work before we get to the proof.

Theorem 42 (FTC-2, Lebesgue, 1904). *Let f be a Lebesgue integrable function on $[a,b]$. Define $F(x) = {}^L\!\int_a^x f$ on $[a,b]$. Then F is absolutely continuous on $[a,b]$ and $F' = f$ a.e. on $[a,b]$.*

Absolute continuity is a strengthening of uniform continuity. It is the critical condition for the Lebesgue integral.

Definition 20 (Absolute continuity). *A function f is **absolutely continuous** on a set A if given any $\varepsilon > 0$ there is a $\delta > 0$ so that $\sum_k |f(x_k) - f(y_k)| < \varepsilon$ for any finite choice of disjoint intervals $\{(x_k, y_k)\}$ from A satisfying $\sum_k |x_k - y_k| < \delta$.*

Intuitively, continuity means that as x approaches x_0, the value of $f(x)$ approaches $f(x_0)$. Absolute continuity requires that, in addition, $f(x)$ does not "wiggle around too much" as it approaches $f(x_0)$. For example, the function

$$f(x) = \begin{cases} x \sin \frac{\pi}{x}, & x \neq 0 \\ 0, & x = 0 \end{cases}$$

is uniformly continuous but not absolutely continuous on $[0, 1]$. (See exercise 84.)

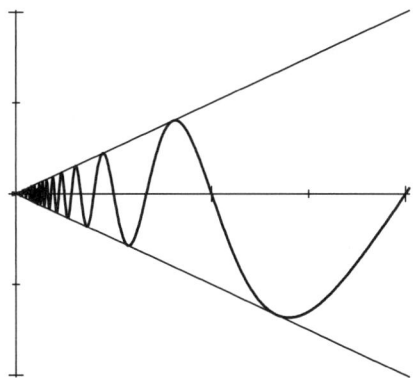

Figure 6.2. $f(x)$ is uniformly but not absolutely continuous

Before turning to the proof of this theorem, notice how it differs from FTC-2 for the Riemann integral (page 40). First, Riemann integrands must be bounded. This is no longer required for Lebesgue integrals. In addition, FTC-2 for the Riemann integral only concludes that $F'(x) = f(x)$ for points where f is continuous. For some Lebesgue integrable functions such as the Dirichlet function, the set of points of continuity is empty. Nevertheless, $F' = f$ a.e. on $[a, b]$. Moreover, Lebesgue's theorem draws the stronger conclusion that F is absolutely continuous.

The proof of FTC-2 for the Lebesgue integral depends on several properties of the integral as well as some facts related to absolute continuity. We will prove the integration-related results here but will only state the results related to absolute continuity, deferring their proofs until Chapter 7.

Lemma 43. *Suppose that f is Lebesgue integrable over $[a, b]$. Then, given any $\varepsilon > 0$, there is a $\delta > 0$ so that $^L\!\!\int_E |f| < \varepsilon$ for any measurable subset E of $[a, b]$ with $\lambda(E) < \delta$.*

Proof. Define

$$f_n(x) = \begin{cases} |f(x)|, & |f(x)| \le n \\ n, & |f(x)| > n. \end{cases}$$

Then $\{f_n\}$ converges pointwise to $|f|$ and, since $0 \le |f| - f_n \le |f|$, we can apply the dominated convergence theorem to conclude that

$$\lim_n {}^L\!\!\int_a^b (|f| - f_n) = 0.$$

Choose n so that $^L\!\!\int_a^b (|f| - f_n) < \varepsilon/2$ and set $\delta = \frac{\varepsilon}{2n}$. Then for any measurable subset E of $[a, b]$ with $\lambda(E) < \delta$,

$$^L\!\!\int_E |f| = {}^L\!\!\int_E (|f| - f_n) + {}^L\!\!\int_E f_n$$

$$\le {}^L\!\!\int_a^b (|f| - f_n) + {}^L\!\!\int_E f_n < \varepsilon. \qquad \square$$

Did you notice how variations of both the divide-and-conquer and truncation techniques were used in the preceding proof?

Lemma 44. *Suppose that f is Lebesgue integrable over $[a, b]$. If $F(x) = {}^L\!\!\int_a^x f = 0$ for all $x \in [a, b]$ then $f = 0$ a.e. on $[a, b]$.*

Proof. Suppose that $\{x \in [a, b] : f(x) > 0\}$ has positive measure. Then for some $n \in \mathbb{N}$, $A = \{x \in [a, b] : f(x) \ge 1/n\}$ also has positive measure.

Since A is measurable, theorem 22 (page 104) guarantees a closed set E contained in A with $\lambda(E) > 0$. Since E is closed, $(a,b) \setminus E$ is open and so can be written as the countable union of disjoint open intervals $\{(a_k, b_k)\}$. (See exercise 74.) By the dominated convergence theorem,

$$
\begin{aligned}
{}^L\!\!\int_{(a,b)\setminus E} f &= {}^L\!\!\int_a^b 1_{(a,b)\setminus E} \cdot f \\
&= {}^L\!\!\int_a^b \sum_k 1_{(a_k,b_k)} \cdot f \\
&= \sum_k {}^L\!\!\int_a^b 1_{(a_k,b_k)} \cdot f \\
&= \sum_k {}^L\!\!\int_{a_k}^{b_k} f \\
&= \sum_k \left({}^L\!\!\int_a^{b_k} f - {}^L\!\!\int_a^{a_k} f \right).
\end{aligned}
$$

This last sum is zero as all its terms are zero by assumption. But then

$$
\begin{aligned}
F(b) &= {}^L\!\!\int_a^b f = {}^L\!\!\int_E f + {}^L\!\!\int_{(a,b)\setminus E} f \\
&= {}^L\!\!\int_E f \geq \frac{1}{n} \lambda(E) > 0
\end{aligned}
$$

contrary to our hypothesis. Thus $f \leq 0$ a.e. on $[a, b]$. By applying the preceding argument to $-f$, we can conclude that $f = 0$ a.e. on $[a, b]$. [9] $\square$

The following three results will be proved in Chapter 7, Section 8.

Lemma 45. *Suppose that f is absolutely continuous on $[a, b]$. Then f can be expressed as $f = f_1 - f_2$ where f_1 and f_2 are increasing functions. (If convenient, we can require f_1 and f_2 to be strictly increasing and/or absolutely continuous.)*

Lemma 46 (Lebesgue, 1904). *If f is increasing on $[a, b]$ then f is differentiable a.e. on $[a, b]$.*

Lemma 47. *If f is absolutely continuous on $[a, b]$ and $f' = 0$ a.e. on $[a, b]$, then f is constant on $[a, b]$.*

---

[9] Exercise 39 outlines an alternate proof of Lemma 44.

With these tools, we are now prepared to prove the derivative form of the fundamental theorem of calculus for Lebesgue integrals which we restate here.

Theorem 42 (FTC-2, Lebesgue, 1904). *Let f be a Lebesgue integrable function on $[a,b]$ and define $F(x) = {}^L\!\int_a^x f$ on $[a,b]$. Then F is absolutely continuous on $[a,b]$ and $F' = f$ a.e. on $[a,b]$.*

Proof. To prove that F is absolutely continuous, let $\varepsilon > 0$ be given. Since f is Lebesgue integrable, lemma 43 tells us that there is a $\delta > 0$ such that ${}^L\!\int_A |f| < \varepsilon$ whenever A is a measurable set with $\lambda(A) < \delta$. Suppose now that $\{(x_k, y_k)\}$ is a finite set of disjoint subintervals from $[a,b]$ with $\sum_k |x_k - y_k| < \delta$. Setting $A = \cup_k (x_k, y_k)$ we have $\lambda(A) = \sum_k |x_k - y_k| < \delta$ so that

$$\sum_k |F(x_k) - F(y_k)| = \sum_k \left| {}^L\!\int_{x_k}^{y_k} f \right|$$

$$\leq \sum_k {}^L\!\int_{x_k}^{y_k} |f|$$

$$= {}^L\!\int_A |f| < \varepsilon.$$

Hence F is absolutely continuous and therefore differentiable almost everywhere.

We first prove that $F' = f$ a.e. under the additional assumption that $|f| \leq B$. Extend F to $[a, b+1]$ by taking $F(x) = F(b)$ for $x \in [b, b+1]$. Then $\lim_n n\left[F\left(x + \frac{1}{n}\right) - F(x)\right] = F'(x)$ for those $x \in (a,b)$ at which F is differentiable. Since

$$\left| n\left[F\left(x + \frac{1}{n}\right) - F(x) \right] \right| = n \left| {}^L\!\int_x^{x+\frac{1}{n}} f \right| \leq n \cdot {}^L\!\int_x^{x+\frac{1}{n}} |f| \leq B,$$

we can follow the proof of theorem 41 (page 141) using the bounded convergence theorem to conclude that

$$ {}^L\!\int_a^x F' = {}^L\!\int_a^x \lim_n n\left[F\left(t + \frac{1}{n}\right) - F(t) \right]$$

$$= F(x) - F(a) = F(x).$$

This means that

$$ {}^L\!\int_a^x (F' - f) = \left({}^L\!\int_a^x F' \right) - F(x) = 0$$

for all $x \in [a,b]$. Hence, by lemma 44 (page 144), $F' = f$ a.e. in $[a,b]$.

If f is not bounded, assume temporarily that $f \geq 0$. Repeat the analysis above using Fatou's lemma (theorem 38, page 139) instead of the bounded convergence theorem to conclude that, for those $x \in (a, b)$ for which F is differentiable,

$$
{}^{L}\!\int_a^x F' = {}^{L}\!\int_a^x \lim_n n \left[F\left(t + \frac{1}{n}\right) - F(t) \right]
$$
$$
\leq \underline{\lim}_n {}^{L}\!\int_a^x n \left[F\left(t + \frac{1}{n}\right) - F(t) \right]
$$
$$
= F(x) - F(a) = F(x).
$$

Hence

$$
{}^{L}\!\int_a^x (F' - f) = \left({}^{L}\!\int_a^x F' \right) - F(x) \leq 0.
$$

For the reverse inequality, define

$$
f_n(x) = \begin{cases} f(x), & f(x) \leq n \\ n, & f(x) > n \end{cases}
$$

and $F_n(x) = {}^{L}\!\int_a^x f_n$ for $x \in [a, b]$. Then $F - F_n$ is nonnegative and increasing on $[a, b]$ so, by lemma 46 (page 145), $(F - F_n)'$ exists and is nonnegative almost everywhere in $[a, b]$. Furthermore, our earlier work shows that $F_n' = f_n$ a.e. on $[a, b]$. Recalling that we are assuming that f, and so f_n, is nonnegative,

$$
0 \leq (F - F_n)' = F' - F_n' = F' - f_n \leq F' \text{ a.e.}
$$

Thus

$$
{}^{L}\!\int_a^x (F' - f_n) \geq 0.
$$

Since the inequality holds for all values of n,

$$
{}^{L}\!\int_a^x (F' - f) \geq 0
$$

by the dominated convergence theorem.

Thus when f is nonnegative, $F' = f$ a.e. The general case now follows by considering $f = f^+ - f^-$. $\qquad\square$

Notice that, even though F and so $n\left[F\left(t + \frac{1}{n}\right) - F(t) \right]$ are continuous functions and therefore Riemann integrable, the Riemann integral does not allow the critical exchanges of the integral with the limit or with $\underline{\lim}$ used in the proofs that $^L\!\int_a^x F' = F(x)$ and $^L\!\int_a^x F' \le F(x)$.

In light of FTC-2, it is tempting to conclude that as long as F is differentiable a.e. on $[a, b]$, then (subtracting $F(a)$ to make the function zero at $x = a$) we have $\int_a^x F' = F(x) - F(a)$. This would be an error. As a counterexample, consider the Cantor function c. Its derivative is $c' = 0$ a.e. on $[a, b]$ so that

$$0 = \int_0^1 c' \ne c(1) - c(0) = 1.$$

Even when F is differentiable everywhere, the conclusion of the theorem may not hold. (See exercise 89.) The conditions under which we can conclude that $\int_a^b F' = F(b) - F(a)$ are given in the following theorem. Notice that, since changing an integrand on a set of measure zero does not change the integral and since antiderivatives may differ by a constant, the theorem is as close to a converse of FTC-2 as is possible.

Theorem 47 (FTC-1, Lebesgue, 1904). *If F is absolutely continuous on $[a, b]$, then F is differentiable a.e., F' is Lebesgue integrable on $[a, b]$, and*

$$^L\!\int_a^x F' = F(x) - F(a)$$

for all $x \in [a, b]$.

Proof. Since F is absolutely continuous, it can be written as $F = F_1 - F_2$ where F_1 and F_2 are increasing functions. Since increasing functions are differentiable a.e., so is F and, as in theorem 41 (page 141), F' is measurable. Moreover, $|F'| \le F_1' + F_2'$ a.e. so that we can use theorem 38 (page 139) as in the proof of FTC-2 to conclude that

$$^L\!\int_a^b |F'| \le {}^L\!\int_a^b F_1' + {}^L\!\int_a^b F_2'$$
$$\le F_1(b) - F_1(a) + F_2(b) - F_2(a).$$

Thus F' is Lebesgue integrable over $[a, b]$.

Now define $G(x) = {}^L\!\int_a^x F'$. Then G, and so $G - F$, are absolutely continuous on $[a, b]$. By theorem 42 (page 146), $(G - F)' = G' - F' = 0$ a.e. on $[a, b]$, so that $G - F$ is constant on $[a, b]$ by lemma 47 (page 145). Since $G(a) = 0$, $G(x) - F(x) = -F(a)$, so that

$$^L\!\int_a^x F' = G(x) = F(x) - F(a). \qquad \square$$

The gauge integral, introduced in the next chapter, allows us to replace the requirement that F be absolutely continuous with simple continuity. On the other hand, F' must be differentiable except at a countable number of points rather than differentiable almost everywhere.

6.5 Exercises

6.1 Variations: filling the gaps

1. Suppose that $f : [a, b] \to (\alpha, \beta)$ is a measurable function. Prove that if $\mathcal{P} = \{[y_{k-1}, y_k]\}_{k=1}^{n}$ is a partition of $[\alpha, \beta]$ and $\mathcal{P}^*$ is a refinement of $\mathcal{P}$, then $S_{\underline{L}}(f, \mathcal{P}) \le S_{\underline{L}}(f, \mathcal{P}^*)$ and $S_{\overline{L}}(f, \mathcal{P}) \ge S_{\overline{L}}(f, \mathcal{P}^*)$. (First show $S_{\underline{L}}(f, \mathcal{P}^*) - S_{\underline{L}}(f, \mathcal{P}) \ge 0$ when a single subinterval is split into two. Then use induction.)

2. Suppose that $f : [a, b] \to (\alpha, \beta)$. Explain why $S_{\underline{L}}(f, \mathcal{P}) \le S_{\overline{L}}(f, \mathcal{Q})$ for any pair of measurable partitions $\mathcal{P}$ and $\mathcal{Q}$ of $[\alpha, \beta]$. Use this fact to prove that $\sup_{\mathcal{P}} S_{\underline{L}}(f, \mathcal{P}) \le \inf_{\mathcal{P}} S_{\overline{L}}(f, \mathcal{P})$ where the supremum and infimum are taken over all partitions of $[\alpha, \beta]$.

3. Explain why the definition of the Lebesgue integral does not mention upper Lebesgue sums.

4. Why does the definition of the simple-Lebesgue integral require the sequence of simple functions converging to f to be increasing?

5. Explain why $^L\!\int_a^b |f| < \infty$ is equivalent to the condition that the integrals of both f^+ and f^- are finite.

6. Verify the standard integral properties for the Lebesgue-style integrals. Chose a different definition for each of the first three properties. (Think about which definition will be most efficacious for proving each of the properties.)

   (a) **Uniqueness.** The value of the integral is unique (if it exists).
   (b) **Linearity.** If f and g are integrable over the interval $[a, b]$ and $c \in \mathbb{R}$, then so are $f + g$ and cf. Moreover, $\int_a^b (f + g) = \int_a^b f + \int_a^b g$ and $\int_a^b cf = c \int_a^b f$.
   (c) **Monotonicity.** If f and g are Lebesgue integrable over the interval $[a, b]$ with $f(x) \le g(x)$ for $x \in [a, b]$, then $\int_a^b f \le \int_a^b g$.
   (d) **Triangle inequality.** If f and $|f|$ are integrable over $[a, b]$, then $\left| \int_a^b f \right| \le \int_a^b |f|$.

6.1 Variations: deeper reflections

7. Suppose that $f : [a, b] \to (\alpha, \beta)$. Even if we cannot use Lebesgue measure because f is not measurable, we can still use Lebesgue outer measure. Let $\mathcal{P} = \{[y_{k-1}, y_k]\}_{k=1}^{n}$ be a partition of $[\alpha, \beta]$ and set $E_k = f^{-1}((y_{k-1}, y_k]), k = 1, 2, \ldots, n$. We can define lower and upper sums by $S_{\underline{L}^*}(f, \mathcal{P}) = \sum_{k=1}^{n} y_{k-1} \cdot \mu^*(E_k)$ and $S_{\overline{L}^*}(f, \mathcal{P}) = \sum_{k=1}^{n} y_k \cdot \mu^*(E_k)$ where $\mu^*(A)$ is the Lebesgue outer measure of the set A. How do things go wrong when we try to define the Lebesgue (outer) integral of f by $\sup_{\mathcal{P}} S_{\underline{L}^*}(f, \mathcal{P})$ or $\inf_{\mathcal{P}} S_{\overline{L}^*}(f, \mathcal{P})$ where the supremum/infimum is taken over all possible partitions of $[\alpha, \beta]$?

8. Suppose that $\{E_k\}_{k=1}^{n}$ are sets of real numbers and that $\phi = \sum_{k=1}^{n} \alpha_k \cdot 1_{E_k}$. How many distinct values could ϕ take on?

9. If f is Lebesgue integrable over $[a, b]$ then so is $|f|$. Does the integrability of $|f|$ imply the integrability of f? Explain.

10. Let f be a nonnegative, measurable function on $[a, b]$. Prove that f is Lebesgue integrable on $[a, b]$ if an only if $\sum_{k=1}^{\infty} k\lambda(E_k)$ converges where $E_k = f^{-1}([k, k+1))$.

6.2 Reconciling the approaches: filling the gaps

11. In the proof of theorem 28 (page 124), why is the set C measurable?

12. Prove that the two alternate statements in Definition 17 (page 125) are equivalent.

13. Explain how Definition 17 (page 125) relates to Definition 3 (page 30).

14. Suppose that $f : [a, b] \to (\alpha, \beta)$. Prove that if $\mathcal{P} = \{E_k\}$ is a measurable partition of $[a, b]$ and $\mathcal{P}^* = \{E_j^*\}$ is a measurable refinement of $\mathcal{P}$, then $S_{\overline{L\text{-}Y}}(f, \mathcal{P}^*) \leq S_{\overline{L\text{-}Y}}(f, \mathcal{P})$. (See lemma 29 on page 126.)

15. Suppose that $f : [a, b] \to (\alpha, \beta)$. Use the previous exercise to prove that $S_{\underline{L\text{-}Y}}(f, \mathcal{P}) \leq S_{\overline{L\text{-}Y}}(f, \mathcal{Q})$ for any measurable partitions $\mathcal{P}$ and $\mathcal{Q}$ of $[a, b]$. Conclude that $\underline{L\text{-}Y}\!\int_a^b f \leq \overline{L\text{-}Y}\!\int_a^b f$.

16. Suppose that ϕ and ψ are measurable simple functions and that c is a real number. Prove that $\psi_1 = \phi + \psi$ and $\psi_2 = c\phi$ are also measurable simple functions.

17. Verify that $^{L}\!\int_a^b \phi = \sum_{k=1}^{n} \alpha_k \cdot \lambda([a, b] \cap E_k)$ for any simple function $\phi = \sum_{k=1}^{n} \alpha_k \cdot 1_{E_k}$.

18. In the proof of theorem 30 (page 126)

    (a) Explain why $\lambda\left(D_k\right) \leq \lambda\left(D_{k,n}\right) = {}^L\!\int_a^b 1_{D_{k,n}} \leq {}^L\!\int_a^b k\left(\phi_n - \psi_n\right)$.
    (b) Fill in the details to explain why f is measurable if ϕ is measurable, $\phi = \psi$ a.e. and $\phi \leq f \leq \psi$.
    (c) Fill in the details at the end of the proof to show that ${}^{L\text{-}Y}\!\int_a^b f = {}^L\!\int_a^b f$.

19. Supply the justifications on page 128 to explain why

$$\underline{D}\int_a^b f \leq \underline{L\text{-}Y}\int_a^b f \leq \overline{L\text{-}Y}\int_a^b f \leq \overline{D}\int_a^b f.$$

20. Suppose that ϕ and ψ are simple functions. Explain why $|\phi - \psi|$ is a simple function.

21. Suppose that ϕ and ψ are simple functions. Prove that $\max\{\phi, \psi\}$ and $\min\{\phi, \psi\}$ are simple functions. Extend this conclusion to finite sets of simple functions.

22. In the proof of theorem 31 (page 128)

    (a) Why does the second half of the proof start with $L < \lim_k s\text{-}L\int_a^b \phi_k$ and an n_0 for which $L < {}^{s\text{-}L}\!\int_a^b \phi_{n_0}$ instead of beginning with $\varepsilon > 0$ and an n_0 for which $\left|\lim_k {}^{s\text{-}L}\!\int_a^b \phi_k - {}^{s\text{-}L}\!\int_a^b \phi_{n_0}\right| < \varepsilon$?
    (b) Explain why $\{\phi_k^*\}$ and $\{\psi_k^*\}$ are increasing sequences of measurable simple functions that converge to f^*.

23. Suppose that $f : [a, b] \to \mathbb{R}$ is Lebesgue integrable over $[a, b]$ and let E be a measurable subset of $[a, b]$. ${}^L\!\int_E f$ could be defined two ways:

    (a) ${}^L\!\int_E f = {}^L\!\int_a^b f \cdot 1_E$ or
    (b) by modifying the standard definition to use
        $E_k = \{x \in E : y_{k-1} < f(x) \leq y_k\}$.

    Prove that the two definitions are equivalent.

24. When proving that the Lebesgue and simple-Lebesgue integrals agree for bounded functions, we used the monotonicity of the simple-Lebesgue integral. Prove this.

25. Suppose that $f : [a, b] \to \mathbb{R}$ is Lebesgue integrable over $[a, b]$. Prove that if E and F are disjoint measurable subsets of $[a, b]$, then ${}^L\!\int_E f + {}^L\!\int_F f = {}^L\!\int_{E \cup F} f$.

26. Fill in the details at the end of the proof of theorem 32 (page 131). Let $\{\phi_n\}$ be a sequence of simple functions that are bounded above by f and converge uniformly to f and for which $^L\!\int_a^b f = \lim_n {}^{s\text{-}L}\!\int_a^b \phi_n$. Define $\widehat{\phi}_n = \max_{k \le n} \phi_k$. Prove that $\{\widehat{\phi}_n\}$ is an increasing sequence of measurable simple functions bounded by f for which $^L\!\int_a^b f = \lim_n {}^{s\text{-}L}\!\int_a^b \widehat{\phi}_n$.

6.2 Reconciling the approaches: deeper reflections

27. Use the ε–δ definition of continuity to prove that if $f : A \to \mathbb{R}$ is continuous and U is an open set of real numbers, then $f^{-1}(U) = V \cap A$ for some open subset V of $\mathbb{R}$.

28. Let $\mathcal{P} = \{[0, 0.2], (0.2, 0.5), [0.5, 0.8), [0.8, 1]\}$ and $\mathcal{Q} = \{[0, 0.1), [0.1, 0.5], (0.5, 0.7), [0.7, 1]\}$. What is $\mathcal{P} \cup \mathcal{Q}$?

29. Approximate $f(x) = x^2$, $x \in [0, 4]$ within 0.25 using a measurable simple function ϕ with $\phi \le f$.

30. Suppose that $f : [a, b] \to (\alpha, \beta)$ and that $\mathcal{P}$ and $\mathcal{Q}$ are measurable partitions of $[a, b]$ with $\mathcal{Q}$ being a refinement of $\mathcal{P}$. Let ϕ and ψ be the lower simple functions corresponding to the partitions $\mathcal{P}$ and $\mathcal{Q}$. In other words, $\phi = \sum_{E \in \mathcal{P}} \inf_E f \cdot 1_E$ and $\psi = \sum_{F \in \mathcal{Q}} \inf_F f \cdot 1_F$. Prove that $\phi \le \psi$. (Given $x \in [a, b]$, how is $\phi(x)$ computed?)

31. Give an example of a pair of sequences $\{a_k\}$ and $\{b_k\}$ for which $\lim_k |a_k - b_k| = 0$ but for which $\lim_k a_k \ne \lim_k b_k$ since neither limit exists (even in the extended real numbers). Explain why this is not an issue in the proof of theorem 31 (page 128).

32. Give an example of a function f and a sequence of partitions $\{\mathcal{P}_n\}$ for which the naturally associated sequence $\{\phi_n\}$ of measurable lower simple functions satisfies $^L\!\int_a^b f = \lim_n S_L(f, \mathcal{P}_n)$ but is not increasing. (See theorem 32 on page 131.)

33. Suppose that f is a measurable function on $[a, b]$ with $0 \le f < \beta$. Use the values $y_{n,k} = \frac{k}{2^n}$ to prove that there is an increasing sequence of simple functions $\{\phi_n\}$ that converges uniformly to f. (To simplify your work, you can assume that $\beta = N$ for some integer N.)

34. Use the sequence of simple functions of the previous exercise to verify that any nonnegative, bounded, measurable function on $[a, b]$ has a finite simple-Lebesgue integral that agrees with the Lebesgue integral.

35. Use the previous exercise to show that any bounded, measurable (not necessarily nonnegative) function has a finite simple-Lebesgue integral that agrees with the Lebesgue integral.

36. Modify exercise 33 to prove if f is a nonnegative, measurable (not necessarily bounded) function on $[a, b]$, there is an increasing sequence of simple functions $\{\phi_n\}$ that converges pointwise to f. (Work with $\left\{ y_{n,k} = \frac{k}{2^n} \right\}_{k=0}^{n \cdot 2^n}$.)

37. Outline the problems one would have to address in order to prove that for a Riemann integrable function f, $^L\!\int_a^b f = {}^R\!\int_a^b f$ using the original Riemann and Lebesgue definitions of the integral.

38. How would the proofs for simple-Lebesgue results change if we modified the definition for nonnegative functions to be

$$^{s\text{-}L}\!\int_a^b f = \sup\left\{ {}^{s\text{-}L}\!\int_a^b \phi : \phi \text{ is a measurable simple function with } \phi \leq f \right\} ?$$

39. Suppose that f is Lebesgue integrable and that $^L\!\int_E f = 0$ for every measurable set $E \subseteq [a, b]$.

   (a) Prove that $f = 0$ a.e. on $[a, b]$. (Begin by showing that $\lambda\left(\{x : f(x) > 1/n\}\right) = 0$.)
   (b) Why is it sufficient to know that $^L\!\int_a^x f = 0$ for all $x \in [a, b]$? (Use sigma algebras.)

6.3 Convergence theorems: filling the gaps

40. Prove that if $f = g$ a.e., then $f^+ = g^+$ a.e. and $f^- = g^-$ a.e.

41. In the proof of theorem 34 (page 135), why are f^+ and g^+ used instead of f and g?

42. In the proof of theorem 35 (page 135), why is $A_{n,k}$ measurable?

43. Suppose that $\{A_n\}$ is a sequence of measurable sets satisfying $A_{n+1} \subseteq A_n$ and $\lambda\left(\cap_n A_n\right) = 0$. Suppose further that $\lambda\left(A_1\right) < +\infty$. Prove that for any $\varepsilon > 0$, there is an integer n so that $\lambda\left(A_n\right) < \varepsilon$. (Express A_1 as the disjoint union of measurable sets $\{B_n\}$ such that $A_n = \left(\cup_{k \geq n} B_k\right) \cup \left(\cap_n A_n\right)$.)

44. Suppose that $\{f_k\}$ is a sequence of functions that converges a.e. to f and $\{f_k\}$ is uniformly bounded by B. Define f^* by

$$f^*(x) = \begin{cases} f(x), & \{f_k(x)\} \text{ converges to } f(x) \\ 0, & \text{otherwise.} \end{cases}$$

    (a) Prove that $|f^*| \le B$.
    (b) Give an example where $|f| \le B$ fails.

45. The proof of theorem 37 (page 138) begins by redefining f and f_k. Provide a specific procedure for this redefinition. Why is the redefinition permissible?

46. In the proof of theorem 37 (page 138), why is $\max_{1 \le k \le n} \phi_{k,n+1} \ge \max_{1 \le k \le n} \phi_{k,n}$?

47. In the proof of theorem 38 (page 139)

    (a) Why is $\{\underline{f}_n\}$ an increasing sequence of measurable functions?
    (b) Why is $^L\!\int_a^b \underline{f}_n \le \inf_{k \ge n} {}^L\!\int_a^b f_k$?
    (c) Why is $\left\{\inf_{k \ge n} {}^L\!\int_a^b f_k\right\}_n$ an increasing sequence?

48. Suppose that $\{a_k\}$ and $\{b_k\}$ are increasing sequences of extended real numbers with $a_k \le b_k$ for all k. Prove that $\lim_k a_k \le \lim_k b_k$.

49. In theorem 39 (page 140)

    (a) Why is f integrable on $[a,b]$?
    (b) Why can we assume that $\{f_k\}$ converges to f and that $|f_k| \le g$ on all of $[a,b]$?
    (c) Why is $\{\underline{f}_k\}$ increasing and why is $\{\overline{f}_k\}$ decreasing?
    (d) Why do $\{\underline{f}_k\}$ and $\{\overline{f}_k\}$ converge to f ?
    (e) Why does $\lim_k {}^L\!\int_a^b \underline{f}_k$ exist?
    (f) Fill in the details to show that $^L\!\int_a^b f = \lim_k {}^L\!\int_a^b \overline{f}_k$.

50. Prove that if $\lambda\left(\{x \in [a,b] : \sum_k |f_k(x)| = \infty\}\right) = m > 0$ then $\sum_k {}^L\!\int_a^b |f_k| = \infty$. (Use exercise 58 to show that for any M, we can find an N so that $\lambda\left(\{x : \sum_{k=1}^N |f_k(x)| \ge M\}\right) \ge m/2$. Explain how this forces $\sum_k {}^L\!\int_a^b |f_k| = \infty$.)

51. Explain how the Monotone and dominated convergence theorems guarantee that $^L\!\int_a^b \sum_k f_k = \sum_k {}^L\!\int_a^b f_k$ in theorem 40 (page 141).

6.3 Convergence theorems: deeper reflections

52. Give an alternate proof of theorem 33 (page 135) based on the definition of the Lebesgue (rather than Lebesgue-Young) integral under the additional assumption that $f \geq 0$.

53. Give an alternate proof of theorem 33 (page 135) based on the definition of the simple-Lebesgue (rather than Lebesgue-Young) integral.

54. Prove that if $\{f_k\}$ is a sequence of bounded functions that converges uniformly to a function f, then $\{f_k\}$ is uniformly bounded. In other words, there is a real number B such that $|f_k| < B$ for all values of k.

55. Suppose that f is measurable and that $g = f$ a.e. Supply an alternate proof that g is measurable by considering the function $f - g$.

56. Give a counterexample to exercise 43 when $\lambda(A_1) = +\infty$.

57. Suppose that $\{A_n\}$ is a sequence of measurable sets. Prove that if $\lambda(\cup_n A_n) > 0$ then $\lambda(A_n) > 0$ for some $n \in \mathbb{N}$.

58. Suppose that $A_1 \subseteq A_2 \subseteq \cdots \subseteq A_n \subseteq A_{n+1} \subseteq \cdots$ is a nested sequence of measurable sets with $\lambda(\cup_n A_n) < \infty$. Prove that there is an integer N for which $\lambda(A_N) \geq \frac{1}{2}\lambda(\cup_n A_n)$. Under what circumstances can you conclude $\lambda(A_N) > \frac{1}{2}\lambda(\cup_n A_n)$?

59. Give an example of a sequence of continuous functions $\{f_k\}$ that converges to f on all of $[0, 1]$ but for which $\lim_k {}^L\!\int_0^1 f_k \neq {}^L\!\int_0^1 f$.

60. State and prove a version of theorem 37 (page 138) for deceasing sequences of functions.

61. Define f on $[0, 1]$ by

$$f(x) = \begin{cases} \frac{1}{\sqrt{x}}, & x \neq 0 \\ 0, & x = 0. \end{cases}$$

Use an increasing sequence of functions to evaluate $^L\!\int_0^1 f$.

62. Define f on $[0, 1]$ by

$$f(x) = \begin{cases} \frac{1}{(x-0.5)^2}, & x \neq 0.5 \\ 0, & x = 0.5. \end{cases}$$

Use an increasing sequence of functions to show that $^L\!\int_0^1 f = +\infty$.

63. Define f on $[0, 1]$ by

$$f(x) = \begin{cases} \frac{1}{(x-0.5)^3}, & x \neq 0.5 \\ 0, & x = 0.5. \end{cases}$$

Show that $^L\!\int_0^1 f$ is not defined.

64. The statement that $^L\!\int_a^b g + {^L\!\int_a^b} f = {^L\!\int_a^b} (g + f)$ in the proof of Theorem 39 (page 140) relies on the assumption that $0 \leq {^L\!\int_a^b} g < \infty$. Give an example of a pair of measurable functions f and g for which $^L\!\int_a^b (g + f) = 0$ but for which $^L\!\int_a^b g + {^L\!\int_a^b} f$ is not defined.

65. Compute the following (provide justifications)

(a) $\sum_{n=0}^{\infty} {^L\!\int_0^1} \frac{x^\alpha}{(1+x^\beta)^n} \, d\lambda\,(x)$ where $\alpha \geq \beta$.

(b) $\sum_{n=1}^{\infty} {^L\!\int_0^1} \frac{x}{(1+(n-1)x)(1+nx)} \, d\lambda\,(x)$.

(c) $\lim_n {^L\!\int_0^1} \frac{nx}{1+n^2x^2} \, d\lambda\,(x)$.

(d) $\lim_n {^L\!\int_{-1}^2} \frac{x^{4n}}{1+x^{4n}} \, d\lambda\,(x)$.

(e) $^L\!\int_0^1 \left(\sum_{n=1}^{\infty} \frac{x^n}{n} \right) d\lambda\,(x)$.

(f) $\lim_n {^L\!\int_0^1} nxe^{-nx^2} \, d\lambda\,(x)$.

(g) $^L\!\int_0^1 g$ where g is defined on $[0, 1]$ by

$$g(x) = \begin{cases} 0, & x \text{ is in the Cantor set} \\ n, & x \text{ is in a removed interval of length } 1/3^n. \end{cases}$$

66. Give an example of a sequence of functions $\{f_n\}$ on $[0, 1]$ for which $\lim_n {^L\!\int_0^1} |f_n| = 0$ but $\{f_n(x)\}$ does not converge to zero for any $x \in (0, 1)$. (Try using characteristic functions.)

6.4 The fundamental theorems: filling the gaps

67. Suppose that F is differentiable almost everywhere on $[a, b]$. Why is F' measurable? (Consider $\lim_n n \left[F \left(t + \frac{1}{n} \right) - F(t) \right]$.)

68. Without using F.T.C., prove that $\lim_n n \cdot {^L\!\int_x^{x+\frac{1}{n}}} f = f(x)$ whenever f is continuous at x.

69. Prove that if f is bounded and measurable on $[a, b+1]$ then

$$^L\!\int_a^b f \left(t + \frac{1}{n} \right) = {^L\!\int_{a+\frac{1}{n}}^{b+\frac{1}{n}}} f(t).$$

70. Prove that if f is nonnegative and A is a measurable subset of $[a,b]$, then $^L\!\int_A f \leq {}^L\!\int_a^b f$.

71. In the proof of lemma 43 (page 144), why is $^L\!\int_E f_n < \varepsilon/2$?

72. Suppose that f is a measurable function for which $\{x \in [a,b] : f(x) > 0\}$ has positive measure. Why must $\{x \in [a,b] : f(x) \geq 1/n\}$ also have positive measure for some n?

73. Suppose that A is measurable with $\lambda(A) > 0$. Why must A contain a closed set E with $\lambda(E) > 0$?

74. Prove that any bounded open set E can be written as a countable union of disjoint open intervals. (The hard part is to make the intervals disjoint. Work inductively. Having chosen $I_1, I_2, \ldots, I_n$, let $E_n = E \setminus \bigcup_{k=1}^n \overline{I}_n$ where $\overline{I}_n$ is the closure of I_n. Given $x \in E_n$, compute $\alpha_x = \inf\{t \in [a,b] : (t,x) \subseteq E\}$ and $\beta_x = \sup\{s \in [a,b] : (x,s) \subseteq E\}$. Take $I_{n+1} = (\alpha_x, \beta_x)$ for some $x \in E_n$. Make sure to choose x in such a way that $\bigcup_n I_n = E$.)

75. Extend the result of the previous exercise to all open sets.

76. Why can the sum and integrals be interchanged in the proof of lemma 44 (page 144)? What is the dominating function?

77. Supply a proof that $f \geq 0$ a.e. on $[a,b]$ for lemma 44 (page 144). (You do not need to mimic what was done to prove $f \leq 0$ a.e. on $[a,b]$.)

78. Prove that if f is a nonnegative function on $[a,b]$, then $F(x) = {}^L\!\int_a^x f$ is increasing.

79. In the proof of theorem 42 (page 146)

    (a) Fill in the details to show that $^L\!\int_a^x F' = F(x) - F(a) = F(x)$ when f is bounded.
    (b) Why is $F - F_n$ nonnegative and increasing on $[a,b]$?
    (c) Why is $F_n' = f_n$ a.e. on $[a,b]$?
    (d) Fill in the details near the middle of the proof to show that $^L\!\int_a^x F' \leq F(x)$ when $f \geq 0$ but unbounded.
    (e) Justify the statement $^L\!\int_a^x (F' - f_n) \geq 0$, near the end of the proof.
    (f) How is the dominated convergence theorem used to conclude that $^L\!\int_a^x (F' - f) \geq 0$? What is the dominating function?
    (g) Complete the proof by showing how $f = f^+ - f^-$ can be used to prove the theorem when f is unbounded but not nonnegative.

80. Supply the details needed to conclude that $^{L}\!\int_{a}^{b}|F'| \leq F_1(b) - F_1(a) + F_2(b) - F_2(a)$ in theorem 47.

6.4 The fundamental theorems: deeper reflections

81. Suppose that f is Riemann integrable on $[a, b]$. Prove that $F(x) = {}^{R}\!\int_{a}^{x} f$ is absolutely continuous on $[a, b]$.

82. Prove that if f and g are absolutely continuous on a set A and c is a scalar, then $f + g$ and cf are also absolutely continuous.

83. Prove that any absolutely continuous function is uniformly continuous.

84. Define
$$f(x) = \begin{cases} x \cos \frac{\pi}{x}, & x \neq 0 \\ 0, & x = 0. \end{cases}$$

    (a) Prove that f is uniformly continuous on $[0, 1]$.
    (b) Prove that f is not absolutely continuous on $[0, 1]$. (Use $(x_k, y_k) = \left(\frac{1}{k+1}, \frac{1}{k}\right)$.)

85. Prove that the Cantor function is not absolutely continuous.

86. Show that continuity is not sufficient in lemma 47 (page 145). In other words, give an example of a nonconstant, continuous function on $[0, 1]$ for which $f' = 0$ a.e. Is uniform continuity sufficient?

87. Suppose that $\{f_n\}$ is a sequence of measurable functions and that
$$\lim_{n} {}^{L}\!\int_{a}^{b} |f_n - f| = 0$$
for some integrable function f. Prove that if $\{f_n\}$ converges a.e. on $[a, b]$, then $\{f_n\}$ converges a.e. to f on $[a, b]$. (See exercise 59.)

88. Let $f(x) = \sqrt{x}$, $x \in [0, 1]$.

    (a) Is $f(x)$ absolutely continuous on $[0, 1]$?
    (b) Is f' Lebesgue integrable on $[0, 1]$?
    (c) Is $^{L}\!\int_{0}^{x} f' = f(x)$ for $x \in [0, 1]$?
    (d) Why are we not worried about the fact that f' does not exist at $x = 0$?

89. Let
$$f(x) = \begin{cases} x^2 \cos \frac{\pi}{x^2}, & x \neq 0 \\ 0, & x = 0. \end{cases}$$

    (a) Prove that f is differentiable on all of $[0, 1]$.

    (b) Is f absolutely continuous? (Consider intervals of the form $\left(\frac{1}{\sqrt{n+1}}, \frac{1}{\sqrt{n}}\right)$.)

    (c) Show that ${}^{L}\!\int_{0}^{x} f' \neq f(x)$ for $x \in [0, 1]$ since ${}^{L}\!\int_{0}^{x} f'$ does not exist. (Again, consider intervals of the form $\left(\frac{1}{\sqrt{n+1}}, \frac{1}{\sqrt{n}}\right)$.)

90. In some situations, the Riemann integral can be extended to handle isolated points near which a function is unbounded. In particular, when a function f is unbounded at 0, we can attempt to define ${}^{R}\!\int_{0}^{x} f$ by

$$ {}^{R}\!\int_{0}^{x} f = \lim_{\alpha \to 0^{+}} {}^{R}\!\int_{\alpha}^{x} f. $$

Show that the extension works for the function f' in exercise 88 and that in this extended sense

$$ {}^{R}\!\int_{0}^{x} f' = f(x) - f(0) $$

for all $x \in (0, 1)$.

91. Give an example of an unbounded function that is Lebesgue integrable but for which it is impossible to use the technique of exercise 90 to assign a value to the Riemann integral.

6.6 References

Bartle, R.G. (1966). *The Elements of Integration*. John Wiley & Sons.

Bressoud, D.M. (2006). *A Radical Approach to Real Analysis* (2nd ed.). Mathematical Association of America.

Burk, F.E. (2007). *A Garden of Integrals*. Mathematical Association of America.

Daniell, P.J. (1918), A general form of integral, *Annals of Mathematics, Second Series* **19** (4): 279–294. JSTOR 1967495.

Hewitt, E. and K. Stromberg (1975). *Real and Abstract Analysis*. Springer.

Taylor, A.E. (2010). *General Theory of Functions and Integration*. Dover.

CHAPTER 7

The Gauge Integral

As we have seen, the Lebesgue integral has far more powerful convergence properties than the Riemann integral. The Lebesgue integral also has significantly more flexible fundamental theorems of calculus. Why?

From the Darboux-integral point of view, expanding the set of partitions of $[a, b]$ to include measurable partitions in addition to partitions by intervals creates much larger sets of upper and lower sums. These larger sets allow the infimum of the upper sums to meet the supremum of the lower sums more often, thus creating a much larger set of integrable functions. More specifically, the use of measurable partitions makes it possible to choose sets on which the supremum and infimum of the integrand f are nearly identical. A bound-and-telescope argument shows that whenever we can ensure that $\sup_E f - \inf_E f < \varepsilon / (b - a)$ for each set E in a partition of $[a, b]$, the upper and lower Lebesgue-Young sums for the partition will be within ε of each other.

The connection between the values of the integrand and the elements of a partition is even stronger in the original conception of the Lebesgue integral. The use of inverse images of small intervals forces the upper and lower bounds on the Lebesgue integral together. Fundamentally, this forcing is possible because measurable sets allow us to guarantee that whenever x and y belong to the same set in the partition of $[a, b]$, $f(x)$ and $f(y)$ are nearly the same.

While the Lebesgue integral is based on a strong connection between the values of the integrand f and the sets used in the partition, the Riemann integral does not support any such connection. At its core, the definition of the Riemann integral is uniform. A single $\delta > 0$ is selected to control the behavior of the entire partition with no accommodation available to account for the local behavior of f. From this point of view, it is not particularly surprising that convergence theorems for the Riemann integral rely on uniform convergence.

What would happen if, instead of employing a single value of $\delta > 0$ for the entire interval of integration, the value of δ is allowed to vary from point to point in a fashion analogous to the way that δ in the ε-δ definition of continuity (as opposed to uniform continuity) depends on both the value of ε and a particular point? Such an approach would move us away from an essentially uniform definition of integration and support a tie between the values of f and a partition of $[a, b]$. Although allowing the value of δ to depend on $x \in [a, b]$ cannot preclude the integrand taking on significantly different values within a subinterval, as we shall see, the contribution such subintervals make to the sum can be controlled by restricting their lengths.

Independently, Ralph Henstock (1923–2007) and Jaroslav Kurzweil (b. 1926) took up this line of inquiry around 1960. The result is an integral that avoids the complications of measure theory while extending the set of integrable functions. In particular, any derivative is integrable in this new setting. (Compare theorem 71 on page 191 with example 47 on page 192.)

We will refer to this type of integral as a gauge integral. The gauge integral maintains the Riemann integral's focus on the integrand's domain, uses only intervals in the partition, and is based on Riemann sums. The power (and complexity) of the gauge integral stems from the localized control a gauge exerts on the size of the intervals in a partition.

7.1   Definition and basic examples

Recall that a function $f : [a, b] \to \mathbb{R}$ is Riemann integrable over $[a, b]$ if there is a real number A such that for any $\varepsilon > 0$ we can find a $\delta > 0$ so that

$$|S_R(f, \mathcal{P}) - A| = \left| \sum_{k=1}^{n} f(t_k)(x_k - x_{k-1}) - A \right| < \varepsilon$$

whenever $\mathcal{P}$ is a tagged partition of $[a, b]$ with $|x_k - x_{k-1}| < \delta$ for all $k = 1, 2, \ldots, n$. The gauge integral modifies this definition slightly by replacing δ with $\delta(t_k)$.

Definition 21 (δ **gauge**). *Let E be a subset of $\mathbb{R}$. A **gauge** on E is a positive function $\delta : E \to \mathbb{R}^+$.*

Think of a gauge $\delta(t)$ as comparable to creating a function by piecing together the required values of δ in the ε-δ definition of continuity. In the case of the gauge, however, we are creating a function to verify integrability rather than continuity.

Definition 22 (δ-fine). *Given a gauge δ, a tagged partition $\mathcal{P} = \left\{\left(t_k, [x_{k-1}, x_k]\right)\right\}_{k=1}^n$ is said to be δ-**fine** if for all $k = 1, 2, \ldots, n$ we have $|x_k - x_{k-1}| < \delta\,(t_k)$.*[1]

Definition 23 (Gauge integrable). *A function $f : [a, b] \to \mathbb{R}$ is **gauge integrable** over $[a, b]$ if there is a real number A satisfying the following condition: For any $\varepsilon > 0$ there is a gauge δ such that*

$$|S_R(f, \mathcal{P}) - A| = \left| \sum_{k=1}^n f\,(t_k)\,(x_k - x_{k-1}) - A \right| < \varepsilon$$

for any δ-fine partition $\mathcal{P}$ of $[a, b]$. In this case, the value of the integral is $^g\!\int_a^b f = A$.

Notice the strong relationship between the gauge integral and the Riemann integral. In particular, we are once again working with Riemann sums. Also take note of the fact that if f is a Riemann integrable function, then f can be shown to be gauge integrable using the uniform gauge $\delta\,(t) = \delta_0$ where $\delta_0 > 0$ is taken from the definition of the Riemann integral. We will ignore this fact in the initial examples in order to investigate the gauge integral based on its definition.

Just as continuity has an alternative characterization that uses open sets, there is an alternative and equivalent (see exercise 1) definition of gauge in terms of open intervals.[2]

Definition 24 (γ gauge). *Let E be a subset of $\mathbb{R}$. A **gauge** on E is a function γ from E to the set of open intervals in $\mathbb{R}$ such that for any $t \in E$, we have $t \in \gamma\,(t)$.*

Definition 25 (γ-fine). *Given a gauge γ, a tagged partition $\mathcal{P} = \{(t_k, I_k)\}$ is said to be γ-**fine** if for all $k = 1, 2, \ldots, n$ we have $I_k \subset \gamma\,(t_k)$.*

While the second definition may appear to be more awkward, in practice it is often simpler and more transparent to use. It is typically (but not always) easier to use the first type of gauge when working with specific functions

---

[1] Some texts require instead that $[x_{k-1}, x_k] \subset (t_k - \delta\,(t_k)\,, t_k + \delta\,(t_k))$. (See Exercise 7.)
[2] One could more closely mimic the open set definition of continuity and take $\gamma\,(t)$ to be an open set containing t. Since we eventually want to use partitions consisting of intervals, there is nothing to be gained from this generality.

and more efficient to use the second type when proving general results. This state of affairs should not surprise us as it is similar to the situation for the ε-δ versus the open-set characterizations of continuity.

We will consistently use δ or γ to indicate which type of gauge we are using.

Example 35 (Constant functions). Let $f(x) = c$ on $[a, b]$. Fix $\varepsilon > 0$ and set $\gamma(t) = (a - 1, b + 1)$. If $\mathcal{P} = \{(t_k, I_k)\}$ is a γ-fine tagged partition of $[a, b]$, then

$$S_R(f, \mathcal{P}) = \sum_{\mathcal{P}} c \cdot \Delta x_k = c(b - a)$$

so that

$$|S_R(f, \mathcal{P}) - c(b - a)| = 0 < \varepsilon.$$

We conclude that the constant function $f(x) = c$ is gauge integrable over $[a, b]$ with

$$^g\!\int_a^b f = c(b - a).$$

Example 36 (Dirichlet). The Dirichlet function

$$d(x) = \begin{cases} 1, & x \in \mathbb{Q} \\ 0, & x \notin \mathbb{Q} \end{cases}$$

is gauge integrable over $[a, b]$.

To see why, suppose that $\varepsilon > 0$ and let $\{r_k\}$ be an enumeration of the rational numbers in $[a, b]$. Define a gauge δ on $[a, b]$ by

$$\delta(t) = \begin{cases} \frac{\varepsilon}{2^k}, & t = r_k \\ 1, & t \notin \mathbb{Q}. \end{cases}$$

If $\mathcal{P}$ is a δ-fine tagged partition of $[a, b]$, we can separate $\mathcal{P}$ into $\mathcal{P}_r$, those tagged intervals with rational tags, and $\mathcal{P}_i$, those with irrational tags. Then

$$|S_R(f, \mathcal{P}) - 0| = \sum_{\mathcal{P}} d(t_k) \cdot \Delta x_k$$

$$= \sum_{\mathcal{P}_r} 1 \cdot \Delta x_k + \sum_{\mathcal{P}_i} 0 \cdot \Delta x_k$$

$$< \sum_n \frac{\varepsilon}{2^n} = \varepsilon.$$

Thus the Dirichlet function is gauge integrable with $^g\!\int_a^b d = 0$.

Example 37 (Identity function). Let $f(x) = x$. We know that ${}^g\!\int_0^2 f = 2$ since ${}^R\!\int_0^2 f = 2$, but we will verify this result directly from the definition of the gauge integral.

Let $\varepsilon > 0$ be given. Define a gauge δ by $\delta(t) = \varepsilon/2$ and suppose that $\mathcal{P} = \{(t_k, [x_{k-1}, x_k])\}_{k=1}^n$ is a δ-fine tagged partition of $[0, 2]$. Because both t_k and $\frac{x_k + x_{k-1}}{2}$ fall in the interval $[x_{k-1}, x_k]$,

$$\left| t_k - \frac{x_k + x_{k-1}}{2} \right| \le x_k - x_{k-1} < \delta(t_k) = \varepsilon/2.$$

Hence

$$|S_R(f, \mathcal{P}_n) - 2| = \left| \sum_{k=1}^n t_k (x_k - x_{k-1}) - \frac{1}{2} \sum_{k=1}^n (x_k^2 - x_{k-1}^2) \right|$$

$$\le \sum_{k=1}^n \left| \left(t_k - \frac{x_k + x_{k-1}}{2} \right) \right| (x_k - x_{k-1})$$

$$< \sum_{k=1}^n \frac{\varepsilon}{2} (x_k - x_{k-1})$$

$$= \frac{\varepsilon}{2} \cdot 2 = \varepsilon.$$

We conclude that the identity function is gauge integrable over $[0, 2]$ with ${}^g\!\int_0^2 f = 2$.

The first and third examples used gauges whose widths do not depend on $x \in [a, b]$. The use of a constant-width gauge reflects the fact that the constant and identity functions are Riemann integrable. This is not the case for the Dirichlet function nor for the next example.

Also note the appearance of the telescoping sum $\frac{1}{2} \sum_{k=1}^n (x_k^2 - x_{k-1}^2) = 2$ in the last example. When proving that a function is integrable, we will often replace the value of the integral by such a telescoping sum.

Example 38. The function

$$f(x) = \begin{cases} \frac{1}{\sqrt{x}}, & x > 0 \\ 0, & x \le 0 \end{cases}$$

is not Riemann integrable over $[0, 1]$ since f is unbounded. However, f is gauge integrable over $[0, 1]$ with ${}^g\!\int_0^1 f = 2$.

To verify this, let $0 < \varepsilon \le 1$ be given. We will use a gauge γ on $[0, 1]$ of the form

$$\gamma(t) = \begin{cases} \left(t \cdot a^2(t), \frac{t}{a^2(t)}\right), & t > 0 \\ \left(-\varepsilon^2, \varepsilon^2\right), & t = 0 \end{cases}$$

where $0 < a(t) < 1$ is yet to be defined. Suppose that $\mathcal{P} = \left\{\left(t_k, [x_{k-1}, x_k]\right)\right\}_{k=1}^n$ is a γ-fine tagged partition of $[0, 1]$. Then, for $k > 1$, $\frac{2}{\sqrt{x_k} + \sqrt{x_{k-1}}}$ and $\frac{1}{\sqrt{t_k}}$ both fall in the interval $\left[\frac{1}{\sqrt{x_k}}, \frac{1}{\sqrt{x_{k-1}}}\right]$ so that

$$\left| \frac{2}{\sqrt{x_k} + \sqrt{x_{k-1}}} - \frac{1}{\sqrt{t_k}} \right| < \frac{1}{\sqrt{x_{k-1}}} - \frac{1}{\sqrt{x_k}}$$

$$< \frac{1}{\sqrt{a^2(t_k) \cdot t_k}} - \frac{1}{\sqrt{t_k/a^2(t_k)}}$$

$$= \frac{1}{\sqrt{t_k}} \left(\frac{1}{a(t_k)} - a(t_k) \right).$$

If we take $a(t) = 1 - \frac{\varepsilon\sqrt{t}}{4}$, then (recalling that $0 < \varepsilon, t \le 1$)

$$\frac{1}{a(t)} - a(t) = \frac{2\varepsilon\sqrt{t} - \frac{1}{4}\varepsilon^2 t}{4 - \varepsilon\sqrt{t}} < \frac{2}{3}\varepsilon\sqrt{t}.$$

Hence

$$\left| \frac{2}{\sqrt{x_k} + \sqrt{x_{k-1}}} - \frac{1}{\sqrt{t_k}} \right| < \varepsilon.$$

Due to the way γ is defined, the tagged interval $(t_0, [x_0, x_1])$ cannot have $x_0 = 0$ unless $t_0 = 0$. Thus when $k = 1$, $f(t_1) = f(0) = 0$ and $0 < x_1 < \varepsilon^2$. Hence

$$|S_R(f, \mathcal{P}_n) - 2| = \left| \sum_{k=2}^n \frac{1}{\sqrt{t_k}} (x_k - x_{k-1}) - 2 \sum_{k=1}^n \left(\sqrt{x_k} - \sqrt{x_{k-1}}\right) \right|$$

$$\le \left| \sum_{k=2}^n \left(\frac{1}{\sqrt{t_k}} - \frac{2}{\sqrt{x_k} + \sqrt{x_{k-1}}} \right) (x_k - x_{k-1}) \right| + 2\sqrt{x_1}$$

$$< \sum_{k=2}^n \varepsilon \cdot (x_k - x_{k-1}) + 2\varepsilon$$

$$< 3\varepsilon.$$

Thus f is gauge integrable with $\displaystyle{}^g\!\int_0^1 f = 2$.

Notice the structure of the last two examples. In both cases, a special value s_k is chosen for the subinterval $[x_{k-1}, x_k]$ in such a way that, when $s_k \cdot (x_k - x_{k-1})$ is expanded, the sum $\sum_{k=1}^{n} s_k \cdot (x_k - x_{k-1})$ simplifies by telescoping. The gauge is selected to control $|f(t_k) - s_k|$ so that we can also bound and telescope $\sum_k |f(t_k) - s_k| (x_k - x_{k-1})$. Though it need not always be the case, $|f(t_k) - s_k|$ often is bounded by controlling $|f(x_k) - f(x_{k-1})|$.

Before turning to more general results, we need to verify that, given a gauge γ, a γ-fine partition exists. If we could construct a gauge γ for which there were no γ-fine tagged partitions, we would be in much the same situation as if the real-valued δ used to bound the mesh of a partition in the Riemann integral were allowed to be negative. The condition for gauge integrability would be satisfied vacuously. In this case, any value of A would satisfy the conditions of the definition.

Theorem 48 (Cousin, 1895). *If γ is a gauge on $[a, b]$, then there is a γ-fine tagged partition of $[a, b]$.*[3]

Proof. Let $S = \{s \in [a, b] : \text{there exists a } \gamma\text{-fine tagged partition of } [a, s]\}$ and let β be the supremum of the bounded and nonempty set S. We will prove that $\beta \in S$ and that $\beta = b$. Hence $b \in S$ which means that there is a γ-fine tagged partition of $[a, b]$.

$\beta \in S$: By the definition of sup S, the open interval $\gamma(\beta) \cap (a, \beta)$ must contain an element c of S. Since $c \in S$, there is a γ-fine partition $\mathcal{P}$ of $[a, c]$. Add $(\beta, [c, \beta])$ to $\mathcal{P}$ to create a γ-fine tagged partition of $[a, \beta]$ thus showing that $\beta \in S$.

$\beta = b$: Since b is an upper bound for S, $\beta \leq b$. Suppose that $\beta < b$. Because $\beta \in S$, there is a γ-fine tagged partition $\mathcal{P}$ of $[a, \beta]$. Choose $d \in \gamma(\beta) \cap (\beta, b)$ and add $(\beta, [\beta, d])$ to $\mathcal{P}$. The result is a γ-fine tagged partition of $[a, d]$. Since $d > \beta$, we have contradicted the fact that $\beta = \sup S$. Hence $\beta = b$. $\square$

7.2   The art of constructing gauges

While the general role of a gauge γ is to ensure that the Riemann sum of any γ-fine tagged partition of $[a, b]$ is close to the posited value of the integral, it is not yet clear how such a task is accomplished. This general objective is usually achieved by accomplishing one or more specific jobs: force a

---

[3] In fact, there are infinitely many γ-fine tagged partitions of $[a, b]$.

particular point to be a tag, ensure that the contribution made by a particular point is small, guarantee that the value of the function throughout a subinterval is close to the functions's value at the interval's tag. In the course of a proof or when working with a particular integral, one often wants to achieve more than one such objective. This is easily accomplished using the intersection of gauges. If γ_1 and γ_2 are two gauges on $[a, b]$ and γ is defined by $\gamma(t) = \gamma_1(t) \cap \gamma_2(t)$, then γ is also a gauge on $[a, b]$ and any γ-fine partition is also γ_1-fine and γ_2-fine. Thus any γ-fine partition will have the properties that both γ_1 and γ_2 were designed to guarantee. The point is that one can deal with individual conditions separately and then easily create a gauge that handles all the conditions at once using intersection. The same goals can be accomplished with delta gauges using $\delta(t) = \min\{\delta_1(t), \delta_2(t)\}$.

The following examples illustrate some of the ways that gauges can be used.

7.2.1   Forced tags and tag splitting

Example 39 (Forced tags). Design a gauge γ on $[0, 1]$ that will force $0, \frac{1}{2}$, and 1 to be tags of any γ-fine partition.

Take

$$
\gamma(t) = \begin{cases} \left(0, \frac{1}{2}\right), & t \in \left(0, \frac{1}{2}\right) \\ \left(\frac{1}{2}, 1\right), & t \in \left(\frac{1}{2}, 1\right) \\ \left(t - \frac{1}{2}, t + \frac{1}{2}\right), & t = 0, \frac{1}{2}, 1. \end{cases}
$$

A tagged interval in a γ-fine partition can only contain $\frac{1}{2}$ if its tag is $\frac{1}{2}$. The same is true for 0 and 1.

In a similar fashion, we can use a gauge γ to force any finite set to be included among the tags of any γ-fine partition. Sometimes it is convenient to assume that these tags are also division points of the partition. This can be accomplished by **tag splitting**: if t_j is not already a division point, replace $(t_j, [x_{j-1}, x_j])$ with $(t_j, [x_{j-1}, t_j])$ and $(t_j, [t_j, x_j])$. The resulting partition is still γ-fine and has the same Riemann sum as the original partition. Tag splitting can be combined with an appropriately chosen gauge γ to force γ-fine tagged partitions to be refinements of a given partition.

7.2.2   Controlling small sets

The next two examples illustrate how gauges can be used to control the contribution of a small set.

Example 40 (Finite set). Let f be a function on $[a, b]$. Given a finite set $S = \{c_k\}_{k=1}^n$ from $[a, b]$ and $\varepsilon > 0$, construct a gauge γ so that if $\mathcal{P}$ is a γ-fine tagged partition of $[a, b]$, then the total contribution to $S_R(f, \mathcal{P})$ made by tagged intervals whose tags come from S is bounded by ε.

Let

$$\gamma(t) = \begin{cases} \left(t - \frac{\varepsilon}{2n(|f(t)|+1)}, t + \frac{\varepsilon}{2n(|f(t)|+1)} \right), & t \in S \\ (t-1, t+1), & t \notin S. \end{cases}$$

If $t_i = c_k$ is a tag in the γ-fine partition $\mathcal{P}$, then the contribution to $S_R(f, \mathcal{P})$ made by that tagged interval[4] satisfies

$$|f(t_i) \Delta x_i| < |f(c_k)| \frac{2\varepsilon}{2n(|f(c_k)|+1)} < \frac{\varepsilon}{n}.$$

Hence the maximal contribution to $S_R(f, \mathcal{P})$ made by tagged intervals with tags in S is less than ε.

Example 41 (Sets of measure zero). Suppose that f is a function on $[a, b]$ with $|f| \leq B$. Given a set Z of measure zero and $\varepsilon > 0$, construct a gauge γ so that for any γ-fine tagged partition of $[a, b]$, the contribution to the Riemann sum made by terms with tags in Z is at most ε.

Since Z has measure zero, it can be covered by a countable set of open intervals $\{I_k\}$ such that $\sum_k l(I_k) < \varepsilon/B$. If $t \in Z$, then $t \in I_k$ for some k. Let $k(t)$ be the smallest such index and define

$$\gamma(t) = \begin{cases} I_{k(t)}, & t \in Z \\ (t-1, t+1), & t \notin Z. \end{cases}$$

Suppose now that $\mathcal{P} = \{(t_k, J_k)\}_{k=1}^n$ is a γ-fine tagged partition of $[a, b]$. If several tags satisfy $k(t_{i_1}) = k(t_{i_2}) = \cdots = k(t_{i_m}) = k_0$, then the corresponding subintervals will all fit inside I_{k_0}. Since the subintervals do not overlap, the total length of the intervals corresponding to those tags can be at most that of I_{k_0}. Consequently, the total contribution of all the terms with tags in Z will be at most

$$\left| \sum_{t_i \in Z} f(t_i) l(J_i) \right| \leq \sum_k B \cdot l(I_k) < \varepsilon.$$

---

[4] If a tag is used for two adjoining intervals, we can use tag joining to treat them as a single interval.

7.2.3 Controlling function differences

As suggested by our previous work, it is often helpful to bound $|f(t_k) - f(w_k)|$ where t_k is the given tag for the interval $[x_{k-1}, x_k]$ and w_k is the tag that we wish we had. Often gauges can be used to bound this difference so that $\sum_k |f(t_k) - f(w_k)|(x_k - x_{k-1})$ can be telescoped.

Example 42. Let

$$f(x) = \begin{cases} \frac{1}{x}, & x > 0 \\ 0, & x = 0 \end{cases}$$

and suppose that $0 < \varepsilon < 1$. Find a gauge δ on $[0, 1]$ so that for any δ-fine tagged partition $\mathcal{P} = \{(t_k, [x_{k-1}, x_k])\}_{k=1}^n$ of $[0, 1]$, we have $|f(x_{k-1}) - f(t_k)| < \varepsilon$ when $t_k > 0$.[5]

Set

$$\delta(t) = \begin{cases} \frac{\varepsilon t^2}{1 + \varepsilon t}, & t > 0 \\ 1, & t = 0. \end{cases}$$

Then when $t_k > 0$, we have $x_{k-1} > t_k - \frac{\varepsilon t_k^2}{1 + \varepsilon t_k}$ so that

$$0 \le f(x_{k-1}) - f(t_k)$$

$$< \frac{1}{t_k - \frac{\varepsilon t_k^2}{1 + \varepsilon t_k}} - \frac{1}{t_k}$$

$$= \frac{1 + \varepsilon t_k}{t_k} - \frac{1}{t_k} = \varepsilon.$$

Of course the point of all these gauge constructions is to gain the capacity to create a gauge to verify that a function is integrable. We close this section with such an example.

Example 43. Show that the unbounded function

$$f(x) = \begin{cases} (-1)^k k, & x \in \left(\frac{1}{k+1}, \frac{1}{k} \right] \\ 0, & x = 0, \end{cases}$$

is gauge integrable over $[0, 1]$ with

$$\,^g\!\!\int_0^1 f = \sum_{k=1}^{\infty} \frac{(-1)^k}{k+1} = \ln 2 - 1.$$

---

[5] Note that this bound does not prove that f is gauge integrable since we have not shown that $\sum_k f_k(x_{k-1})(x_k - x_{k-1})$ telescopes or otherwise simplifies.

Begin by noting that f is constant on the interval $\left(\frac{1}{k+1}, \frac{1}{k}\right]$ and that $\int_{\frac{1}{k+1}}^{\frac{1}{k}} f = (-1)^k k \left(\frac{1}{k} - \frac{1}{k+1}\right) = \frac{(-1)^k}{k+1}$ (see exercise 13). With this fact in mind, we want to construct a gauge so that

1. the union of the intervals with tags in $\left(\frac{1}{k+1}, \frac{1}{k}\right]$ approximates $\left(\frac{1}{k+1}, \frac{1}{k}\right]$ and

2. a sufficient number of the intervals $\left(\frac{1}{k+1}, \frac{1}{k}\right]$ are considered to get a good approximation to $\sum_{k=1}^{\infty} \frac{(-1)^k}{k+1}$.

To this end, let $\varepsilon > 0$ and define our gauge by

$$\gamma(t) = \begin{cases} \left(\frac{1}{k+1}, \frac{1}{k} + \frac{\varepsilon}{2k \cdot 2^k}\right), & t \in \left(\frac{1}{k+1}, \frac{1}{k}\right], \\ \left(-\frac{\varepsilon}{2}, \frac{\varepsilon}{2}\right), & t = 0. \end{cases}$$

Now suppose that $\mathcal{P} = \left\{(t_k, [x_{k-1}, x_k])\right\}_{k=1}^{n}$ is a γ-fine partition of $[0, 1]$. Let m be the natural number satisfying $\frac{1}{m+1} < x_1 \le \frac{1}{m}$. Note that the tag for $[0, x_1]$ must be $t_1 = 0$ because no other value can tag an interval with a left endpoint of 0. If we express the union of the intervals from $\mathcal{P}$ with tags in $\left(\frac{1}{k+1}, \frac{1}{k}\right]$ as $[z_{k+1}, z_k]$, then $S_R(f, \mathcal{P})$ can be expressed as

$$S_R(f, \mathcal{P}) = \sum_{k=1}^{m} (-1)^k k (z_k - z_{k+1}).$$

Since $z_1 = 1$ and $\frac{1}{k+1} < z_{k+1}$ for $1 \le k \le m$,

$$\frac{1}{k} \le z_k < \frac{1}{k} + \frac{\varepsilon}{2k \cdot 2^k} \quad \text{for } 1 \le k \le m+1.$$

Thus for $1 \le k \le m$ we can use the fact that $k\left(\frac{1}{k} - \frac{1}{k+1}\right) = \frac{1}{k+1}$ to see that

$$\frac{1}{k+1} - \frac{\varepsilon}{2} \frac{1}{2^k} < k\left(\frac{1}{k} - \left(\frac{1}{k+1} + \frac{\varepsilon}{2(k+1) \cdot 2^{k+1}}\right)\right)$$

$$< k(z_k - z_{k+1})$$

$$< k\left(\frac{1}{k} + \frac{\varepsilon}{2k \cdot 2^k} - \frac{1}{k+1}\right) = \frac{1}{k+1} + \frac{\varepsilon}{2} \frac{1}{2^k}.$$

Hence

$$\left| S_R(f, \mathcal{P}) - \sum_{k=1}^{m} \frac{(-1)^k}{k+1} \right| \le \sum_{k=1}^{m} \left| k(z_k - z_{k+1}) - \frac{1}{k+1} \right|$$

$$< \sum_{k=1}^{m} \frac{\varepsilon}{2} \frac{1}{2^k} < \frac{1}{2}\varepsilon.$$

As $\frac{1}{m+1} < x_1 < \frac{\varepsilon}{2}$, the properties of alternating series tell us that

$$\left| \sum_{k=1}^{\infty} \frac{(-1)^k}{k+1} - \sum_{k=1}^{m} \frac{(-1)^k}{k+1} \right| < \frac{\varepsilon}{2}.$$

Hence

$$\begin{aligned} |S_R(f, \mathcal{P}) - (\ln 2 - 1)| &\leq \left| S_R(f, \mathcal{P}) - \sum_{k=1}^{m} \frac{(-1)^k}{k+1} \right| \\ &\quad + \left| \sum_{k=1}^{m} \frac{(-1)^k}{k+1} - \sum_{k=1}^{\infty} \frac{(-1)^k}{k+1} \right| < \varepsilon. \end{aligned}$$

We conclude that f is gauge integrable over $[0, 1]$ with

$$ {}^g\!\!\int_0^1 f = \sum_{k=1}^{\infty} \frac{(-1)^k}{k+1} = \ln 2 - 1.$$

Note that f, being unbounded, is not Riemann integrable.

7.3   Basic integrability results

We begin by formally recording our previously noted connection between the Riemann and gauge integrals.

Theorem 49 (Riemann $\Longrightarrow$ gauge). *Any Riemann integrable function is gauge integrable and the integrals agree.*

Proof. Exercise 2. □

In many contexts we do not need to identify the value of an integral. We simply want to verify the integrability of a function. As with the Darboux integral there is a form of the Cauchy criterion that guarantees a gauge integral exists without requiring us to identify a specific value for the integral.

Theorem 50 (Cauchy criterion). *Let f be a function on $[a, b]$. Then f is gauge integrable over $[a, b]$ if and only if for each $\varepsilon > 0$ there is a gauge γ on $[a, b]$ such that*

$$|S_R(f, \mathcal{P}_1) - S_R(f, \mathcal{P}_2)| < \varepsilon$$

for any pair of γ-fine tagged partitions $\mathcal{P}_1$ and $\mathcal{P}_2$ of $[a, b]$.

Proof. The necessity of the condition is left as an exercise (exercise 37).

To prove sufficiency, let $\{\gamma_n\}$ be a sequence of gauges on $[a,b]$ such that

$$|S_R(f,\mathcal{P}_1) - S_R(f,\mathcal{P}_2)| < \frac{1}{n}$$

whenever $\mathcal{P}_1$ and $\mathcal{P}_2$ are two γ_n-fine tagged partitions of $[a,b]$. Define a second sequence of gauges by $\widehat{\gamma}_n(t) = \cap_{i=1}^{n}\gamma_i(t)$ and let $\{\mathcal{P}_n\}$ be a corresponding sequence of $\widehat{\gamma}_n$-fine tagged partitions of $[a,b]$. By construction, if $m > n$ then $\mathcal{P}_m$ is also $\widehat{\gamma}_n$-fine. Hence

$$|S_R(f,\mathcal{P}_m) - S_R(f,\mathcal{P}_n)| < \frac{1}{N}$$

whenever $n, m > N$. Thus $\{S_R(f,\mathcal{P}_n)\}$ is a Cauchy sequence and so converges to some value A.

Now let $\varepsilon > 0$ and select a natural number n_0 so that $\frac{1}{n_0} < \frac{\varepsilon}{2}$ and $\left|S_R\left(f,\mathcal{P}_{n_0}\right) - A\right| < \varepsilon/2$. If $\mathcal{P}$ is a $\widehat{\gamma}_{n_0}$-fine tagged partition, then

$$|S_R(f,\mathcal{P}) - A| \leq \left|S_R(f,\mathcal{P}) - S_R(f,\mathcal{P}_{n_0})\right| + \left|S_R(f,\mathcal{P}_{n_0}) - A\right| < \varepsilon.$$

Hence f is gauge integrable over $[a,b]$ with $^g\!\int_a^b f = A$.   □

The next theorem is an example of the type of circumstance in which the existence but not the value of a gauge integral is important.

Theorem 51 (Subintervals 1). *Suppose that the function f is gauge integrable over $[a,b]$ and $[c,d] \subset [a,b]$. Then f is gauge integrable over $[c,d]$.*

Proof. For definiteness, consider the case where $a < c < d < b$, the cases when $a = c$ or $d = b$ being simpler.

Let $\varepsilon > 0$ and let γ be a gauge on $[a,b]$ such that

$$|S_R(f,\mathcal{P}_1) - S_R(f,\mathcal{P}_2)| < \varepsilon$$

whenever $\mathcal{P}_1$ and $\mathcal{P}_2$ are two γ-fine tagged partitions of $[a,b]$. Let $\widehat{\gamma}$ be the restriction of γ to $[c,d]$ and suppose that $\mathcal{Q}_1$ and $\mathcal{Q}_2$ are two $\widehat{\gamma}$-fine partitions of $[c,d]$. Let $\widehat{\gamma}_a$ and $\widehat{\gamma}_b$ be the restrictions of γ to $[a,c]$ and $[d,b]$ respectively and let $\mathcal{R}_a$ and $\mathcal{R}_b$ be $\widehat{\gamma}_a$-fine and $\widehat{\gamma}_b$-fine tagged partitions of $[a,c]$ and $[d,b]$ respectively. Then $\mathcal{S}_1 = \mathcal{R}_a \cup \mathcal{Q}_1 \cup \mathcal{R}_b$ and $\mathcal{S}_2 = \mathcal{R}_a \cup \mathcal{Q}_2 \cup \mathcal{R}_b$ are γ-fine tagged partitions of $[a,b]$.[6] Hence

$$|S_R(f,\mathcal{Q}_1) - S_R(f,\mathcal{Q}_2)| = |S_R(f,\mathcal{S}_1) - S_R(f,\mathcal{S}_2)| < \varepsilon.$$

---

[6] Note that, since the tagged partitions are on non-overlapping intervals, $\cup$ in this case is the usual union of sets. If we were working with tagged partitions of a common interval, then $\cup$ would refer to a refinement.

We conclude that f is gauge integrable over $[c,d]$ by the Cauchy criterion.

$\square$

The next theorem serves as a converse.

Theorem 52 (Subintervals 2). *Suppose that the function f is gauge integrable on $[a,c]$ and $[c,b]$. Then f is gauge integrable on $[a,b]$ with $\,^g\!\int_a^b f = \,^g\!\int_a^c f + \,^g\!\int_c^b f$.*

Proof. Exercise 39.

$\square$

In the future, rather than assigning a new symbol when a gauge γ is restricted to a subinterval, we will continue to use γ. This slight abuse of notation will cause no harm and will make our notation cleaner. Following this convention, we would refer to $\mathcal{R}_a$, $\mathcal{R}_b$, and $\mathcal{Q}_1$ in the above proof as γ-fine tagged partitions of $[a,b]$, $[d,b]$, and $[c,d]$ respectively.

As with the Lebesgue integral, we can safely ignore changes to a function when they occur on sets of measure zero.

Theorem 53 (Zero a.e.). *If $f = 0$ a.e. on $[a,b]$, then f is gauge integrable with $\,^g\!\int_a^b f = 0$.*

Proof. Let $\varepsilon > 0$ be given and define

$$E_n = \{x \in [a,b] : n-1 < |f(x)| \le n\}, n = 0,1,2,\dots.$$

When $n > 0$, $\lambda(E_n) = 0$ so (using the union of a collection of open intervals) there is an open set G_n with $E_n \subseteq G_n$ and $\lambda(G_n) < \varepsilon/n2^n$. As G_n is an open set, for each $x \in E_n$ there is an $r_x > 0$ so that $y \in G_n$ whenever $|y - x| < r_x$. Set $E = \cup_n E_n$ and define a gauge δ by

$$\delta(t) = \begin{cases} r_t, & t \in E \\ 1, & t \notin E. \end{cases}$$

Now suppose that $\mathcal{P} = \{(t_i, I_i)\}$ is a δ-fine tagged partition of $[a,b]$. Let $\mathcal{P}_n$ be the set of intervals with tags in E_n. Then for $n > 0$, $\cup_{\mathcal{P}_n} I_i \subseteq G_n$ and, since the intervals in $\mathcal{P}_n$ do not overlap,

$$\sum_{\mathcal{P}_n} l(I_i) = \lambda(\cup_{\mathcal{P}_n} I_i) \le \lambda(G_n) < \varepsilon/n2^n.$$

Thus

$$|S_R(f,\mathcal{P})| = \left| \sum_{n=0}^\infty \sum_{\mathcal{P}_n} f(t_i) l(I_i) \right| \le \sum_{n=1}^\infty n \sum_{\mathcal{P}_n} l(I_i) < \sum_{n=1}^\infty \frac{\varepsilon}{2^n} = \varepsilon.$$

Hence f is gauge integrable with $\,^g\!\int_a^b f = 0$.

$\square$

Corollary 54 (Equal a.e.). *If f is gauge integrable on $[a, b]$ and $f = g$ a.e. on $[a, b]$, then g is gauge integrable on $[a, b]$ with $^g\!\int_a^b f = {}^g\!\int_a^b g$.*

Proof. Exercise.   □

The proof of theorem 53 uses sums taken over subsets of a tagged partition. We have used similar tactics before, but this technique plays a more central role in proofs related to the gauge integral.

On a seemingly unrelated note, one of the standard techniques when working with Riemann integrals is to create a common refinement. When this is done, however, the tags are lost. We can use a gauge and tag splitting to make sure that the tags are preserved, but then the intervals may change and the new intervals will have additional tags. Consequently, the common refinement approach generally is not effective when working with the gauge integral.

Both of these issues are addressed by relating partial Riemann sums to sums of integrals over subintervals. The key observation is that sums of integrals over subintervals do not depend on a choice of tags. The next definition and the two theorems that follow provide us with the needed vocabulary and tools to make effective use of this strategy.

Definition 26 (Partial tagged partition). *A **partial tagged partition** of $[a, b]$ is a finite set of tagged, closed intervals $\{(t_i, I_i)\}$ where*

1. $t_i \in I_i \subseteq [a, b]$, and

2. I_i and I_j are non-overlapping when $i \neq j$.

A partial tagged partition is like a tagged partition of $[a, b]$ except that the union of the intervals in a partial tagged partition need not (but may) cover all $[a, b]$. Typically a partial tagged partition arises from considering a subset of a tagged partition.

Lemma 55 (Henstock). *Let f be a gauge integrable function on $[a, b]$. Fix $\varepsilon > 0$ and suppose further that γ is a gauge on $[a, b]$ for which any γ-fine partition $\mathcal{P}$ satisfies*

$$\left| S_R(f, \mathcal{P}) - {}^g\!\int_a^b f \right| < \varepsilon.$$

If $\widehat{\mathcal{P}} = \{(t_i, I_i)\}$ is a γ-fine partial tagged partition of $[a, b]$, then

$$\left| \sum_{\widehat{\mathcal{P}}} \left\{ f(t_i) \cdot \Delta x_i - {}^g\!\int_{I_i} f \right\} \right| \leq \varepsilon.$$

Alternatively, we can express the last inequality as

$$\left| S_R(f, \widehat{\mathcal{P}}) - \sum_{\widehat{\mathcal{P}}} {}^g\!\!\int_{I_i} f \right| \leq \varepsilon.$$

Proof. The set $[a, b] \setminus \cup_i I_i$ consists of a finite number (perhaps zero) of disjoint intervals. Add their endpoints and label the resulting closed intervals as $J_1, J_2, \ldots, J_m$. By theorem 51, f is gauge integrable over J_k, $k = 1, 2, \ldots, m$. Thus for any $\xi > 0$ and for each $k = 1, 2, \ldots, m$, we can find a γ-fine partition $\mathcal{P}_k$ of J_k for which

$$\left| S_R(f, \mathcal{P}_k) - {}^g\!\!\int_{J_k} f \right| < \frac{\xi}{m}.$$

Putting the partial tagged partitions together, $\mathcal{Q} = \widehat{\mathcal{P}} \cup \mathcal{P}_1 \cup \mathcal{P}_2 \cup \cdots \cup \mathcal{P}_m$ forms a γ-fine tagged partition of $[a, b]$. Hence

$$\left| S_R(f, \widehat{\mathcal{P}}) - \sum_{\widehat{\mathcal{P}}} {}^g\!\!\int_{I_i} f \right|$$

$$= \left| S_R(f, \mathcal{Q}) - {}^g\!\!\int_a^b f - \sum_{k=1}^m \left(S_R(f, \mathcal{P}_k) - {}^g\!\!\int_{J_k} f \right) \right|$$

$$\leq \left| S_R(f, \mathcal{Q}) - {}^g\!\!\int_a^b f \right| + \sum_{k=1}^m \left| S_R(f, \mathcal{P}_k) - {}^g\!\!\int_{J_k} f \right|$$

$$< \varepsilon + \xi.$$

Since $\xi > 0$ was arbitrary, we conclude that

$$\left| S_R(f, \widehat{\mathcal{P}}) - \sum_{\widehat{\mathcal{P}}} {}^g\!\!\int_{I_i} f \right| \leq \varepsilon. \qquad \square$$

Corollary 56. *Under the hypotheses of Henstock's lemma,*

$$\sum_{\widehat{\mathcal{P}}} \left| f(t_i) \cdot \Delta x_i - {}^g\!\!\int_{I_i} f \right| \leq 2\varepsilon$$

and

$$\left| \sum_{\widehat{\mathcal{P}}} \left(|f(t_i)| \cdot \Delta x_i - \left| {}^g\!\!\int_{I_i} f \right| \right) \right| \leq 2\varepsilon$$

or, equivalently,

$$\left| S_R(|f|, \widehat{\mathcal{P}}) - \sum_{\widehat{\mathcal{P}}} \left| {}^g\!\!\int_{I_i} f \right| \right| \leq 2\varepsilon.$$

Proof. Let $\widehat{\mathcal{P}}_+$ be those intervals from $\widehat{\mathcal{P}}$ for which $f(t_i) \cdot \Delta x_i - {}^g\!\!\int_{I_i} f \geq 0$. Define $\widehat{\mathcal{P}}_-$ analogously. Then both $\widehat{\mathcal{P}}_+$ and $\widehat{\mathcal{P}}_-$ are partial tagged partitions satisfying the conditions of Henstock's lemma. Thus

$$\sum_{\widehat{\mathcal{P}}} \left| f(t_i) \cdot \Delta x_i - {}^g\!\!\int_{I_i} f \right|$$

$$= \left| \sum_{\widehat{\mathcal{P}}_+} \left\{ f(t_i) \cdot \Delta x_i - {}^g\!\!\int_{I_i} f \right\} \right| + \left| \sum_{\widehat{\mathcal{P}}_-} \left\{ f(t_i) \cdot \Delta x_i - {}^g\!\!\int_{I_i} f \right\} \right|$$

$$\leq 2\varepsilon.$$

The second inequality is a consequence of the fact that for $a, b \in \mathbb{R}$, $||a| - |b|| \leq |a - b|$ so that

$$\left| \sum_{\widehat{\mathcal{P}}} \left(|f(t_i)| \cdot \Delta x_i - \left| {}^g\!\!\int_{I_i} f \right| \right) \right| \leq \sum_{\widehat{\mathcal{P}}} \left| |f(t_i)| \cdot \Delta x_i - \left| {}^g\!\!\int_{I_i} f \right| \right|$$

$$\leq \sum_{\widehat{\mathcal{P}}} \left| f(t_i) \cdot \Delta x_i - {}^g\!\!\int_{I_i} f \right| \leq 2\varepsilon. \quad \square$$

7.4   Absolute integrability

At the end of Section 7.2, we showed that the function

$$f(x) = \begin{cases} (-1)^k k, & x \in \left(\frac{1}{k+1}, \frac{1}{k} \right] \\ 0, & x = 0 \end{cases}$$

is gauge integrable. The function $|f|$ is not gauge integrable, however. To see why, suppose that $|f|$ is gauge integrable over $[0, 1]$. Then by theorems 51 and 52 (page 173),

$$ {}^g\!\!\int_0^1 |f| = {}^g\!\!\int_0^{\frac{1}{n}} |f| + {}^g\!\!\int_{\frac{1}{n}}^1 |f| > {}^g\!\!\int_{\frac{1}{n}}^1 |f|.$$

But $|f|$ is constant on each of the intervals $\left(\frac{1}{k+1}, \frac{1}{k}\right]$, $1 \leq k < n$, so that

$$^g\!\!\int_{\frac{1}{n}}^{1} |f| = \sum_{k=1}^{n-1} {}^g\!\!\int_{\frac{1}{k+1}}^{\frac{1}{k}} |f| = \sum_{k=1}^{n-1} \frac{1}{k+1}.$$

Thus $^g\!\!\int_0^1 |f|$ cannot be finite and $|f|$ is not gauge integrable.

We introduce the terms **absolutely integrable** to describe those functions f for which f and $|f|$ are integrable and **conditionally integrable** to describe those cases when f is integrable but $|f|$ is not. For the Lebesgue integral, the integrability of a measurable function f is equivalent to the integrability of $|f|$ and $\left| {}^L\!\!\int_a^b f \right| \leq {}^L\!\!\int_a^b |f|$. As just illustrated, this equivalence is not true for the gauge integral. A function can be gauge integrable without being absolutely gauge integrable because a gauge can force a Riemann sum to consider both positive and negative contributions in a balanced way.

Initially, the fact that some functions are conditionally gauge integrable appears to be interesting but not particularly consequential. In fact, the distinction between integrable and absolutely integrable complicates the development of the theory of convergence for the gauge integral. For example, many of the convergence proofs for the Lebesgue integral rely on the fact that measurable functions are closed under taking linear combinations, maximums, minimums, supremums, and infimums. Functions constructed with those operations can be used freely in Lebesgue integral proofs. This is not the case for the gauge integral. Given a pair of gauge integrable functions f and g and a scalar c, $f+g$ and cf are gauge integrable, but $h_1 = \max\{f, g\}$ and $h_2 = \min\{f, g\}$ may not be. However, as we shall see, $h_1 = \max\{f, g\}$ and $h_2 = \min\{f, g\}$ are gauge integrable when f and g are absolutely gauge integrable.

So when is a gauge integrable function f absolutely integrable over $[a, b]$? It is relatively straightforward to prove that a gauge integrable function f is absolutely integrable if and only if both f^+ and f^- are gauge integrable (see exercise 51). This characterization suffices for some, but certainly not all, of our purposes. For example, it is tempting to try to prove that the maximum of two absolutely gauge integrable functions is gauge integrable using $\max\{f, g\} = \frac{1}{2}(f + g + |f - g|)$, but we cannot easily prove that the sum or difference of two absolutely gauge integrable functions is absolutely gauge integrable since $(f + g)^+$ is not $f^+ + g^+$.

For a measurable function, the only requirement for absolute Lebesgue integrability is that the Lebesgue sums be bounded. We will eventually prove a related but rather more subtle result for the gauge integral. However, unlike the situation for the Lebesgue integral, the gauge theorem does not follow directly or easily from the definition of the gauge integral. The pathway runs through the concept of bounded variation.

Definition 27 (Variation). *Let f be a function on $[a, b]$. Given a partition $\mathcal{P} = \{[x_{k-1}, x_k]\}$ of $[a, b]$, the **variation of** f **with respect to** $\mathcal{P}$ is*

$$V(f, \mathcal{P}) = \sum_k |f(x_k) - f(x_{k-1})|.$$

*The **variation of** f **over** $[a, b]$ is*

$$V_a^b f = \sup_P V(f, \mathcal{P})$$

*where the supremum is taken over all partitions $\mathcal{P}$ of $[a, b]$. If $V_a^b f$ is finite, then f is said to be of **bounded variation on** $[a, b]$. The set of all such functions is denoted by $BV([a, b])$.*

Example 44. The gauge integrable function

$$f(x) = \begin{cases} (-1)^k k, & x \in \left(\frac{1}{k+1}, \frac{1}{k}\right] \\ 0, & x = 0 \end{cases}$$

(see example 43, page 170) does not belong to $BV([0, 1])$.

To see why, use the partition $\mathcal{P}_n = \left\{\left[0, \frac{1}{n}\right], \left[\frac{1}{n}, 1\right]\right\}$. Then

$$V(f, \mathcal{P}_n) = \sum_{k=1}^{2} |f(x_k) - f(x_{k-1})|$$
$$= \left|(-1)^n n - 0\right| + \left|1 - (-1)^n n\right| \geq 2n - 1.$$

Since $2n - 1$ is not bounded, $f \notin BV([0, 1])$.

Theorem 57 (Monotone functions). *If f is a monotone function on $[a, b]$, then $f \in BV([a, b])$ with $V_a^b f = |f(b) - f(a)|$.*

Proof. For definiteness, assume that f is monotone increasing. Let $\mathcal{P} = \{[x_{k-1}, x_k]\}$ be a partition of $[a, b]$. Then

$$V(f, \mathcal{P}) = \sum_k |f(x_k) - f(x_{k-1})|$$
$$= \sum_k (f(x_k) - f(x_{k-1})) = f(b) - f(a).$$

Hence $V_a^b f = |f(b) - f(a)|$.

The verification for monotone decreasing functions is similar. □

Step functions form another class of functions with bounded variation. Step functions have appeared in the exercises (see exercises 13, 14, and 31) and will play an important role in Section 7. However, we have not yet formally defined them. We do so now.

Definition 28 (Step function). *A function f defined on $[a,b]$ is a step function if there is a partition $\mathcal{P} = \{[x_{i-1}, x_i]\}$ of $[a,b]$ such that f is constant on each open interval (x_{i-1}, x_i).*

At first glance, step functions appear deceptively similar to measurable simple functions. In a sense, step functions are the gauge integral's analog to the Lebesgue integral's measurable simple functions. Step functions are based on partitions (by intervals) and simple functions are based on measurable partitions. Despite their apparent similarities, you should not assume that step functions and measurable simple functions have the same properties. For example, the Dirichlet function is a measurable simple function whose variation is definitely not bounded while all step functions have bounded variation. Both simple and step functions take on a finite number of values, but while simple functions can have an infinite number of points of discontinuity, step functions have only finitely many discontinuities.

Example 45. The function

$$f(x) = \begin{cases} 1, & 0 \le x < \frac{1}{2} \\ 3, & x = \frac{1}{2} \\ 2, & \frac{1}{2} < x < \frac{3}{4} \\ 5, & x = \frac{3}{4} \\ 4, & \frac{3}{4} < x < 1 \\ 0, & x = 1 \end{cases}$$

is a step function with three points of discontinuity: $x = \frac{1}{2}$, $x = \frac{3}{4}$, and $x = 1$.

Theorem 58 (Step functions). *If f is a step function on $[a,b]$, then f has bounded variation.*

Proof. Since a step function f can only take on a finite number of values, f is bounded by some value B. Suppose that f has m points of discontinuity. If $\mathcal{P} = \{[x_{k-1}, x_k]\}$ is an arbitrary partition of $[a,b]$, then at most $2m$ of the intervals of $\mathcal{P}$ can produce different values for $f(x_{k-1})$ and $f(x_k)$ and the difference can be at most $2B$. Thus $V(f, \mathcal{P}) = \sum_k |f(x_k) - f(x_{k-1})|$ is bounded by $4mB$. $\square$

Before developing the relationship between functions of bounded variation and functions that are absolutely gauge integrable, we investigate the properties of $BV([a, b])$ and $V_a^b f$.

Theorem 59 ($BV([a, b])$ is a vector space). *Suppose that $f, g \in BV([a, b])$ and that $c \in \mathbb{R}$. Then $f + g$ and cf are both in $BV[(a, b)]$.*

Proof. Exercise. $\square$

Theorem 60 (Properties of $V_a^b f$). *Suppose that f is a function on $[a, b]$. Then*

1. *$V(f, \mathcal{P}) \leq V(f, \mathcal{Q})$ whenever $\mathcal{P}$ and $\mathcal{Q}$ are partitions of $[a, b]$ and $\mathcal{Q}$ is a refinement of $\mathcal{P}$.*
   If, in addition, $f \in BV([a, b])$, then

2. *$V_a^b f = V_a^c f + V_c^b f$ for any $c \in (a, b)$ and*

3. *$V_a^x f$ and $V_a^x f - f(x)$ are increasing functions of x on $[a, b]$.*

Proof. We leave (1) as an exercise.

To prove (2), suppose that $f \in BV([a, b])$ and $c \in (a, b)$. Then $f \in BV([a, c])$ and $f \in BV([c, b])$. Fix $\varepsilon > 0$ and choose partitions $\mathcal{P}_a$ and $\mathcal{P}_b$ of $[a, c]$ and $[c, b]$ such that

$$V_a^c f - \frac{\varepsilon}{2} < V(f, \mathcal{P}_a) \leq V_a^c f$$

and

$$V_c^b f - \frac{\varepsilon}{2} < V(f, \mathcal{P}_b) \leq V_c^b f.$$

Then $\mathcal{P} = \mathcal{P}_a \cup \mathcal{P}_b$ is a partition of $[a, b]$ for which

$$V_a^c f + V_c^b f - \varepsilon < V(f, \mathcal{P}_a) + V(f, \mathcal{P}_b)$$
$$= V(f, \mathcal{P})$$
$$\leq V_a^b f.$$

Hence

$$V_a^c f + V_c^b f \leq V_a^b f.$$

For the reverse inequality, select a partition $\mathcal{Q}$ of $[a, b]$ satisfying

$$V_a^b f - \varepsilon < V(f, \mathcal{Q}) \leq V_a^b f.$$

Create a new partition $\mathcal{Q}^*$ by inserting c as a division point and let $\mathcal{Q}_a$ and $\mathcal{Q}_b$ consist of the intervals from $\mathcal{Q}^*$ to the left and right of c respectively. Then by (1),

$$
\begin{aligned}
V_a^b f - \varepsilon &< V(f, \mathcal{Q}) \\
&\leq V(f, \mathcal{Q}^*) \\
&= V(f, \mathcal{Q}_a) + V(f, \mathcal{Q}_b) \\
&\leq V_a^c f + V_c^b f.
\end{aligned}
$$

Hence

$$
V_a^b f \leq V_a^c f + V_c^b f
$$

and equality follows.

To prove (3), suppose that $a \leq x < y \leq b$. From (2),

$$
V_a^y f - V_a^x f = V_x^y f \geq 0.
$$

Thus $V_a^x f$ is increasing. Moreover,

$$
V_a^y f - V_a^x f = V_x^y f \geq |f(y) - f(x)| \geq f(y) - f(x).
$$

Hence $V_a^x f - f(x)$ is also increasing. $\square$

The previous theorem has a corollary that characterizes functions of bounded variation.

Corollary 61 (BV = difference of increasing functions). *Let f be a function defined on $[a, b]$. Then $f \in BV([a, b])$ if and only if f can be written as the difference of a pair of increasing functions on $[a, b]$.*

Proof. Since monotone functions on $[a, b]$ have bounded variation and $BV[(a, b)]$ is a vector space, any function on $[a, b]$ that can be expressed as the difference of increasing functions is in $BV[(a, b)]$.

For the converse, note that $f(x) = V_a^x f - \left(V_a^x f - f(x) \right)$. $\square$

By writing $f(x) = \left(V_a^x f + x \right) - \left(V_a^x f - f(x) - x \right)$ we can replace "increasing" with "strictly increasing" in the corollary.

We now come to the key theorem that connects $BV([a, b])$ with absolutely integrable functions on $[a, b]$. In the proof, watch for the use of a gauge and tag cutting to force a tagged partition to be a refinement of a given partition.

Theorem 62 (Absolute integrability and $BV([a, b])$). *Suppose that f is a gauge integrable function on $[a, b]$. Then f is absolutely gauge integrable on $[a, b]$ if and only if $F(x) = {}^g\!\int_a^x f$ has bounded variation on $[a, b]$. In this case,*

$$V_a^b F = {}^g\!\int_a^b |f|.$$

Proof. Begin by noting that if $\mathcal{P} = \{[x_{k-1}, x_k]\}$ is a partition of $[a, b]$, then

$$V(F, \mathcal{P}) = \sum_k |F(x_k) - F(x_{k-1})| = \sum_k \left| {}^g\!\int_{x_{k-1}}^{x_k} f \right|. \qquad (7.1)$$

Suppose that f is absolutely integrable and let $\mathcal{P} = \{[x_{k-1}, x_k]\}$ be a partition of $[a, b]$. Then

$$V(F, \mathcal{P}) = \sum_k \left| {}^g\!\int_{x_{k-1}}^{x_k} f \right| \le \sum_k {}^g\!\int_{x_{k-1}}^{x_k} |f| = {}^g\!\int_a^b |f|.$$

Hence $V_a^b F \le {}^g\!\int_a^b |f|$.

For the converse, suppose that $F \in BV([a, b])$. Given $\varepsilon > 0$, choose a partition $\mathcal{P}_0 = \{[x_{k-1}, x_k]\}_{k=1}^n$ such that

$$V_a^b F - \varepsilon/2 < V(f, \mathcal{P}_0) \le V_a^b F.$$

Set $x_{-1} = a - 1$ and $x_{n+1} = b + 1$ and define γ_0 on $[a, b]$ by

$$\gamma_0(t) = \begin{cases} (x_{k-1}, x_k), & t \in (x_{k-1}, x_k) \\ (x_{k-1}, x_{k+1}), & t = x_k. \end{cases}$$

Then any γ_0-fine tagged partition $\mathcal{P}$ will include $\{x_0, x_1, \ldots, x_n\}$ among the tags so that, after tag cutting, $\mathcal{P}$ is a refinement of $\mathcal{P}_0$. Hence, by theorem 60 (page 181),

$$V_a^b F - \varepsilon/2 < V(f, \mathcal{P}_0) \le V(f, \mathcal{P}) \le V_a^b F.$$

Because f is integrable, we can find a gauge γ_1 such that any γ_1-fine tagged partition $\mathcal{P} = \{(t_j, [y_{j-1}, y_j])\}$ satisfies

$$\left| S_R(f, \mathcal{P}) - {}^g\!\int_a^b f \right| < \frac{\varepsilon}{4}.$$

Using (7.1) and Corollary 56 of Henstock's lemma (page 176) applied to all of $\mathcal{P}$ we conclude that

$$|S_R(|f|,\mathcal{P}) - V(F,\mathcal{P})| = \left|S_R(|f|,\mathcal{P}) - \sum_k \left|{}^g\!\!\int_{x_{k-1}}^{x_k} f\right|\right| \leq \frac{\varepsilon}{2}.$$

Set $\gamma(t) = \gamma_0(t) \cap \gamma_1(t)$ and suppose that $\mathcal{P}$ is any γ-fine tagged partition of $[a,b]$. Then

$$\left|S_R(|f|,\mathcal{P}) - V_a^b F\right|$$
$$\leq |S_R(|f|,\mathcal{P}) - V(F,\mathcal{P})| + \left|V(F,\mathcal{P}) - V_a^b F\right|$$
$$< \varepsilon.$$

Hence $|f|$ is gauge integrable with

$${}^g\!\!\int_a^b |f| = V_a^b F$$

as claimed. $\qquad\qquad\square$

While theorem 62 provides a complete characterization of absolutely gauge integrable functions, it is rather awkward to apply. The next three results (the goals of this section) provide tools that are much more straightforward to use.

Theorem 63 (Comparison test). *Let f and g be gauge integrable functions on $[a,b]$ with $|f| \leq g$. Then f is absolutely integrable with*

$$\left|{}^g\!\!\int_a^b f\right| \leq {}^g\!\!\int_a^b |f| \leq {}^g\!\!\int_a^b g.$$

Proof. Set $F(x) = {}^g\!\!\int_a^x f$, $x \in [a,b]$. If $\mathcal{P} = \{[x_{k-1}, x_k]\}$ is a partition of $[a,b]$, then

$$V(F,\mathcal{P}) = \sum_k |F(x_k) - F(x_{k-1})|$$
$$= \sum_k \left|{}^g\!\!\int_{x_{k-1}}^{x_k} f\right| \leq \sum_k {}^g\!\!\int_{x_{k-1}}^{x_k} g = {}^g\!\!\int_a^b g.$$

Thus

$$V_a^b F \leq {}^g\!\!\int_a^b g$$

so that f is absolutely integrable with

$$ {}^{\mathrm{g}}\!\int_a^b |f| = V_a^b F \leq {}^{\mathrm{g}}\!\int_a^b g. \qquad \square $$

Corollary 64 (Absolutely integrable functions form a vector space). *Let f and g be absolutely gauge integrable functions on $[a,b]$ and let c be a scalar. Then $f + g$ and cf are also absolutely gauge integrable.*

Proof. Exercise. $\square$

Theorem 65 (Max and min of absolutely integrable functions). *Suppose that f and g are absolutely gauge integrable functions. Then*

1. f^+ and f^- are gauge integrable and

2. $\max\{f, g\}$ and $\min\{f, g\}$ are gauge integrable.

Proof. 1. Since f and $|f|$ are gauge integrable, so are $f^+ = \frac{1}{2}(|f| + f)$ and $f^- = \frac{1}{2}(|f| - f)$.

2. Since $f - g$, $|f|$, and $|g|$ are gauge integrable and $|f - g| \leq |f| + |g|$, $|f - g|$ is gauge integrable by the comparison test. Hence so are $\min\{f, g\} = \frac{1}{2}(f + g - |f - g|)$ and $\max\{f, g\} = \frac{1}{2}(f + g + |f - g|)$. $\square$

The condition that f and g are absolutely gauge integrable can be relaxed a bit.

Theorem 66 (Max and min of dominated functions). *Suppose that f, g, and h are gauge integrable functions on $[a,b]$.*

1. If $f \leq h$ and $g \leq h$ then $\max\{f, g\}$ and $\min\{f, g\}$ are gauge integrable.

2. If $f \geq h$ and $g \geq h$ then $\max\{f, g\}$ and $\min\{f, g\}$ are gauge integrable.

Proof. 1. Suppose that $f, g \leq h$. Observe that

$$ |f - g| = 2\max\{f, g\} - f - g \leq 2h - f - g. $$

By the comparison test, $|f - g|$ is gauge integrable. Consequently, so are

$$ \max\{f, g\} = \frac{1}{2}(f + g + |f - g|) $$

and

$$\min\{f, g\} = \frac{1}{2}\left(f + g - |f - g|\right).$$

Negate the functions to prove 2. □

This result will be extended to apply to supremums and infimums in the next section once we have established some convergence results.

7.5   Convergence theorems

The convergence results for the gauge integral are similar to those for the Lebesgue integral. However, the proofs have a rather different flavor.

Theorem 67 (Monotone convergence). *Suppose that $\{f_k\}$ is a monotone sequence of gauge integrable functions that converges pointwise to f on $[a, b]$. Then f is gauge integrable if and only if $\left\{ {}^g\!\int_a^b f_k \right\}$ is bounded. In this case,*

$$\lim_k {}^g\!\int_a^b f_k = {}^g\!\int_a^b f.$$

Before turning to the proof, consider how it will need to differ from the proof of the corresponding result for the Lebesgue integral. The measurability of the limit function was never an issue. All we needed to worry about was the value of the integral. We proved the monotone convergence theorem for the Lebesgue integral (theorem 37 on page 138) by approximating each of the functions f_k with an increasing sequence of measurable simple functions. A new increasing sequence of measurable simple functions converging to f was constructed using maximums and the conclusion of the monotone convergence theorem followed from the monotonicity of the Lebesgue integral.

We know that the gauge integral is monotone and that, with some care, maximums of gauge integrable functions are again gauge integrable. However, we don't know that measurable simple functions are gauge integrable nor do we know that a gauge integrable function is the limit of a monotone sequence of measurable simple functions. Moreover, proving that the limit of the gauge integrals of a such a sequence is the gauge integral of the limit function requires the theorem we are trying to prove. We need to work from scratch here and construct an appropriate gauge.

Since we are dealing with pointwise convergence, different values of x will require different values of k in order to force $|f(x) - f_k(x)|$ to be appropriately small. Thus the gauge we seek will be defined in terms of a gauge associated with f_k where k depends on the particular value of x. Henstock's lemma (page 175) will be used to provide the critical connection for the triangle inequality.

Proof. For definiteness, assume that $\{f_k\}$ is monotone increasing.

If f is gauge integrable, then the increasing sequence $\left\{ {}^{g}\!\int_a^b f_k \right\}$ is bounded below by ${}^{g}\!\int_a^b f_1$ and above by ${}^{g}\!\int_a^b f$.

For the converse, suppose that $\left\{ {}^{g}\!\int_a^b f_k \right\}$ is bounded above. Since $\left\{ {}^{g}\!\int_a^b f_k \right\}$ is increasing, the sequence converges to a real value A. We will show that f is gauge integrable with ${}^{g}\!\int_a^b f = A$.

To that end, let $\varepsilon > 0$ be given. Then we can find a natural number N so that

$$\left| {}^{g}\!\int_a^b f_k - A \right| < \varepsilon/3$$

for all $k \geq N$. Since $\{f_k\}$ converges pointwise to f, we can associate to each x in $[a, b]$ a natural number $n(x) \geq N$ such that $|f(x) - f_k(x)| < \frac{\varepsilon}{3(b-a)}$ for all $k \geq n(x)$. Finally, we can use the gauge integrability of f_k to find a gauge γ_k on $[a, b]$ such that

$$\left| S_R(f_k, \mathcal{P}) - {}^{g}\!\int_a^b f_k \right| < \frac{\varepsilon}{3 \cdot 2^k}$$

for any γ_k-fine tagged partition $\mathcal{P}$ of $[a, b]$.

Now define a new gauge on $[a, b]$ by $\gamma(x) = \gamma_{n(x)}(x)$ and suppose that $\mathcal{P} = \{(t_k, I_k)\}_{k=1}^n$ is a γ-fine tagged partition of $[a, b]$. By the triangle inequality,

$$\begin{aligned}
|S_R(f, \mathcal{P}) - A| \leq &\left| \sum_{k=1}^n f(t_k)\,\Delta x_k - \sum_{k=1}^n f_{n(t_k)}(t_k)\,\Delta x_k \right| \\
&+ \left| \sum_{k=1}^n f_{n(t_k)}(t_k)\,\Delta x_k - \sum_{k=1}^n {}^{g}\!\int_{I_k} f_{n(t_k)} \right| \\
&+ \left| \sum_{k=1}^n {}^{g}\!\int_{I_k} f_{n(t_k)} - A \right|.
\end{aligned}$$

We will show that each of the terms on the right side is less than $\varepsilon/3$.

For the first term, note that $n(x)$ was chosen so that $|f(t_k) - f_{n(t_k)}(t_k)| < \frac{\varepsilon}{3(b-a)}$. Thus

$$\left| \sum_{k=1}^{n} f(t_k) \Delta x_k - \sum_{k=1}^{n} f_{n(t_k)}(t_k) \Delta x_k \right| \leq \sum_{k=1}^{n} \left| f(t_k) - f_{n(t_k)}(t_k) \right| \Delta x_k$$

$$< \frac{\varepsilon}{3(b-a)} \sum_{k=1}^{n} \Delta x_k = \frac{\varepsilon}{3}.$$

For the second term, let $\mathcal{P}_j$ be the partial tagged partition consisting of those tagged intervals from $\mathcal{P}$ for which $n(t_k) = j$. Set $M = \max_{1 \leq k \leq n} n(t_k)$. Using Henstock's lemma (page 175) with the γ_j-fine partial tagged partition $\mathcal{P}_j$, we conclude that

$$\left| \sum_{k=1}^{n} f_{n(t_k)}(t_k) \Delta x_k - \sum_{k=1}^{n} {}^g\!\!\int_{I_k} f_{n(t_k)} \right|$$

$$\leq \sum_{j=1}^{M} \left| \sum_{\mathcal{P}_j} \left[f_j(t_k) \Delta x_k - {}^g\!\!\int_{I_k} f_j \right] \right|$$

$$\leq \sum_{j=1}^{M} \frac{\varepsilon}{3 \cdot 2^j} < \frac{\varepsilon}{3}.$$

Keeping the same value of M for the third term, use the facts that $\{f_k\}$ is an increasing sequence and that $N \leq n(t_k) \leq M$ to see that

$${}^g\!\!\int_a^b f_N = \sum_{k=1}^{n} {}^g\!\!\int_{I_k} f_N \leq \sum_{k=1}^{n} {}^g\!\!\int_{I_k} f_{n(t_k)} \leq \sum_{k=1}^{n} {}^g\!\!\int_{I_k} f_M = {}^g\!\!\int_a^b f_M \leq A.$$

Hence

$$\left| \sum_{k=1}^{n} {}^g\!\!\int_{I_k} f_{n(t_k)} - A \right| \leq \left| {}^g\!\!\int_a^b f_N - A \right| < \frac{\varepsilon}{3}.$$

Putting the three terms together, we find that $|S_R(f, \mathcal{P}) - A| < \varepsilon$ for any γ-fine tagged partition $\mathcal{P}$ of $[a, b]$. Thus f is gauge integrable with

$${}^g\!\!\int_a^b f = A = \lim_k {}^g\!\!\int_a^b f_k. \qquad \square$$

We are now in a position to extend theorem 66 to apply to infimums and supremums. We need this result for the proof of the dominated convergence theorem.

Theorem 68 (Dominated supremums). *Suppose that* $\{f_k\}$ *is a sequence of gauge integrable functions on* $[a, b]$ *and that* g *is a gauge integrable function on* $[a, b]$.

 1. If $f_k \leq g$ *for all* $k \in \mathbb{N}$, *then* $\sup_k f_k$ *is gauge integrable.*

 2. If $f_k \geq g$ *for all* $k \in \mathbb{N}$, *then* $\inf_k f_k$ *is gauge integrable.*

Proof. Suppose that $f_k \leq g$ for all $k \in \mathbb{N}$. If we define $g_n = \max_{1 \leq k \leq n} f_k$, then theorem 66 tells us that g_n is a gauge integrable function. By its definition, $\{g_n\}$ is an increasing sequence satisfying

$$\sideset{^g}{}\int_a^b g_1 \leq \sideset{^g}{}\int_a^b g_n \leq \sideset{^g}{}\int_a^b g.$$

Whence $\sup_k f_k = \lim_n \max_{k \leq n} f_k = \lim_n g_n$ is gauge integrable.

Negate the functions to establish (2).   □

With infimums and supremums now available, the proof of the gauge integral form of the dominated convergence theorem is very similar to the proof in the Lebesgue context.

Theorem 69 (Dominated convergence). *Suppose that* $\{f_k\}$ *is a sequence of gauge integrable functions that converges pointwise to* f *on* $[a, b]$. *If there exist gauge integrable functions* g_1 *and* g_2 *on* $[a, b]$ *satisfying* $g_1 \leq f_k \leq g_2$ *for all* $k \in \mathbb{N}$, *then* f *is gauge integrable and*

$$\sideset{^g}{}\int_a^b f = \lim_k \sideset{^g}{}\int_a^b f_k.$$

Proof. Define the sequences $\{\underline{f}_k\}$ and $\{\overline{f}_k\}$ by $\underline{f}_k = \inf_{j \geq k} f_j$ and $\overline{f}_k = \sup_{j \geq k} f_j$. By construction, $\{\underline{f}_k\}$ and $\{\overline{f}_k\}$ are monotone sequences converging to f and satisfying

$$g_1 \leq \underline{f}_k \leq \overline{f}_k \leq g_2.$$

By theorem 68, $\underline{f}_k$ and $\overline{f}_k$ are gauge integrable for $k \in \mathbb{N}$. The monotonicity of the gauge integral allows us to bound their integrals by

$$\sideset{^g}{}\int_a^b g_1 \leq \sideset{^g}{}\int_a^b \underline{f}_k \leq \sideset{^g}{}\int_a^b \overline{f}_k \leq \sideset{^g}{}\int_a^b g_2.$$

By the monotone convergence theorem, f is gauge integrable with

$$\lim_k \sideset{^g}{}\int_a^b \underline{f}_k = \sideset{^g}{}\int_a^b f = \lim_k \sideset{^g}{}\int_a^b \overline{f}_k.$$

Moreover,

$$\underline{f}_k \le f_k \le \overline{f}_k$$

so that

$$^g\!\int_a^b \underline{f}_k \le {}^g\!\int_a^b f_k \le {}^g\!\int_a^b \overline{f}_k.$$

Hence

$$^g\!\int_a^b f = \lim_k {}^g\!\int_a^b f_k$$

as claimed.   □

Example 46. Use the dominated convergence theorem to show that

$$f(x) = \begin{cases} (-1)^k, & x \in \left(\frac{1}{k+1}, \frac{1}{k}\right] \\ 0, & x = 0 \end{cases}$$

is gauge integrable over $[0, 1]$ with $^g\!\int_0^1 f = 2\sum_{k=2}^\infty \frac{(-1)^k}{k} - 1 = 1 - 2\ln 2$.
Define

$$f_n(x) = \begin{cases} (-1)^k, & x \in \left(\frac{1}{k+1}, \frac{1}{k}\right], \ k < n \\ 0, & x \in \left[0, \frac{1}{n}\right]. \end{cases}$$

Then $\{f_n\}$ converges to f on $[0, 1]$ with $-1 \le f_n \le 1$. Hence, by the
dominated convergence theorem,

$$^g\!\int_0^1 f = \lim_n {}^g\!\int_0^1 f_n = \lim_n \sum_{k=1}^{n-1} {}^g\!\int_{\frac{1}{k+1}}^{\frac{1}{k}} (-1)^k$$

$$= \lim_n \sum_{k=1}^{n-1} (-1)^k \left(\frac{1}{k} - \frac{1}{k+1}\right) = 2\sum_{k=2}^\infty \frac{(-1)^k}{k} - 1 = 1 - 2\ln 2.$$

Compare this example to example 43 on page 170.

7.6 The fundamental theorems

Given the close relationship between the definitions of the gauge and Riemann integrals, we should expect that the proofs of the fundamental theorems for the gauge integral should have a structure similar to those for the
Riemann integral. Review the proof of FTC-1 for the Riemann integral (theorem 9, page 39) and you will find the key to the proof: Given a partition

of $[a, b]$, use the mean value theorem to select the tag for each subinterval so that the resulting Riemann sum telescopes. This is not possible for the gauge integral since a subinterval and its tag are not selected independently. So the critical step in the proof of FTC-1 for the gauge integral is to show that there is a gauge γ that will guarantee that, given any γ-fine tagged partition of $[a, b]$, the term $f'(t_k)(x_k - x_{k-1})$ in the Riemann sum and the term $f(x_k) - f(x_{k-1})$ in the telescoping sum are essentially the same. This result is known as the straddle lemma since the endpoints of the subinterval straddle the tag. In contrast to the mean value theorem which starts with an interval and selects a point in that interval, the straddle lemma begins with a point and identifies a containing interval.

Lemma 70 (Straddle). *Let f be defined on $[a, b]$ and differentiable at $z \in [a, b]$. Then for each $\varepsilon > 0$ there is an open interval I_z containing z so that*

$$\left| f(v) - f(u) - f'(z)(v - u) \right| < \varepsilon(v - u)$$

for all u, v satisfying $z \in [u, v] \subseteq [a, b] \cap I_z$.

Proof. Fix $\varepsilon > 0$. Since f is differentiable at z, there is a $\delta > 0$ such that for all $x \in [a, b] \cap (z - \delta, z + \delta)$

$$\left| \frac{f(x) - f(z)}{x - z} - f'(z) \right| < \varepsilon.$$

Equivalently,

$$\left| f(x) - f(z) - f'(z)(x - z) \right| < \varepsilon |x - z|.$$

Set $I_z = (z - \delta, z + \delta)$ and suppose that $z \in [u, v] \subseteq [a, b] \cap I_z$. If $z = u$ or $z = v$, the conclusion follows. So suppose that $u < z < v$. Then

$$\left| f(v) - f(u) - f'(z)(v - u) \right|$$
$$\leq \left| f(v) - f(z) - f'(z)(v - z) \right| + \left| f(z) - f(u) - f'(z)(z - u) \right|$$
$$< \varepsilon(v - z) + \varepsilon(z - u) = \varepsilon(v - u)$$

as claimed.   □

Theorem 71 (FTC-1). *Suppose that F is differentiable on $[a, b]$. Then F' is gauge integrable on $[a, b]$ and*

$$ {}^g\!\!\int_a^y F' = F(y) - F(a)$$

for all $y \in [a, b]$.

Proof. Let $\varepsilon > 0$ be given and define a gauge γ on $[a, b]$ by taking $\gamma(t)$ to be the open interval I_t guaranteed by the straddle lemma when applied to F and $\frac{\varepsilon}{(b-a)} > 0$. Suppose that $\mathcal{P} = \{(t_i, [x_{i-1}, x_i])\}$ is a γ-fine tagged partition of $[a, y]$. Then

$$\left| S_R\left(F', \mathcal{P}\right) - [F(y) - F(a)] \right|$$

$$= \left| \sum_{\mathcal{P}} F'(t_i)(x_i - x_{i-1}) - \sum_{\mathcal{P}} [F(x_i) - F(x_{i-1})] \right|$$

$$\leq \sum_{\mathcal{P}} \left| F'(t_i)(x_i - x_{i-1}) - [F(x_i) - F(x_{i-1})] \right|$$

$$< \sum_{\mathcal{P}} \frac{\varepsilon}{b-a}(x_i - x_{i-1}) = \varepsilon.$$

$\square$

Compare theorem 71 to the statements of FTC-1 for the Riemann (theorem 9, page 39) and Lebesgue (theorem 41, page 141) integrals. Where the Riemann integral requires the derivative to be continuous (at least a.e.) and the Lebesgue integral requires the derivative to be bounded a.e., the gauge integral has no constraints on the derivative.

The gauge-integral version of FTC-1 (theorem 71) points us toward functions that are gauge integrable but not Lebesgue integrable.

Example 47. Let

$$F(x) = \begin{cases} x^2 \cos \frac{\pi}{x^2}, & x \neq 0 \\ 0, & x = 0 \end{cases}$$

and set

$$f(x) = F'(x) = \begin{cases} 2x \cos \frac{\pi}{x^2} + \frac{1}{x} 2\pi \sin \frac{\pi}{x^2}, & x \neq 0 \\ 0, & x = 0. \end{cases}$$

FTC-1 for the gauge integral (theorem 71) tells us that f is gauge integrable. However, $^L\!\int_0^x f$ does not exist when $x > 0$ since both $^L\!\int_0^x f^+$ and $^L\!\int_0^x f^-$ are infinite.

To see why, let $I_n = [\frac{1}{\sqrt{n+1}}, \frac{1}{\sqrt{n}}]$, choose m so that $\frac{1}{\sqrt{m}} < x$, and define $g_n = \left(\sum_{k=m}^n 1_{I_{2k}}\right) \cdot f$. Since f is bounded on I_k, we can apply FTC-1 for the Lebesgue integral to conclude that

$$
{}^{L}\!\int_0^x g_n = \sum_{k=m}^n {}^{L}\!\int_{\frac{1}{\sqrt{2k+1}}}^{\frac{1}{\sqrt{2k}}} f = \sum_{k=m}^n \left[F\left(\frac{1}{\sqrt{2k}}\right) - F\left(\frac{1}{\sqrt{2k+1}}\right) \right]
$$

$$
= \sum_{k=m}^n \left[\frac{1}{2k} + \frac{1}{2k+1} \right].
$$

But $g_n \le f^+$ so that

$$
\sum_{k=m}^n \left[\frac{1}{2k} + \frac{1}{2k+1} \right] \le {}^{L}\!\int_0^x f^+.
$$

As n increases, the summation will grow without bound forcing ${}^{L}\!\int_0^x f^+ = +\infty$.

The verification that ${}^{L}\!\int_0^x f^- = +\infty$ is similar.

Similarly to the way that FTC-1 for the Lebesgue integral can be extended to allow F to be differentiable almost everywhere as long as F is also absolutely continuous, FTC-1 for the gauge integral can be extended to allow F to be nondifferentiable at a countable number of points. For the gauge integral, ordinary continuity, the minimal condition on F, still suffices.

Theorem 72 (FTC-1). *Suppose that F is a continuous function on $[a,b]$ that is differentiable except at a countable number of points. Then F' is gauge integrable on $[a,b]$ with*

$$
{}^{g}\!\int_a^y F' = F(y) - F(a)
$$

for all $y \in [a,b]$.

Proof. Denote the countable exceptional set on which F is not differentiable by $\{z_k\}$. By Corollary 54 (page 175), we can set $F'(z_k) = 0$ for all k.

Given $\varepsilon > 0$, define a gauge γ on $[a,b]$ in the following manner. If F is differentiable at t, take $\gamma(t)$ to be the open interval I_t guaranteed by the straddle lemma when applied to F and $\frac{\varepsilon}{2(b-a)}$. When $t = z_k$, use the continuity of F to choose an open interval $\gamma(z_k)$ that contains z_k and such that

$$
|F(x) - F(z_k)| < \frac{\varepsilon}{2^{k+2}}
$$

for all $x \in [a,b] \cap \gamma(z_k)$.

Let $\mathcal{P} = \{(t_i, [x_{i-1}, x_i])\}_{i=1}^n$ be a γ-fine tagged partition of $[a, y]$. Suppose that $t_i = z_k$ and compare the term associated with z_k in the telescoping sum $\sum_i (F(x_i) - F(x_{i-1}))$ with the corresponding term from the Riemann sum $\sum_i F'(t_i)(x_i - x_{i-1})$.

$$\left| F(x_i) - F(x_{i-1}) - F'(t_i)(x_i - x_{i-1}) \right|$$
$$\leq \left| F(x_i) - F(z_k) \right| + \left| F(z_k) - F(x_{i-1}) \right| + \left| F'(z_k)(x_i - x_{i-1}) \right|$$
$$< 2\frac{\varepsilon}{2^{k+2}} + 0 = \frac{\varepsilon}{2^{k+1}}.$$

It could be that z_k is the tag for two adjacent subintervals ($z_k = x_j = t_j = t_{j+1}$). In this case, an additional term in the second line will be zero and the total difference between the corresponding terms in the telescoping sum and the Riemann sum is still bounded by $\frac{\varepsilon}{2^{k+1}}$.

Now separate $\mathcal{P}$ into two partial tagged partitions $\mathcal{P}_e$ and $\mathcal{P}_r$ corresponding to the intervals whose tags belong to the exceptional set $\{z_k\}$ and the intervals whose tags do not. Then

$$\left| \left[F(y) - F(a) \right] - S_R(f, \mathcal{P}) \right|$$
$$\leq \left| \sum_{\mathcal{P}_e} \left[F(x_i) - F(x_{i-1}) - F'(t_i)(x_i - x_{i-1}) \right] \right|$$
$$+ \left| \sum_{\mathcal{P}_r} \left[F(x_i) - F(x_{i-1}) - F'(t_i)(x_i - x_{i-1}) \right] \right|$$
$$< \sum_{k=1}^{\infty} \frac{\varepsilon}{2^{k+1}} + \frac{\varepsilon}{2(b-a)} \sum_{i=1}^{n} (x_i - x_{i-1}) = \varepsilon.$$

Hence F' is gauge integrable with

$$ {}^g\!\int_a^y F' = F(y) - F(a). \qquad \qquad \square$$

You of course recognized the by now familiar divide-and-conquer move in the proof.

For the derivative form of the fundamental theorem of calculus, we begin with a simple form that is closely related to the corresponding theorem for the Riemann integral.

Theorem 73 (FTC-2). *Suppose that f is a gauge integrable function on $[a, b]$. Then the function $F(x) = {}^g\!\int_a^x f$ is continuous on $[a, b]$ and differentiable with $F' = f$ at those points where f is continuous.*

Proof. We will show that F is continuous at $x_0 \in (a, b)$, the analysis at the endpoints being similar. Fix $\varepsilon > 0$ and choose a gauge δ on $[a, b]$ so that any δ-fine tagged partition $\mathcal{P}$ satisfies

$$\left| S_R(f, \mathcal{P}) - {}^g\!\!\int_a^b f \right| < \frac{\varepsilon}{2}.$$

Let δ' be the positive value $\delta' = \min\left\{ \delta(x_0), \frac{\varepsilon}{2(|f(x_0)|+1)} \right\}$ and suppose that $x \in [a, b] \cap (x_0, x_0 + \delta')$. Then $\{(x_0, [x_0, x])\}$ is a δ-fine partial tagged partition of $[a, b]$. Using the triangle inequality followed by an application of Henstock's lemma, we find that

$$|F(x) - F(x_0)| \leq \left| {}^g\!\!\int_{x_0}^x f - f(x_0)(x - x_0) \right| + |f(x_0)(x - x_0)|$$

$$< \frac{\varepsilon}{2} + \left| f(x_0) \frac{\varepsilon}{2(|f(x_0)| + 1)} \right| < \varepsilon.$$

The analysis when $x_0 - \delta' < x < x_0$ is similar. Thus F is continuous at x_0.

The proof that $F'(x_0) = f(x_0)$ when f is continuous at x_0 closely follows that for the Riemann integral and is left as an exercise. □

As with FTC-2 for the Lebesgue integral, we can remove the condition that f is continuous at x_0 and still conclude that F is differentiable almost everywhere. For the gauge integral, the Vitali covering theorem (below) plays a role analogous to the roles of lemmas 45, 46, and 47 (page 145) in the Lebesgue context. Not coincidentally, the Vitali covering theorem will also do a lot of heavy lifting in Section 8 where we provide the deferred proofs of these lemmas.

Definition 29 (Vitali covering). *Let E be a bounded set. A **Vitali covering** of E is a collection $\mathcal{V}$ of nondegenerate, closed intervals such that, given any $x \in E$ and $\varepsilon > 0$, there is an interval $I \in \mathcal{V}$ with $x \in I$ and $l(I) < \varepsilon$. We do not assume that $\mathcal{V}$ is countable.*

Example 48. Let C be the Cantor set. Then $\mathcal{V} = \left\{ \left[x - \frac{1}{n}, x \right] : x \in C, n \in \mathbb{N} \right\}$ is a Vitali covering of C. Note that $\mathcal{V}^* = \left\{ \left[x - \frac{1}{2}, x \right] : x \in C \right\}$ is not a Vitali covering of C because there is no interval I in $\mathcal{V}^*$ for which $\frac{1}{3} \in I$ with $l(I) < \frac{1}{4}$.

The Vitali covering theorem, next, captures an idea similar to the concept of compactness. Any open cover of a compact set has a finite subcover.

Given a Vitali covering, there is a finite subset of disjoint sets that is "almost" a subcover.

Theorem 74 (Vitali Covering Theorem). *Let $E \subseteq [a, b]$ and let $\mathcal{V}$ be a Vitali covering of E. Then given any $\varepsilon > 0$, there is a finite set of disjoint intervals $\{I_k\}_{k=1}^n$ from $\mathcal{V}$ with $\mu^* \left(E \backslash \cup_{k=1}^n I_k \right) < \varepsilon$.*

The following proof will use a slightly unusual construction. Given a non-degenerate interval $I = [\alpha, \beta]$, let $\hat{I} = [\alpha - 2l, \beta + 2l]$ where $l = \beta - \alpha$ is the length of I. The construction extends the interval I by twice its length on each end. The important features of $\hat{I}$ are the length of $\hat{I}$ is 5 times that of I and any point within $2l$ of I is an element of $\hat{I}$.

Proof. If all the intervals of length greater than 1 are removed from $\mathcal{V}$, the result is still a Vitali covering of E. Hence we can assume that the intervals in $\mathcal{V}$ have length at most 1.

Choose any interval from $\mathcal{V}$ to be I_1. Suppose that $I_1, \ldots, I_m$ have been chosen. If $E \subseteq \cup_{k=1}^m I_k$, then we are done. Otherwise, let $\mathcal{V}_m$ be the set of intervals in $\mathcal{V}$ that are disjoint from $\cup_{k=1}^m I_k$ and that contain at least one element of E. Let λ_m be the supremum of the lengths of the intervals in $\mathcal{V}_m$ and choose an interval from $\mathcal{V}_m$ with length greater than $\lambda_m/2$ to be I_{m+1}.

Suppose that the process does not terminate with a finite number of intervals and that $x \in E \backslash \left(\cup_{k=1}^\infty I_k \right)$. Since the intervals $\{I_k\}_{k=1}^\infty$ are disjoint and contained in $[a - 1, b + 1]$, $\sum_{k=1}^\infty l(I_k)$ converges. Hence we can select $N \in \mathbb{N}$ so that $\sum_{k=N+1}^\infty l(I_k) < \varepsilon/5$. Pick any interval J from $\mathcal{V}_N$ that contains x and let λ be the length of J. Since $\lambda_m \geq \lambda$ as long as $J \in \mathcal{V}_m$ and since $\lim_n \lambda_n = 0$, we see that $J \notin \mathcal{V}_m$ for sufficiently large values of m. Let M be the smallest integer for which $J \notin \mathcal{V}_M$. As J is disjoint from $\cup_{k<M} I_k$, J must intersect I_M which implies that x is within λ of I_M. Since the length of I_M is at least $\lambda_M/2 \geq \lambda/2$, we know that $x \in \hat{I}_M$.

Because $x \in E \backslash \left(\cup_{k=1}^\infty I_k \right)$ was arbitrary, we can conclude that $E \backslash \left(\cup_{k=1}^\infty I_k \right) \subseteq \cup_{k=N+1}^\infty \hat{I}_k$. Consequently,

$$\mu^* \left(E \backslash \cup_{k=1}^n I_k \right) \leq \mu^* \left(\cup_{k=N+1}^\infty \hat{I}_k \right) \leq \sum_{k=N+1}^\infty l \left(\hat{I}_k \right)$$

$$= 5 \sum_{k=N+1}^\infty l(I_k) < \varepsilon$$

as claimed.   □

In the proof of the next theorem, the Vitali covering theorem is used to prove that a set Z_+ has measure zero. By constructing a Vitali covering of Z_+, the theorem allows us "essentially" to cover Z_+ with a union of a finite set $\{I_k\}_{k=1}^n$ of disjoint intervals. Henstock's lemma is used to prove that the size of $\cup_{k=1}^n I_k$ is arbitrarily small. This divide-and-conquer use of the Vitali covering theorem is typical.

Theorem 75 (FTC-2). *Suppose that f is a gauge integrable function on $[a, b]$. Then the function $F(x) = {}^g\!\int_a^x f$ is continuous on $[a, b]$ and, except on a set of measure zero, F is differentiable with $F' = f$.*

Proof. A proof of the continuity of F appears in theorem 73 (page 194).

To prove that F' exists and is equal to f almost everywhere, let Z_+ be the set of points t where

$$F'_+(t) = \lim_{x \to t^+} \frac{F(x) - F(t)}{x - t}$$

fails to exist or is not equal to $f(t)$. Z_- is defined analogously. If $t \in Z_+$ then, negating the definition of a limit, there is some $\varepsilon_t > 0$ such that for every $s > 0$ there is an $x_{t,s} \in [a, b] \cap (t, t + s)$ for which

$$\left| \frac{F(x_{t,s}) - F(t)}{x_{t,s} - t} - f(t) \right| > \varepsilon_t.$$

Equivalently,

$$|F(x_{t,s}) - F(t) - f(t)(x_{t,s} - t)| > \varepsilon_t (x_{t,s} - t). \qquad (7.2)$$

Set $E_n = \left\{ t \in Z_+ : \varepsilon_t \geq \frac{1}{n} \right\}$ and let $\varepsilon > 0$ be given. Since f is gauge integrable, there is a gauge δ such that

$$\left| S_R(f, \mathcal{P}) - {}^g\!\int_a^b f \right| < \frac{\varepsilon}{3n}$$

for any δ-fine tagged partition $\mathcal{P}$ of $[a, b]$.

Observe that $\mathcal{V}_n = \{[t, x_{t,s}] : t \in E_n, 0 < s < \delta(t)\}$ is a Vitali covering of E_n. By the Vitali covering theorem we can find a finite set of disjoint intervals $\{I_k\} = \{[t_k, x_k]\}$ from $\mathcal{V}_n$ for which $\mu^*(E_n \setminus \cup_k I_k) < \varepsilon/3$. Since the elements of $\{I_k\}$ are disjoint, $\mathcal{P}^* = \{(t_k, [t_k, x_k])\}$ is a δ-fine partial tagged partition of $[a, b]$. Using the corollary to Henstock's lemma (page 176) and then inequality (7.2), we can conclude that

$$2\frac{\varepsilon}{3n} \geq \sum_{\mathcal{P}*} \left| f\left(t_k\right) \cdot \left(x_k - t_k\right) - {}^g\!\int_{t_k}^{x_k} f \right|$$

$$= \sum_{\mathcal{P}*} \left| f\left(t_k\right)\left(x_k - t_k\right) - \left[F\left(t_k\right) - F\left(x_k\right)\right] \right|$$

$$> \sum_{\mathcal{P}*} \varepsilon_t \left(x_k - t_k\right)$$

$$\geq \sum_{\mathcal{P}*} \frac{1}{n} \left(x_k - t_k\right).$$

Hence

$$\mu^* \left(\cup_k I_k\right) = \sum_{\mathcal{P}*} \left(x_k - t_k\right) \leq \frac{2}{3}\varepsilon.$$

Thus

$$\mu^* \left(E_n\right) \leq \mu^* \left(\cup_k I_k\right) + \mu^* \left(E_n \backslash \cup_k I_k\right) < \varepsilon.$$

Because $\varepsilon > 0$ was arbitrary, $\lambda\left(E_n\right) = 0$ and since $Z_+ = \cup_n E_n$, we conclude that Z_+ is a set of measure zero. A similar argument shows that $\lambda\left(Z_-\right) = 0$ so $Z_+ \cup Z_-$, the set of points where F fails to have a derivative equal to f, has measure zero. □

We observe that the conclusion of theorem 75 is somewhat weaker than the corresponding theorem for the Lebesgue and hence the Riemann-Darboux integrals. For Lebesgue integrable functions, the function $F(x) = {}^L\!\int_a^x f$ is absolutely continuous. This need not be true for the gauge integral. To see why not, consider the functions of example 47 (page 192) and the intervals $\{[\frac{1}{\sqrt{2k+1}}, \frac{1}{\sqrt{2k}}]\}_{k=m}^n$. Given any $\delta > 0$, we can choose m and n so that the total length of the intervals is

$$\sum_{k=m}^n \left(\frac{1}{\sqrt{2k}} - \frac{1}{\sqrt{2k+1}}\right) < \frac{1}{\sqrt{2m}} < \delta$$

while

$$\sum_{k=m}^n \left[F\left(\frac{1}{\sqrt{2k}}\right) - F\left(\frac{1}{\sqrt{2k+1}}\right)\right] = \sum_{k=m}^n \left[\frac{1}{2k} + \frac{1}{2k+1}\right] > 1.$$

7.7   Integral relationships

We have already seen (exercise 2) that any Riemann integrable function is gauge integrable and that the two integrals agree. We have also encountered a function that is gauge but not Lebesgue integrable (example 47 on page 192). Our purpose in this section is to more closely examine the relationship between Lebesgue and gauge integrability. The nature of the Lebesgue integral dictates that the discussion must focus on the measurability of functions and, consequently, on measurable sets.

We begin our investigation by considering characteristic functions. First note that when I is an interval then 1_I is a step function and so is gauge integrable over $[a, b]$ with

$$ {}^g\!\int_a^b 1_I = l\,(I \cap [a, b]) = \lambda\,(I \cap [a, b]) = {}^L\!\int_a^b 1_I. $$

Let G be an open set. Then G can be expressed as a countable union of disjoint, open intervals, $G = \cup_k I_k$. If the union is finite, then the linearity of the Lebesgue and gauge integrals implies that 1_G is gauge integrable over $[a, b]$ with

$$ {}^g\!\int_a^b 1_G = {}^g\!\int_a^b \sum_k 1_{I_k} = \sum_k {}^g\!\int_a^b 1_{I_k} = \sum_k {}^L\!\int_a^b 1_{I_k} = {}^L\!\int_a^b 1_G. $$

The result for countably infinite unions follows from the monotone convergence theorems for the gauge and Lebesgue integrals. If F is a closed set, then $G = F^c$ is open and $1_F = 1 - 1_G$. Hence 1_F is gauge integrable over $[a, b]$ with

$$ {}^g\!\int_a^b 1_F = {}^L\!\int_a^b 1_F = \lambda\,(F \cap [a, b])\,. $$

In fact, the conclusion extends to all measurable sets.

Theorem 76 (Measurable sets have gauge integrable characteristic functions). *Suppose that E is a (Lebesgue) measurable subset of $\mathbb{R}$. Then 1_E is gauge integrable over $[a, b]$ with*

$$ {}^g\!\int_a^b 1_E = {}^L\!\int_a^b 1_E = \lambda\,(E \cap [a, b])\,. $$

Proof. Since E is measurable, we can use theorem 22 (page 104) to find a sequence of closed sets $\{F_n\}$ contained in E such that $\lambda\,(E \backslash F_n) < \frac{1}{n}$. Replacing F_n by $\cup_{k=1}^n F_k$, we may assume that the sequence of closed set is increasing ($F_n \subseteq F_{n+1}$). Set $E^* = \cup_n F_n$. Again applying the monotone convergence theorems, we see that 1_{E^*} is gauge integrable over $[a, b]$ with

$$^g\!\!\int_a^b 1_{E^*} = {}^L\!\!\int_a^b 1_{E^*} = \lambda \left(E^* \cap [a, b] \right).$$

Now for any $n \in \mathbb{N}$, we have $F_n \subseteq E^* \subseteq E$ so $\lambda (E \backslash E^*) \leq \lambda (E \backslash F_n) < \frac{1}{n}$. Hence $E \backslash E^*$ has measure zero so that $1_E = 1_{E^*}$ a.e. Therefore 1_E is gauge integrable with

$$^g\!\!\int_a^b 1_E = {}^L\!\!\int_a^b 1_E = \lambda \left(E \cap [a, b] \right). \qquad \square$$

This is a satisfying result that begs the question: Can we use $\lambda_g (E) = {}^g\!\!\int_a^b 1_E$ to define a gauge measure that extends Lebesgue measure to be defined on a larger collection of subsets of $\mathbb{R}$? One suspects not, but the question should be investigated. In any case, the previous theorem provides us with exactly the tools we need to verify that any Lebesgue integrable function is gauge integrable.

Theorem 77 (Lebesgue $\Longrightarrow$ gauge). *Suppose that f is a Lebesgue integrable function over $[a, b]$. Then f is also gauge integrable over $[a, b]$ and the two integrals agree.*

Proof. First assume that $f \geq 0$. Then there is an increasing sequence of measurable simple functions $\{\phi_n\}$ that converges to f. By linearity of the integrals, the previous theorem implies that each ϕ_n is gauge integrable over $[a, b]$ with

$$^g\!\!\int_a^b \phi_n = {}^L\!\!\int_a^b \phi_n.$$

The conclusion now follows from the monotone convergence theorems for the Lebesgue and gauge integrals. When f is not nonnegative, consider $f = f^+ - f^-$. $\qquad \square$

We already know from example 47 (page 192) that the converse of theorem 77 is false. Can we say anything about those functions that are gauge but not Lebesgue integrable? Indeed we can. And the proof of the result features a cameo appearance of a familiar sequence of functions.

Theorem 78 (Gauge $\Longrightarrow$ measurable). *Let f be a gauge integrable function on $[a, b]$. Then f is measurable.*

Proof. Define $F(x) = {}^g\!\int_a^x f$ for $x \in [a,b]$. Then by the FTC-2 (theorem 75 on page 197), F is continuous on $[a,b]$ and $F' = f$ a.e. on $[a,b]$. Extend F to $[a, b+1]$ by taking $F(x) = F(b)$ for $x \in [b, b+1]$ and define g_n on $[a,b]$ by $g_n(x) = n\left[F\left(x + \frac{1}{n}\right) - F(x)\right]$. Then $\{g_n\}$ is a sequence of continuous and therefore measurable functions that converges a.e. to f. Hence f is measurable. □

Note that we have settled the question of whether or not the gauge integral can be used to extend the Lebesgue measure. If E is a set for which 1_E is gauge integrable, then E is Lebesgue measurable. The gauge integral cannot be used to extend Lebesgue measure.

So if a gauge integrable function f is measurable, how can it fail to be Lebesgue integrable? Only if at least one of ${}^L\!\int_a^b f^+$ or ${}^L\!\int_a^b f^-$ is infinite. This observation allows us to completely characterize the relationship between the gauge and Lebesgue integrals.

Theorem 79 (Lebesgue = absolutely gauge). *Let f be a real-valued function on $[a,b]$.*

1. *If $f \geq 0$, then f is Lebesgue integrable over $[a,b]$ if and only if f is gauge integrable over $[a,b]$.*

2. *In general, f is Lebesgue integrable over $[a,b]$ if and only if f is absolutely gauge integrable over $[a,b]$.*

Proof. Exercise. □

7.8 Loose ends and Dini derivatives

The techniques we have developed for the gauge integral can be used to provide the proofs that were deferred from the previous chapter. Specifically, bounded variation and the Vitali covering theorem are the appropriate tools to prove lemmas 45, 46, and 47 (page 145).

You no doubt noticed the similarity between the definitions of bounded variation and of absolute continuity. Absolute continuity appears to be a strengthening of the condition of bounded variation. This is indeed the case.

Theorem 80 (Absolutely continuous $\implies$ BV). *If f is an absolutely continuous function on $[a,b]$, then $f \in BV([a,b])$.*

Proof. Since f is absolutely continuous, we can find a $\delta > 0$ so that whenever $\{(x_k, y_k)\}$ is a finite set of disjoint intervals satisfying $\sum_k |y_k - x_k| < \delta$, we have $\sum_k |f(y_k) - f(x_k)| < 1$. Choose n so that $\frac{b-a}{n} < \delta$ and set $z_k = a + k(\frac{b-a}{n})$, $k = 0, 1, 2, \ldots, n$. Suppose that $\mathcal{P}$ is a partition of $[z_{k-1}, z_k]$. Then the total length of the intervals in $\mathcal{P}$ is $\frac{b-a}{n} < \delta$ so that $V(f, \mathcal{P}) < 1$. This inequality holds for all partitions of $[z_{k-1}, z_k]$ so that $V_{z_{k-1}}^{z_k} f \leq 1$. By theorem 60 (page 181) we conclude that

$$V_a^b f = \sum_{k=1}^{n} V_{z_{k-1}}^{z_k} f \leq n.$$

Hence $f \in BV([a,b])$. $\qquad\square$

The previous theorem when combined with Corollary 61 (page 182) provides the deferred proof of lemma 45 on page 145.

Lemma 45. *Suppose that f is absolutely continuous on $[a,b]$. Then f can be expressed as $f = f_1 - f_2$ where f_1 and f_2 are increasing functions. (If convenient, we can require f_1 and f_2 to be strictly increasing or absolutely continuous.)*

Proof. If f is absolutely continuous, then $f \in BV([a,b])$. By theorem 60 (page 60), $V_a^x f$ and $V_a^x f - f$ are increasing functions. Thus we can express $f(x) = V_a^x f - (V_a^x f - f(x))$ as the difference of increasing functions. To make the functions strictly increasing write $f(x) = V_a^x f + x - (V_a^x - f(x) + x)$. Since $V_a^x f$ is absolutely continuous, so are all the functions in the decompositions.

To see that $V_a^x f$ is absolutely continuous, let $\varepsilon > 0$ be given. By the absolute continuity of f, there is a $\delta > 0$ so that whenever $\{(x_k, y_k)\}$ is a finite set of disjoint intervals satisfying $\sum_k |y_k - x_k| < \delta$, we have $\sum_k |f(y_k) - f(x_k)| < \varepsilon$. Let $\{(x_k, y_k)\}_{k=1}^{n}$ be such a set of intervals and suppose that for each k, $\mathcal{P}_k = \{[x_{k,j-1}, x_{k,j}]\}_{j=1}^{n_k}$ is a partition of $[x_k, y_k]$ for which

$$V_{x_k}^{y_k} f - \frac{\varepsilon}{n} < V(f, \mathcal{P}_k).$$

Then $\{(x_{k,j-1}, x_{k,j})\}_{k,j}$ is a finite set of disjoint intervals satisfying

$$\sum_{k=1}^{n}\sum_{j=1}^{n_k} |x_{k,j} - x_{k,j-1}| = \sum_{k=1}^{n} |y_k - x_k| < \delta$$

so that

$$\sum_{k=1}^{n} V(f, \mathcal{P}_k) = \sum_{k=1}^{n} \sum_{j=1}^{n_k} \left| f\left(x_{k,j}\right) - f\left(x_{k,j-1}\right) \right| < \varepsilon.$$

Thus

$$\sum_{k} |V_a^{y_k} f - V_a^{x_k} f| = \sum_{k} V_{x_k}^{y_k} f < \sum_{k} V(f, \mathcal{P}_k) + \varepsilon < 2\varepsilon.$$

We conclude that $V_a^x f$ is absolutely continuous. □

The proofs of the other two lemmas make heavy use of the Vitali covering theorem. The next proof uses the Vitali covering theorem to divide-and-conquer. We split (a, b) into two parts: a collection of intervals on which f cannot change much because the derivative is zero at one endpoint and a collection of intervals with very small total length on which the absolutely continuous function f can change very little.

Lemma 47. *If f is absolutely continuous on $[a, b]$ and $f' = 0$ a.e. on $[a, b]$, then f is constant on $[a, b]$.*

Proof. Using lemma 45, it is sufficient to assume that f is also monotone increasing and prove that $f(a) = f(b)$.

Fix $\varepsilon > 0$ and let δ be the corresponding δ of absolute continuity. Set $E = \{x \in (a, b) : f'(x) = 0\}$. Then if $x \in E$,

$$\lim_{t \to x} \frac{f(t) - f(x)}{t - x} = 0$$

so that

$$\mathcal{V} = \{[x, t] : x \in E, t \in (x, b), |f(t) - f(x)| < \varepsilon(t - x)\}$$

is a Vitali covering of E. The Vitali covering theorem implies that there is a finite set of disjoint, closed intervals $\{I_k\} = \{[x_k, y_k]\}_{k=1}^{n}$ from $\mathcal{V}$ such that $\mu^*(E \setminus \cup_k I_k) < \delta$ and $|f(y_k) - f(x_k)| < \varepsilon |y_k - x_k|$.

Set $y_0 = a$ and $x_{n+1} = b$. Then (a, b) is the disjoint union of the $2n + 1$ intervals $\{[x_k, y_k]\}_{k=1}^{n}$ and $\{(y_k, x_{k+1})\}_{k=0}^{n}$. Since

$$E \setminus \cup_k I_k \subseteq (a, b) \setminus \cup_k I_k \subseteq ((a, b) \setminus E) \cup (E \setminus \cup_k I_k)$$

and $(a, b) \setminus E$ has measure zero, we conclude that

$$\sum_{k=0}^{n} |x_{k+1} - y_k| = \mu^*((a, b) \setminus \cup_k I_k) = \mu^*(E \setminus \cup_k I_k) < \delta$$

so that

$$\sum_{k=0}^{n} |f(x_{k+1}) - f(y_k)| < \varepsilon.$$

Thus

$$
\begin{aligned}
|f(b) - f(a)| &= \left| \sum_{k=1}^{n} (f(y_k) - f(x_k)) + \sum_{k=0}^{n} (f(x_{k+1}) - f(y_k)) \right| \\
&\leq \sum_{k=1}^{n} |f(y_k) - f(x_k)| + \sum_{k=0}^{n} |f(x_{k+1}) - f(y_k)| \\
&< \varepsilon \sum_{k=1}^{n} |y_k - x_k| + \varepsilon \leq \varepsilon(b - a + 1).
\end{aligned}
$$

As $\varepsilon > 0$ was arbitrary, we conclude that $f(b) = f(a)$. Hence f, being monotone increasing, is constant on $[a, b]$.   □

To prove lemma 46, we employ a more general notion of the derivative. A function f is not differentiable at $x = c$ when $\lim_{x \to x_0} \frac{f(x) - f(c)}{x - c}$ fails to exist. However, there are four related limits that always exist in the extended real numbers. These are the Dini derivatives. A function is differentiable at c exactly when its four Dini derivatives at c are finite and equal.

Definition 30 (Dini derivatives). *Let f be a function defined on a neighborhood of c. The four Dini derivatives of f at c are*

$$
D_- f(c) = \varliminf_{x \to c^-} \frac{f(x) - f(c)}{x - c}, \qquad D^+ f(c) = \varlimsup_{x \to c^+} \frac{f(x) - f(c)}{x - c},
$$

$$
D^- f(c) = \varlimsup_{x \to c^-} \frac{f(x) - f(c)}{x - c}, \qquad D_+ f(c) = \varliminf_{x \to c^+} \frac{f(x) - f(c)}{x - c}.
$$

The limits may take values in the extended real numbers.

The positions of the four Dini derivatives in the definition reflect the relative positions of rays having the corresponding slopes. Figure 7.1 displays a function with its four Dini derivatives. In this case, the four Dini derivatives are all different.

Example 49. The four Dini derivatives of the absolute value function $f(x) = |x|$ at zero are

$$
\begin{aligned}
D_- f(0) &= -1, & D^+ f(0) &= 1, \\
D^- f(0) &= -1, & D_+ f(0) &= 1.
\end{aligned}
$$

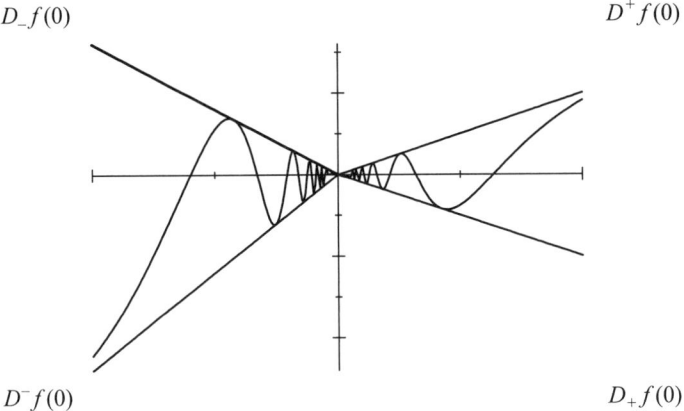

$D_- f(0)$

$D^+ f(0)$

$D^- f(0)$

$D_+ f(0)$

Figure 7.1. f with rays for the four Dini derivatives at $x = 0$

Example 50. The four Dini derivatives of

$$f(x) = \begin{cases} x \sin \frac{1}{x}, & x \neq 0 \\ 0, & x = 0 \end{cases}$$

at zero are

$$D_- f(0) = -1, \quad D^+ f(0) = 1,$$
$$D^- f(0) = 1, \quad D_+ f(0) = -1.$$

From the definition of the Dini derivatives, we see that $D^+ f \geq D_+ f$ and $D^- f \geq D_- f$. Moreover, when f is an increasing function, all four derivatives are nonnegative. Our goal is to prove that when f is an increasing function on $[a, b]$, then f is differentiable a.e. on $[a, b]$. This will involve showing that the four Dini derivatives are equal and finite a.e. on $[a, b]$. We begin by showing that $D^+ f$ is finite a.e. on $[a, b]$.

Theorem 81. *Let f be an increasing function on $[a, b]$. Then $D^+ f$ is finite a.e. on $[a, b]$.*

Proof. Let $E = \{x \in (a, b) : D^+ f(x) = \infty\}$ and set $m = \mu^*(E)$. If $x \in E$, then

$$\varlimsup_{t \to x^+} \frac{f(t) - f(x)}{t - x} = +\infty$$

so that, for any constant B,

$$\mathcal{V} = \{[x, t] : x \in E,\ t \in (x, b),\ f(t) - f(x) > B(t - x)\}$$

is a Vitali covering of E. By the Vitali covering theorem, there is a finite, disjoint collection of intervals $\{I_k\} = \{[x_k, t_k]\}$ from $\mathcal{V}$ so that $\mu^* (E \backslash \cup_k I_k) < \frac{m}{2}$ and hence $\sum_k (t_k - x_k) = \mu^* (\cup_k I_k) > \frac{m}{2}$. Because f is increasing,

$$f (b) - f (a) \geq \sum_k (f (t_k) - f (x_k)) > B \sum_k (t_k - x_k) > B \frac{m}{2}.$$

Since B is arbitrary, we must have $\mu^* (E) = m = 0$ as claimed.   □

By reflecting f, we can interchange the Dini derivatives.

Lemma 82. *Suppose that f is a function on $[a, b]$. Define h on $[-b, -a]$ by $h (x) = -f (-x)$. Then for any $c \in (a, b)$*

1. $D_+ f (c) = D_- h (-c)$,

2. $D^- f (c) = D^+ h (-c)$, and

3. if f is increasing, so is h.

Proof. Exercise.   □

We are now prepared to provide the last of the deferred proofs. The proof of lemma 46 proceeds by carefully approximating the set of points where f is not differentiable in three stages: by an open set, by a Vitali approximation[7] associated with left limits, and a second Vitali approximation related to right limits. The open set contains the set of points where f is not differentiable and each subsequent approximating set will fit inside the previous one.

Lemma 46 (Lebesgue, 1904). *If f is increasing on $[a, b]$ then f is differentiable a.e. on $[a, b]$.*

Proof. Since we have already established that the $D^+ f$ is finite a.e. when f is an increasing function, our task is to show the Dini derivatives are equal a.e. on $[a, b]$. If we can show that $D_- f \geq D^+ f$ and $D_+ f \geq D^- f$ a.e. on $[a, b]$, then our conclusion will follow from the chain of inequalities

$$D_- f \geq D^+ f \geq D_+ f \geq D^- f \geq D_- f.$$

In fact, it suffices to show that $D_- f \geq D^+ f$ for any increasing function f, since lemma 82 then can be used to conclude that $D_+ f \geq D^- f$.

---

[7] The union of the finite set of closed disjoint intervals guaranteed by the Vitali covering theorem.

Suppose then that f is an increasing function on $[a, b]$. Let

$$E = \{x \in (a, b) : D_- f(x) < D^+ f(x)\}$$

and note that $E = \cup_{p < q \in \mathbb{Q}^+} E_p^q$ where

$$E_p^q = \{x \in (a, b) : D_- f(x) < p < q < D^+ f(x)\}.$$

Fix $p, q \in \mathbb{Q}^+$ with $p < q$ and set $m = \mu^*(E_p^q)$. Then given $\varepsilon > 0$, there is an open set G with $E \subseteq G \subseteq (a, b)$ and $\mu^*(G) < m + \varepsilon$.
If $x \in E_p^q$, then

$$\lim_{t \to x^-} \frac{f(t) - f(x)}{t - x} < p$$

so that

$$V = \{[t, x] : x \in E_p^q, t < x, [t, x] \subset G, f(x) - f(t) < p(x - t)\}$$

is a Vitali covering of E_p^q. Use the Vitali covering theorem to find a finite set of disjoint, closed intervals $\{I_k\} = \{[x_k, y_k]\}$ such that $\mu^*(E_p^q \setminus \cup_k I_k) < \varepsilon$ and

$$\sum_k (f(y_k) - f(x_k)) < p \sum_k (y_k - x_k)$$
$$< p\mu^*(G) < p(m + \varepsilon). \tag{7.3}$$

Without affecting the outer measure of E_p^q, remove any endpoints of the $\{I_k\}$ from E_p^q and consider the set $F_p^q = E_p^q \cap \cup_k (x_k, y_k)$. Note that our modified E_p^q is the disjoint union of F_p^q and $E_p^q \setminus \cup_k I_k$. Thus $\mu^*(F_p^q) > m - \varepsilon$.
If $x \in F_p^q$ then $x \in E_p^q$ and x belongs to exactly one interval I_{k_x}. Since $x \in E_p^q$,

$$\overline{\lim_{t \to x^+}} \frac{f(t) - f(x)}{t - x} > q$$

so that

$$V = \{[x, t] : x \in F_p^q, t > x, [x, t] \subset I_{k_x}, f(t) - f(x) > q(t - x)\}$$

is a Vitali covering of F_p^q. Again, the Vitali covering theorem asserts the existence of a finite number of disjoint closed intervals $\{J_i\} = \{[u_i, v_i]\}$ such that $\mu^*(F_p^q \setminus \cup_i J_i) < \varepsilon$ and

$$\sum_i (f(v_i) - f(u_i)) > q \sum_i (v_i - u_i). \tag{7.4}$$

Since F_p^q is contained in the disjoint union of $F_p^q \setminus \cup_i J_i$ and $\cup_i J_i$, we see that

$$\mu^* \left(\cup_i J_i \right) > \mu^* \left(F_p^q \right) - \varepsilon > m - 2\varepsilon. \tag{7.5}$$

As f is increasing and every J_i is contained in some I_k,

$$\sum_i \left(f\left(v_i\right) - f\left(u_i\right) \right) \le \sum_k \left(f\left(y_k\right) - f\left(x_k\right) \right). \tag{7.6}$$

Hence from (7.5), (7.4), (7.6), and (7.3),

$$m < \mu^* \left(\cup_i J_i \right) + 2\varepsilon$$

$$= \sum_i \left(v_i - u_i \right) + 2\varepsilon$$

$$< \frac{1}{q} \sum_i \left(f\left(v_i\right) - f\left(u_i\right) \right) + 2\varepsilon$$

$$\le \frac{1}{q} \sum_k \left(f\left(y_k\right) - f\left(x_k\right) \right) + 2\varepsilon$$

$$< \frac{p}{q} \left(m + \varepsilon \right) + 2\varepsilon.$$

Because $\frac{p}{q} < 1$ and $\varepsilon > 0$ was arbitrary, we conclude that $\mu^* \left(E_p^q \right) = m = 0$. Hence $E = \cup_{p<q \in \mathbb{Q}^+} E_p^q$, the countable union of sets of measure zero, is itself a set of measure zero. Since $D_- f \ge D^+ f$ a.e. on $[a, b]$, we conclude that the increasing function f is differentiable a.e. on $[a, b]$. $\square$

Since functions of bounded variation can be expressed as the difference of increasing functions, we have actually proved a slightly stronger result.

Theorem 83 (BV $\implies$ differentiable a.e). *If $f \in BV\left([a,b]\right)$ then f is differentiable a.e. on $[a,b]$.*

7.9   Some reflections on the gauge integral

It generally takes students a while to become comfortable working with the gauge integral. Once the basic ideas are well grasped, however, the development of the theory of the gauge integral seems to proceed more smoothly than that of the Lebesgue integral which requires the development of the theory of Lebesgue-measurable sets and functions. Our proof of FTC-2 for the Lebesgue integral required five lemmas, three of which were postponed until

Section 7.8 in the current chapter on the gauge integral. Since the three postponed proofs use ideas related to the gauge integral (bounded variation and the Vitali covering theorem) they have been placed at the end of the chapter on the gauge integral. In terms of logical complexity, the gauge integral comes out a winner.

In addition to having a simpler development, the gauge integral strictly extends the Lebesgue integral and has a more satisfying version of FTC-1. When a function is differentiable on an interval, its derivative is always gauge integrable over any subinterval. Why then is the gauge integral relatively unknown compared to the Lebesgue integral? Let me suggest three reasons.

First, Lebesgue published his integral in 1902 while the gauge integral was not introduced until around 1960. Consequently, the Lebesgue integral had over 50 years to become established in the structure of mathematics and to be extended before the gauge integral was developed. In some sense, the ecological niche was already occupied by the time the gauge integral was developed. Second, the mathematical community took note of the Lebesgue integral since it solved a problem that mathematicians were struggling with at the time. In 1960, the goal of extending FTC-1 to all derivatives did not have the same kind of priority. Third and most enduring, the Lebesgue integral can more naturally be extended to nonscalar-valued integrals of functions over non-Euclidean domains. We will see a hint of these extensions in the next chapter.

7.10 Exercises

7.1 Definition and basic examples: filling the gaps

1. Prove that the two definitions of gauge on page 162 are equivalent in the following sense.

   (a) Given any gauge δ of the first type on $[a, b]$, there is a gauge γ of the second type so that any γ-fine tagged partition of $[a, b]$ is also δ-fine.

   (b) Given any gauge γ of the second type on $[a, b]$, there is a gauge δ of the first type such that any δ-fine tagged partition of $[a, b]$ is also γ-fine.

2. Prove that any Riemann integrable function is gauge integrable and that the integrals agree.

3. Prove that if f is gauge integrable over $[a, b]$ then for each $\varepsilon > 0$ there is a gauge γ on $[a, b]$ such that

$$|S_R(f, \mathcal{P}_1) - S_R(f, \mathcal{P}_2)| < \varepsilon$$

for any pair of γ-fine tagged partitions $\mathcal{P}_1$ and $\mathcal{P}_2$ of $[a, b]$.

4. In example 38 (page 165)

   (a) What is it about the calculations related to the function

   $$f(x) = \begin{cases} \frac{1}{\sqrt{x}}, & x > 0 \\ 0, & x \le 0 \end{cases}$$

   that would lead one to think that using a gauge of the form

   $$\gamma(t) = \begin{cases} \left(t \cdot a^2(t), \frac{t}{a^2(t)}\right), & t > 0 \\ \left(-\varepsilon^2, \varepsilon^2\right), & t = 0 \end{cases}$$

   might be an effective line of attack?

   (b) Why was $\frac{2}{\sqrt{x_k} + \sqrt{x_{k-1}}}$ used?

   (c) Explain how $a(t) = 1 - \frac{\varepsilon\sqrt{t}}{4}$ was probably derived.

5. Explain what is meant by the phrase "satisfied vacuously" on page 167.

6. In the proof of Cousin's theorem (page 167),

   (a) Why is S nonempty and bounded?

   (b) Why must there be a $c \in S$ with $c \in \gamma(\beta) \cap (a, \beta)$?

   (c) Why is $\mathcal{P} \cup \{(\beta, [c, \beta])\}$ γ-fine?

7.1 Definition and basic examples: deeper reflections

7. Some texts use an alternative condition for a tagged partition $\mathcal{P} = \{(t_k, I_k)\}$ to be δ-fine: $t_k \in I_k \subset (t_k - \delta(t_k), t_k + \delta(t_k))$ for all tagged intervals. Prove that this condition is equivalent to the condition given in Definition 21 (page 162) in the sense that given a gauge δ_1 using one of the conditions, there is a second gauge δ_2 using the other condition such that any partition that is δ_2-fine is δ_1-fine.

8. How would you change γ in example 38 (page 165) if the interval of integration was $[0, 3]$?

9. In examples 37 and 38 (page 165), the size of $|f(t_k) - f(s_k)|$ was controlled by bounding $|f(x_k) - f(x_{k-1})|$. Under what circumstances can this approach be effective?

10. Let $\{r_k\}$ be the enumeration of the rational numbers in $[0, 1]$ given by $\{0, 1, \frac{1}{2}, \frac{1}{3}, \frac{2}{3}, \frac{1}{4}, \frac{3}{4}, \frac{1}{5}, \frac{2}{5}, \frac{3}{5}, \ldots\}$ and define a gauge δ on $[0, 1]$ by

$$\delta(t) = \begin{cases} \frac{1}{2^k}, & t = r_k \\ 0.1, & t \notin \mathbb{Q}. \end{cases}$$

Construct two δ-fine tagged partitions of $[0, 1]$.

11. Let

$$\gamma(t) = \begin{cases} \left(\frac{1}{4}t, 2t\right), & t > 0 \\ (-0.1, 0.1), & t = 0. \end{cases}$$

Construct two γ-fine tagged partitions of $[0, 1]$.

12. Using the definition of the gauge integral, prove that $f(x) = x^2$ is gauge integrable over $[0, 2]$ and that $\int_0^2 f = \frac{8}{3}$. (Note that $x_{i-1}^2 < \frac{x_i^2 + x_i x_{i-1} + x_{i-1}^2}{3} < x_i^2$.)

13. Prove from the definition that

$$f(x) = \begin{cases} \beta_1, & x = a \\ \alpha, & a < x < b \\ \beta_2, & x = b \end{cases}$$

is gauge integrable on $[a, b]$ and find the value of its integral.

14. Prove from the definition that

$$f(x) = \begin{cases} \alpha_1, & a \le x < c \\ \beta, & x = c \\ \alpha_2, & c < x \le b \end{cases}$$

is gauge integrable on $[a, b]$ and find the value of its integral.

15. Explain why Cousin's theorem (page 167) also implies the existence of a δ-fine partition.

16. Construct an alternative proof of Cousin's theorem (page 167) based on nested subintervals. Suppose that there is no γ-fine tagged partition of $[a, b]$.

    (a) Let c be the midpoint of $[a, b]$. Can both $[a, c]$ and $[c, b]$ have γ-fine tagged partitions?

    (b) Create a nested sequence of closed subintervals whose lengths go to zero. Explain how the intersection point can be used to generate a contradiction.

17. Since $\{\gamma(t)\}_{t\in[a,b]}$ is an open cover of the compact set $[a,b]$, it is tempting to try to prove Cousin's theorem (page 167) by constructing a tagged partition based on a finite subcover $\{\gamma(t_k)\}_{k=1}^n$ of $[a,b]$. What goes wrong when this approach is attempted?

7.2 The art of constructing gauges: filling the gaps

18. Suppose that δ_1 and δ_2 are two gauges on $[a,b]$ with $\delta_1 \le \delta_2$. Prove that any δ_1-fine tagged partition is also δ_2-fine.

19. Suppose that γ_1 and γ_2 are two gauges on $[a,b]$ with $\gamma_1(t) \subseteq \gamma_2(t)$. Prove that any γ_1-fine tagged partition is also γ_2-fine.

20. Given a finite set of points $A = \{a_1, a_2, \ldots, a_m\}$ from $[a,b]$, explain how a gauge γ on $[a,b]$ can be constructed so that the elements of A must be tags of any γ-fine partition. (You may find it convenient to add ghost points $a_0 = a - 1$ and $a_{n+1} = b + 1$.)

21. In example 40 (page 169), why was $\frac{\varepsilon}{2n(|f(t)|+1)}$ used instead of $\frac{\varepsilon}{2n(|f(t)|)}$?

22. Suppose that δ_1 and δ_2 are two gauges on $[a,b]$. Prove that $\delta_3 = \min\{\delta_1, \delta_2\}$ is also a gauge on $[a,b]$ and that any δ_3-fine partition is both δ_1-fine and δ_2-fine.

23. Suppose that γ_1 and γ_2 are two gauges on $[a,b]$. Prove that γ_3 defined by $\gamma_3(t) = \gamma_1(t) \cap \gamma_2(t)$ is also a gauge on $[a,b]$ and that any any γ_3-fine partition is both γ_1-fine and γ_2-fine.

24. Fill in the details to explain why we can assume that the tags of any partition are also division points of the partition without changing the associated Riemann sum.

25. Given a gauge γ on $[a,b]$ and a γ-fine partition $\mathcal{P}$ of $[a,b]$, explain how a second gauge γ' can be combined with tag splitting to force subsequent γ'-fine tagged partitions of $[a,b]$ to be γ-fine refinements of $\mathcal{P}$.

7.2 The art of constructing gauges: deeper reflections

26. Design a gauge δ on $[0,1]$ that that will force 0, $\frac{1}{2}$, and 1 to be tags of any δ-fine partition.

27. Let $\mathcal{P} = \{(0.2, [0, 0.25]), (0.3, [0.25, 0.5]), (0.6, [0.5, 0.75]), (0.8, [0.75, 1])\}$.

Identify an alternative tagged partition $\mathcal{P}^*$ so that all the tags of $\mathcal{P}$ are division points of $\mathcal{P}^*$ and $S_R(f, \mathcal{P}^*) = S_R(f, \mathcal{P})$ for any function f defined on $[0, 1]$.

28. Explain how δ probably was derived in example 42 (page 170).

29. Explain why it is impossible to construct a gauge γ that will prevent a predetermined point from being a tag of a γ-fine partition.

30. Let f be defined on $[a, b]$ and let $\varepsilon > 0$.

(a) Design a gauge δ on $[a, b]$ so that for any δ-fine tagged partition $\mathcal{P} = \{(t_k, J_k)\}_{k=1}^{n}$ of $[a, b]$, the contribution of any one term in the Riemann sum will be less than ε.

(b) Part (a) may appear to bound the value of $S_R(f, \mathcal{P})$. Explain why this is not the case.

31. Find a gauge that shows that any step function of the form

$$f(x) = \begin{cases} \alpha_1, & a \le x < c_1 \\ \beta_1, & x = c_1 \\ \alpha_2, & c_1 < x < c_2 \\ \beta_2, & x = c_2 \\ \alpha_3, & c_2 < x < c_3 \\ \beta_3, & x = c_3 \\ \alpha_4, & c_3 < x < c_4 \\ \cdots \\ \alpha_n, & c_{n-1} < x \le b \end{cases}$$

is gauge integrable on $[a, b]$. What is the value of $g\!\int_a^b f$?

32. Let

$$f(x) = \begin{cases} n, & x = \frac{m}{n} \text{ with } n, m \in \mathbb{N} \text{ and } \frac{m}{n} \text{ in lowest terms,} \\ 0, & x \notin \mathbb{Q}, \end{cases}$$

and let $\varepsilon > 0$ be given. Design a gauge γ so that for any γ-fine partition $\mathcal{P}$ of $[0, 1]$, $S_R(f, \mathcal{P}) < \varepsilon$. What can you conclude?

33. Let f be defined on $[a, b]$ and let $\varepsilon > 0$. Given a countable set S, design a gauge γ on $[a, b]$ so that for any γ-fine tagged partition $\mathcal{P}$ of $[a, b]$, the total contribution to $S_R(f, \mathcal{P})$ by tagged intervals with tags in S is at most ε.

34. Suppose that f is gauge integrable over $[a, b]$ and that $g = f$ except at a countable set of points. Prove that g is gauge integrable and that ${}^g\!\int_a^b g = {}^g\!\int_a^b f$.

35. Suppose that f is gauge integrable over $[a, b]$ and that $g = f$ except on a set Z of measure zero. Suppose further that the difference between f and g is bounded.

    (a) Prove that g is gauge integrable and that ${}^g\!\int_a^b g = {}^g\!\int_a^b f$.
    (b) What goes wrong with your proof when you try to extend it by removing the condition that the difference between f and g is bounded? Can you find a way around this difficulty?

7.3 Basic integrability results: filling the gaps

36. Verify the standard integral properties for the gauge integral.

    (a) **Uniqueness.** The value of the gauge integral is unique (if it exists).
    (b) **Linearity.** Let $c \in \mathbb{R}$. If f and g are gauge integrable over the interval $[a, b]$, then so are $f + g$ and cf. Moreover, ${}^g\!\int_a^b (f + g) = {}^g\!\int_a^b f + {}^g\!\int_a^b g$ and ${}^g\!\int_a^b cf = c\, {}^g\!\int_a^b f$.
    (c) **Monotonicity.** If f and g are gauge integrable over the interval $[a, b]$ with $f(x) \le g(x)$ for $x \in [a, b]$, then ${}^g\!\int_a^b f \le {}^g\!\int_a^b g$.
    (d) **Triangle inequality.** If f and $|f|$ are gauge integrable over $[a, b]$, then $\left| {}^g\!\int_a^b f \right| \le {}^g\!\int_a^b |f|$.

37. Supply a proof for the forward direction of the Cauchy criterion (theorem 50 on page 172): If f is gauge integrable over $[a, b]$ then for each $\varepsilon > 0$ there is a gauge γ on $[a, b]$ such that

$$|S_R(f, \mathcal{P}_1) - S_R(f, \mathcal{P}_2)| < \varepsilon$$

for any pair of γ-fine tagged partitions $\mathcal{P}_1$ and $\mathcal{P}_2$ of $[a, b]$.

38. In the proof of theorem 51 (page 173)

    (a) Why is $|S_R(f, \mathcal{Q}_1) - S_R(f, \mathcal{Q}_2)| = |S_R(f, \mathcal{S}_1) - S_R(f, \mathcal{S}_2)|$?
    (b) How should the proof be modified when $a = c < d < b$?

39. Prove that if the function f is gauge integrable on $[a, c]$ and $[c, b]$ then f is gauge integrable on $[a, b]$ and that ${}^g\!\int_a^b f = {}^g\!\int_a^c f + {}^g\!\int_c^b f$.

(Using gauges γ_a and γ_b on $[a, c]$ and $[c, b]$, define

$$\gamma(t) = \begin{cases} \gamma_a(t) \cap (-\infty, c), & t < c \\ \gamma_a(t) \cap \gamma_b(t), & t = c \\ \gamma_b(t) \cap (c, +\infty), & t > c \end{cases}$$

and split the tag at $t = c$.)

40. Let γ be a gauge on $[a, b]$ and suppose that $\mathcal{P}_1$ is a γ-fine tagged partition of $[a, c]$ where $a < c < b$. Prove that there is another γ-fine tagged partition $\mathcal{P}_2$ of $[c, b]$ so that $\mathcal{P}_1 \cup \mathcal{P}_2$ is a γ-fine partition of $[a, b]$.

41. Use theorem 53 (page 174) to prove Corollary 54 (page 175).

42. In the proof of Henstock's lemma (page 175)

(a) The gauge integrability of f over J_k ensures that there is a gauge γ_k on J_k such that any γ_k-fine tagged partition $\mathcal{P}$ of J_k satisfies

$$\left| S_R(f, \mathcal{P}) - {}^g\!\!\int_{J_k} f \right| < \frac{\xi}{m}.$$

So why is there a γ-fine (as opposed to γ_k-fine) tagged partition $\mathcal{P}_k$ such that

$$\left| S_R(f, \mathcal{P}_k) - {}^g\!\!\int_{J_k} f \right| < \frac{\xi}{m}?$$

(b) Why is

$$\left| S_R(f, \widehat{\mathcal{P}}) - \sum_i {}^g\!\!\int_{I_i} f \right|$$
$$= \left| S_R(f, Q) - {}^g\!\!\int_a^b f - \sum_{k=1}^m \left(S_R(f, P_k) - {}^g\!\!\int_{J_k} f \right) \right|?$$

7.3 Basic integrability results: deeper reflections

43. The Darboux and gauge integrals both have a Cauchy criterion. Though not presented in this text, Cauchy first developed the criterion for the Riemann integral. Of all the integrals we have considered, only the Lebesgue integral does not have a Cauchy criterion. Why not?

44. (Alternative approach for exercise 31) Use exercises 39 and 13 to prove that any step function (see Definition 28, page 180) is gauge integrable. Describe the value of the integral.

45. Show that

$$f(x) = \begin{cases} (-1)^k, & x \in \left(\frac{1}{k+1}, \frac{1}{k}\right] \\ 0, & x = 0 \end{cases}$$

is gauge integrable.

46. Prove that

$$f(x) = \begin{cases} k, & x \in \left(\frac{1}{k+1}, \frac{1}{k}\right] \\ 0, & x = 0 \end{cases}$$

is not gauge integrable.

47. Define f by

$$f(x) = \begin{cases} 1, & x \in [0, 1] \\ -1, & x \in (1, 2] \\ 2, & x \in (2, 13] \end{cases}$$

Define a gauge γ on $[0, 3]$ by

$$\gamma(t) = \begin{cases} (-1, 1), & t \in [0, 1) \\ (1 - \varepsilon, 1 + \varepsilon), & t = 1 \\ (1, 2), & t \in (1, 2) \\ (2 - \varepsilon, 2 + \varepsilon), & t = 2 \\ (2, 3) & t \in (2, 3] \end{cases}$$

and suppose that $\mathcal{P} = \{(t_i, [x_{i-1}, x_i])\}_{i=1}^n$ is a γ-fine tagged partition of $[a, b]$.

(a) Compute $\left| S_R(f, \mathcal{P}) - {}^g\!\int_a^b f \right|$ and give a bound in terms of ε.

(b) Why must 1 be a tag in the partition $\mathcal{P}$?

Let be k the minimal integer for which $t_k = 1$ and set $\mathcal{P}_0 = \{(t_i, [x_{i-1}, x_i])\}_{i=1}^k$ and $\mathcal{P}_3 = \{(t_i, [x_{i-1}, x_i])\}_{i=k+1}^n$.

(c) Compute $\left| S_R(f, \mathcal{P}_0) - \sum_{i=1}^k {}^g\!\int_{x_{i-1}}^{x_i} f \right|$ and identify a bound in terms of ε.

(d) Compute and bound $\left| S_R(f, \mathcal{P}_3) - \sum_{i=k+1}^n {}^g\!\int_{x_{i-1}}^{x_i} f \right|$.

(e) Compute and bound $\sum_i \left| f(t_i) \cdot \Delta x_i - {}^g\!\int_{I_i} f \right|$.

(f) Compute and bound $S_R(|f|, \mathcal{P}_3) - \sum_i \left| {}^g\!\int_{I_i} f \right|$.

(g) Relate the previous parts of this exercise to Henstock's lemma and its corollary.

48. Suppose that f is defined on $[a, b]$ and gauge integrable over $[c, b]$ for all $c \in (a, b)$. Prove that f is gauge integrable over $[a, b]$ if and only

if $\lim_{c \to a+} \int_c^b f$ exists. In this case,

$$ {}^g\!\int_a^b f = \lim_{c \to a+} {}^g\!\int_c^b f. $$

(For one direction, define a gauge on $[a, b]$ by patching together gauges on subintervals like $\left[a + \frac{1}{n}, b\right]$. For the other direction, use theorem 51 (page 173) and Henstock's lemma (page 175) applied to a single interval with tag $t = a$.)

49. Use exercise 48 to give an alternative proof that the function of example 43 (page 170) is gauge integrable over $[0, 1]$ with ${}^g\!\int_0^1 f = \sum_{k=1}^{\infty} \frac{(-1)^k}{k+1} = \ln 2 - 1$. (First show that $\lim_{n \to \infty} {}^g\!\int_{\frac{1}{n}}^1 f = \ln 2 - 1$. Then use the triangle inequality to show that $\lim_{a \to 0+} {}^g\!\int_a^1 f = \ln 2 - 1$.)

50. Suppose that f is a gauge integrable function on $[a, b]$ and that $s > 0$. Define f_s on $[sa, sb]$ by $f_s(x) = f(x/s)$. Prove that f_s is gauge integrable on $[sa, sb]$ with

$$ {}^g\!\int_{sa}^{sb} f_s = s \, {}^g\!\int_a^b f. $$

7.4 Absolute integrability: filling the gaps

51. Let f be a gauge integrable function on $[a, b]$. Prove that f is absolutely gauge integrable over $[a, b]$ if and only if f^+ and f^- are both gauge integrable over $[a, b]$. (Express f^+ as a linear combination of f and $|f|$.)

52. Prove that the Dirichlet function does not have bounded variation.

53. Suppose that $\mathcal{P}$ and $\mathcal{Q}$ are partitions of $[a, b]$ and that $\mathcal{Q}$ is a refinement of $\mathcal{P}$. Prove that

$$ V(f, \mathcal{P}) \leq V(f, \mathcal{Q}). $$

54. Suppose that $f, g \in BV([a, b])$ and that $c \in \mathbb{R}$. Prove that $f + g$ and cf are both in $BV([a, b])$.

55. Suppose that $f \in BV([a, b])$ and that $c \in \mathbb{R}$. Explain why $f \in BV([a, c])$ and $f \in BV([c, b])$.

56. Why is $V_a^b f \geq |f(b) - f(a)|$?

57. Near the end of the proof of theorem 60 (page 181), why does $V_x^y f \geq f(y) - f(x)$ imply that $V_a^x f - f(x)$ is increasing?

58. When proving the first half of theorem 62 (page 183), we only proved $V_a^b F \leq {}^g\!\int_a^b |f|$ rather than $V_a^b F = {}^g\!\int_a^b |f|$. Why is this OK?

59. Why is $\sum_k \left| {}^g\!\int_{x_{k-1}}^{x_k} f \right| \leq \sum_k {}^g\!\int_{x_{k-1}}^{x_k} g$ in the proof of theorem 63 (page 184)?

60. Prove that if f and g are absolutely gauge integrable functions on $[a, b]$ and c is a scalar, then $f + g$ and cf are also absolutely gauge integrable.

61. Suppose that f, g, and h are gauge integrable functions with $f \geq h$ and $g \geq h$. Prove that $\max\{f, g\}$ and $\min\{f, g\}$ are gauge integrable.

7.4 Absolute integrability: deeper reflections

For the next two exercises, take

$$f(x) = \begin{cases} (-1)^k k, & x \in \left(\frac{1}{k+1}, \frac{1}{k}\right] \\ 0, & x = 0. \end{cases}$$

62. Show that given any partition $\mathcal{P}$ of $[0, 1]$ and $B > 0$, it is possible to assign one set of tags so that $S_R(f, \mathcal{P}_+) > B$ and another set of tags so that $S_R(f, \mathcal{P}_-) < -B$. What does this say about the Riemann integrability of f?

63. Show that ${}^{s\text{-}L}\!\int_0^1 f^+$ and ${}^{s\text{-}L}\!\int_0^1 f^-$ are both infinite so that f is not Lebesgue integrable.

64. Prove that

$$f(x) = \begin{cases} (-1)^k, & x \in \left(\frac{1}{k+1}, \frac{1}{k}\right] \\ 0, & x = 0 \end{cases}$$

is not in $BV([0, 1])$.

65. Is

$$f(x) = \begin{cases} \frac{(-1)^k}{k}, & x \in \left(\frac{1}{k+1}, \frac{1}{k}\right] \\ 0, & x = 0 \end{cases}$$

in $BV([0, 1])$?

66. Is

$$f(x) = \begin{cases} \sin\left(\frac{\pi}{x}\right), & x \in (0, 1] \\ 0, & x = 0 \end{cases}$$

in $BV([0, 1])$?

67. Is

$$f(x) = \begin{cases} x^2 \sin\left(\frac{\pi}{x}\right), & x \in (0, 1] \\ 0, & x = 0 \end{cases}$$

in $BV([0, 1])$?

68. Prove that if $f \in BV([a, b])$ then f is bounded on $[a, b]$. Give an example to show that the converse is not true.

69. Give two examples of functions to show that continuous functions need not have bounded variation and that functions of bounded variation need not be continuous.

70. Give an example of a pair of gauge integrable functions f and g for which $h_1 = \max\{f, g\}$ and $h_2 = \min\{f, g\}$ are not gauge integrable.

71. Give an example of a step function f on $[0, 1]$ that has m points of discontinuity, satisfies $|f| \le B$, and attains the maximum value of $V_a^b f = 4mB$. (See the proof of theorem 58 on page 180.)

72. Prove that if $f \in BV([a, b])$, then $|f| \in BV([a, b])$.

73. Prove that if $f, g \in BV([a, b])$, then $\max\{f, g\}$ and $\min\{f, g\}$ have bounded variation on $[a, b]$. (Use exercise 72.)

74. Prove that if f is differentiable with a bounded derivative on $[a, b]$, then $f \in BV([a, b])$.

75. Express $\sin x$ and $\cos^2 x$ on $[0, 2\pi]$ as the difference of two increasing functions.

76. Suppose that $f \in BV([a, b])$.

   (a) Show that f can be written as the difference of two increasing functions in an infinite number of ways.
   (b) Define $g_0 = \frac{1}{2}\left[V_a^x f + f(x)\right]$ and $h_0(x) = \frac{1}{2}\left[V_a^x f - f(x)\right]$ for $x \in [a, b]$. Prove that g_0 and h_0 are increasing functions with $f = g_0 - h_0$ and $V_a^x f = g_0 + h_0$.
   (c) Prove that g_0 and h_0 are most efficient in the sense that any decomposition of $f = g - h$, where g and h are increasing functions, satisfies $g_0(b) - g_0(a) \le g(b) - g(a)$ and $h_0(b) - h_0(a) \le h(b) - h(a)$.

7.5 Convergence theorems: filling the gaps

77. In the proof of the monotone convergence theorem (page 67), we used Henstock's lemma with a sum of the form

$$\sum_{j=1}^{M} \left| \left\{ \sum_{\mathcal{P}_j} \left[f_j(t_k) \Delta x_k - {}^g\!\!\int_{I_k} f_j \right] \right\} \right|.$$

Explain why we could not just use

$$\sum_{k=1}^{n} \left| f_{n(t_k)}(t_k) \Delta x_k - {}^g\!\!\int_{I_k} f_{n(t_k)} \right| \leq \sum_{j=1}^{n} \frac{\varepsilon}{3 \cdot 2^j} < \frac{\varepsilon}{3}.$$

78. Prove theorem 67 (page 186) for monotone decreasing sequences of functions.

79. Prove part (2) of theorem 68 (page 189): If $\{f_k\}$ is a sequence of gauge integrable functions on $[a, b]$ and g is a gauge integrable function on $[a, b]$ satisfying $f_k \geq g$, then $\inf_k f_k$ is gauge integrable.

80. Explain why $\{\underline{f}_k\}$ and $\{\overline{f}_k\}$ in the proof of theorem 69 (page 189) are monotone sequences converging to f.

7.5 Convergence theorems: deeper reflections

81. Suppose that $\{f_k\}$ is a sequence of gauge integrable functions that converges uniformly to f on $[a, b]$. Without using the theorems of Section 5, prove that f is gauge integrable with

$$ {}^g\!\!\int_a^b f = \lim_k {}^g\!\!\int_a^b f_k.$$

82. Let f be a continuous function on $[0, 1]$. Prove that $\lim_k {}^g\!\!\int_0^1 f(x^k) = f(0)$.

83. Suppose that $\{f_k\}$ is a sequence of nonnegative functions on $[0, 1]$.

    (a) Prove that if $\{f_k\}$ is monotone, then $\lim_k {}^g\!\!\int_0^1 f_k = 0$ if and only if $\{f_k\}$ converges to 0 a.e. on $[0, 1]$.
    (b) Find an example of a sequence of nonnegative functions $\{f_k\}$ on $[0, 1]$ such that $\lim_k {}^g\!\!\int_0^1 f_k = 0$ but $\{f_k(x)\}$ does not converge to 0 for any $x \in [0, 1]$. (Work with intervals like $\left[\frac{j}{2^i}, \frac{j+1}{2^i} \right]$.)

84. Suppose that $\{f_k\}$ is a sequence of gauge integrable functions on $[a, b]$ with $n \leq {}^g\!\!\int_a^b f_n$ and that f is a positive function on $[a, b]$ satisfying $f_k \leq f$ for all k. Prove that f is not gauge integrable.

85. State and prove a version of Fatou's lemma (theorem 38, page 139) for the gauge integral.

86. Explain how the hypotheses in the convergence theorems can be modified with the phrase "almost everywhere."

7.6 Fundamental theorems: filling the gaps

87. In the proof of the straddle lemma (page 191), explain why the result holds if $z = u$ or $z = v$.

88. Let
$$f(x) = \begin{cases} x^2 \cos \frac{\pi}{x^2}, & x \neq 0 \\ 0, & x = 0. \end{cases}$$

Prove that
$$f'(x) = \begin{cases} 2x \cos \frac{\pi}{x^2} + \frac{1}{x} 2\pi \sin \frac{\pi}{x^2}, & x \neq 0 \\ 0, & x = 0. \end{cases}$$

In particular, verify that $f'(0) = 0$.

89. Let
$$f(x) = \begin{cases} 2x \cos \frac{\pi}{x^2} + \frac{1}{x} 2\pi \sin \frac{\pi}{x^2}, & x \neq 0 \\ 0, & x = 0. \end{cases}$$

Prove that for any $x > 0$, $\,^L\!\int_0^x f^- = +\infty$.

90. Let f be a function on $[a, b]$ and let E be a subset of $[a, b]$. Prove that $1_E \cdot f \leq f^+$ and $-1_E \cdot f \leq f^-$.

91. In the proof of FTC-1 (theorem 72 on page 193)

    (a) Explain why we can set $F'(z_k) = 0$ for all k.
    (b) Explain why the sum of the differences between the terms associated with z_k in the telescoping sum and the Riemann sum is bounded by $\frac{\varepsilon}{2^{k+1}}$ when $z_k = x_i = t_{i+1} = t_i$.

92. Modify the proof of theorem 10 on page 40 to complete the proof of FTC-2 (page 194).

93. Let I be an interval of length l. Explain why

    (a) the length of $\hat{I}$ is 5 times that of I and
    (b) any point within $2l$ of I is an element of $\hat{I}$.

94. In the proof of the Vitali covering theorem (page 74),

    (a) Why is λ_m finite?
    (b) Why does $\lim_n \lambda_n = 0$?
    (c) Explain why I_M intersects J and the length of I_M is at least $\lambda/2$.

95. Explain how the proof of theorem 75 (page 197) can be modified to show that the set of points x_0 where the left derivative of F, $\lim_{x \to x_0^-} \frac{F(x) - F(x_0)}{x - x_0}$, does not exist or is not equal to $f(x_0)$ has measure zero.

7.6 Fundamental theorems: deeper reflections

96. Use an example to explain the statement that the continuity of F is "the minimal condition on F" on page 191.

97. Let f be a gauge integrable function on $[a, b]$ and define $F(x) = {}^g\!\int_a^x f$ for $x \in [a, b]$. Suppose that f is bounded on an open interval containing $c \in (a, b)$.

    (a) Prove that F is continuous at c.
    (b) Give an example to show that F need not be differentiable at c.

98. Suppose that F and G are two continuous functions on $[a, b]$ such that $F' = G'$ except at a countable number of points. Prove that $F - G$ is constant.

99. Since F in FTC-1 (page 193) is continuous on the compact set $[a, b]$, F is uniformly continuous on $[a, b]$. One can use the uniform continuity of F together with a tight, countable, open-interval cover of the set where F is not differentiable to create a gauge to play the role of γ in the proof. It seems like this proof should work for F that are differentiable except on a set of measure zero. What goes wrong with this approach? Under what circumstances will this tactic succeed?

100. Prove that $V = \left\{ \left[r - \frac{1}{n}, r + \frac{1}{n} \right] : r \in [a, b] \cap \mathbb{Q}, n \in \mathbb{N} \right\}$ is a Vitali covering of $[a, b]$.

101. Suppose that V is a Vitali covering of E and that $E \subset I$ where I is an open interval. Prove that $V_I = \{ J \in V : J \subset I \}$ is also a Vitali covering of E.

102. The Vitali covering of the Cantor set in example 48 (page 195) is uncountable. Find a countable Vitali covering of the Cantor set.

103. Given E and $\{I_k\}$ as constructed in the proof of the Vitali covering theorem, prove that $E \setminus \bigcup_{k=1}^{\infty} I_k$ has measure zero.

104. Give an example of a construction of $\{I_k\}$ as in the proof of the Vitali covering theorem where $E \not\subseteq \bigcup_{k=1}^{\infty} I_k$. ($E = \{\frac{1}{n} : n \in \mathbb{N}\} \cup \{0\}$ is one possible place to start.)

105. Give an example of a function f defined on $[0, 1]$ that is not continuous at $x = \frac{1}{2}$ but for which $F(x) = {}^{g}\!\int_0^x f$ is differentiable at $x = \frac{1}{2}$.

106. FTC-1 (page 193) tells us that when F is a continuous function on $[a, b]$ and has at most a countable set of points of nondifferentiability, then F' is gauge integrable with ${}^{g}\!\int_a^y F' = F(y) - F(a)$ for $y \in [a, b]$. FTC-2 (page 197) tells us that when f is a gauge integrable function on $[a, b]$, the function $F(x) = {}^{g}\!\int_a^x f$ is differentiable with $F' = f$ a.e. on $[a, b]$.

   (a) Give an example to show that we cannot extend FTC-1 (page 193) by allowing the set of points where F is not differentiable to have measure zero instead of being countable.

   (b) Give an example to show that we cannot strengthen FTC-2 (page 197) to conclude that $F(x) = {}^{g}\!\int_a^x f$ has only a countable number of points where F' fails to exist or does not equal f.

7.7 Integral relationships: filling the gaps

107. How are the monotone convergence theorems used to prove that the characteristic function of any open set G is gauge integrable over $[a, b]$ with ${}^{g}\!\int_a^b 1_G = {}^{L}\!\int_a^b 1_G$?

108. In the proof of theorem 76 (page 199)

   (a) How are the monotone convergence theorems used to conclude that

   $$ {}^{g}\!\int_a^b 1_{E^*} = {}^{L}\!\int_a^b 1_{E^*} = \lambda\left(E^* \cap [a, b]\right)? $$

   (b) Fill in the details to conclude that 1_E is gauge integrable with

   $$ {}^{g}\!\int_a^b 1_E = {}^{L}\!\int_a^b 1_E = \lambda\left(E \cap [a, b]\right). $$

109. In the proof of theorem 77 (page 200)

   (a) How does linearity imply that

   $$ {}^g\!\int_a^b \phi_n = {}^L\!\int_a^b \phi_n $$

   for all $n \in \mathbb{N}$?

   (b) How is the monotone convergence theorem applied to conclude that f is gauge integrable over $[a, b]$ and

   $$ {}^g\!\int_a^b f = {}^L\!\int_a^b f? $$

110. Prove theorem 79 (page 201).

7.7 Integral Relationships: deeper reflections

111. Prove directly that the collection of sets whose characteristic function is gauge integrable over every closed and bounded interval $[a, b]$ is a sigma algebra. Explain why every Borel set has a gauge integrable characteristic function.

112. Give an example of a function that is gauge integrable on $[a, b]$ and a set $E \subset [a, b]$ for which

   $$ g(x) = \begin{cases} f(x), & x \in E \\ 0, & x \notin E \end{cases} $$

   is not gauge integrable. What hypotheses can you add so that g is gauge integrable?

113. Give an example of two functions that are not gauge integrable over $[0, 1]$ but whose sum is.

114. What are the advantages and disadvantages of the gauge integral when compared to the Lebesgue integral?

7.8 Loose ends: filling the gaps

115. In the proof of lemma 47 (page 145),

   (a) Why it is sufficient to assume that f is also monotone increasing and prove that $f(a) = f(b)$?

   (b) Is is possible to have degenerate intervals in $\{(y_k, x_{k+1})\}$? Explain.

116. In the proof of theorem 81 (page 205)

   (a) Explain why $(a, b) \setminus \cup_k I_k \subseteq ((a, b) \setminus E) \cup (E \setminus \cup_k I_k)$.
   (b) Explain why $\mathcal{V} = \{[x, t] : t \in (x, b), f(t) - f(x) > B(t - x)\}$
       is a Vitali covering of E.
   (c) Why is $\sum_k (t_k - x_k) = \mu^*(\cup_k I_k) > \frac{m}{2}$?
   (d) Justify each of the inequalities in

   $$f(b) - f(a) \geq \sum_k (f(t_k) - f(x_k)) > B \sum_k (t_k - x_k) > B\frac{m}{2}.$$

117. Suppose that f is a function on $[a, b]$. Define h on $[-b, -a]$ by $h(x) = -f(-x)$. Prove that for any $c \in (a, b)$

   (a) $D_+ f(c) = D_- h(-c)$,
   (b) $D^- f(c) = D^+ h(-c)$, and
   (c) if f is increasing, so is h.

118. Suppose we know that $D_- f \geq D^+ f$ for any for increasing function f. Explain how lemma 82 (page 206) can be used to show that $D_+ f \geq D^- f$ as well.

119. Given a set $E \subseteq (a, b)$ and $\varepsilon > 0$, explain why there is an open set G with $E \subseteq G \subseteq (a, b)$ and $\mu^*(G) < \mu^*(E) + \varepsilon$.

120. In the proof of lemma 46 (page 145)

   (a) Explain why $E = \cup_{p,q \in \mathbb{Q}^+} E_p^q$.
   (b) Explain why G is introduced into the proof.
   (c) Explain why

   $$\mathcal{V} = \{[t, x] : x \in E_p^q, t < x, [t, x] \subset G, f(x) - f(t) < p(x - t)\}$$

   is a Vitali covering of E_p^q.
   (d) Justify each of the inequalities in

   $$\sum_k (f(y_k) - f(x_k)) < p \sum_k (y_k - x_k) < p\mu^*(G) < p(m + \varepsilon).$$

   (e) Why did we remove the endpoints of $\{I_k\}$ from E_p^q instead of defining $F_p^q = E_p^q \cap \cup_k I_k$?
   (f) Why is $\mu^*(F_p^q) > m - \varepsilon$?

7.8 Loose ends: deeper reflections
121. Compute the Dini derivatives of $f(x) = x \sin \frac{1}{x} + 2(x - |x|)$ at $x = 0$.

122. Given $a, b, c, d \in \mathbb{R}$ with $a < b$ and $c < d$, define a function f for which $D_- f(0) = a$, $D^- f(0) = b$, $D_+ f(0) = c$, and $D^+ f(0) = d$.

7.11   References

Gordon, R.A. (1994). *The integrals of Lebesgue, Denjoy, Perron, and Henstock*. Graduate Studies in Mathematics **4**. American Mathematical Society.

Thomson, B.S. (2012). *Theory of the Integral*. ClassicalRealAnalysis.com.

Burk, F.E. (2007). *A Garden of Integrals*. Mathematical Association of America.

DePree, J. and C. Swartz. (1988). *Introduction to Real Analysis*. John Wiley & Sons.

Henstock, R. (1988). *Lectures on the Theory of Integration*. Series in Real Analysis **1**. World Scientific Publishing Company.

Kurzweil, J. (2000). *Henstock-Kurzweil Integration: Its Relation to Topological Vector Spaces*. Series in Real Analysis **7**. World Scientific Publishing Company.

Vitali, G. (1908). Sui gruppi di punti e sulle funzioni di variabili reali, *Atti dell'Accademia delle Scienze di Torino* (in Italian) **43**: 229–246. archive.org/stream/attidellarealeac43real#page/229.

CHAPTER 8

Stieltjes-type Integrals and Extensions

Thus far, all the integrals we have studied have treated intervals according to their length. In 1894, Thomas Joannes Stieltjes published the definition of a new type of integral that did not treat all intervals of the same length equally. Stieltjes' motivation came from the problem of computing moments of inertia. If a mass m lies at a distance l from a fulcrum, then it exerts a force of $\mathbf{g}lm$ on the lever where $\mathbf{g}$ is the constant for the acceleration of gravity. Of course, objects typically do not place all their mass at a single point. Rather, an object's mass is distributed along a range of distances from the fulcrum. The total force exerted on the lever comes from the cumulative contribution of bits of mass at these various distances. In a similar fashion, while a point mass's resistance to rotation is $l^2 m$ where l is the distance from the center of rotation, the resistance of a typical object must be computed from bits of mass spread across an interval.

Instead of focusing on the mass at point x, Stieltjes represented the mass to the left of a point x by $g(x)$. Then $g(x_i) - g(x_{i-1})$ is the mass between x_{i-1} and x_i. Since we will want to allow point masses (weights concentrated at a single point), we need to be a bit more precise. Typically, the function $g(x)$ represents the mass in the interval $(-\infty, x]$ so that $g(x_i) - g(x_{i-1})$ represents the mass in $(x_{i-1}, x_i]$.

Definition 31 (Riemann-Stieltjes Integral). *Let f and g be functions on $[a, b]$ and let $\mathcal{P} = \{(t_i, [x_{i-1}, x_i])\}$ be a tagged partition of $[a, b]$. Then the **Riemann-Stieltjes** sum of f with respect to g and $\mathcal{P}$ is*

$$S_{R\text{-}S}(f, g, \mathcal{P}) = \sum_{i=1}^{n} f(t_i)[g(x_i) - g(x_{i-1})] = \sum_{\mathcal{P}} f \, \Delta g.$$

*The function f is **Riemann-Stieltjes integrable** with respect to g over [a, b] if there is a real value A such that for every ε > 0 there is an associated δ > 0 so that*

$$|S_{R\text{-}S}(f, g, \mathcal{P}) - A| < \varepsilon$$

whenever $\mathcal{P}$ is a tagged partition of [a, b] with $\|\mathcal{P}\| < \delta$. In this case, we write $^{R\text{-}S}\!\int_a^b f \, dg = A$.

Note that when $g(x) = x$, the Riemann-Stieltjes integral becomes the familiar Riemann integral. In Stieltjes' applications referenced above, f would be either $f(x) = \mathbf{g}x$ or $f(x) = x^2$. However, we need not and will not limit ourselves to these options.

8.1   Examples and counterexamples

Example 51 (Constant functions I). Let $f(x) = c$ on $[a, b]$ and let g be an arbitrary function on $[a, b]$. Then for any tagged partition $\mathcal{P} = \{(t_i, [x_{i-1}, x_i])\}$ of $[a, b]$,

$$S_{R\text{-}S}(f, g, \mathcal{P}) = \sum_{i=1}^{n} c \, \Delta g = c \, (g(b) - g(a)).$$

Hence the constant function $f(x) = c$ is Riemann-Stieltjes integrable with respect to g over $[a, b]$ and

$$^{R\text{-}S}\!\int_a^b f \, dg = c \, (g(b) - g(a)).$$

Example 52 (Constant functions II). Let f be an arbitrary function on $[a, b]$ and let $g(x) = c$. If $\mathcal{P} = \{(t_i, [x_{i-1}, x_i])\}$ is any tagged partition of $[a, b]$, then

$$S_{R\text{-}S}(f, g, \mathcal{P}) = \sum_{i=1}^{n} f(t_i)[c - c] = 0.$$

Thus the function f is Riemann-Stieltjes integrable with respect to the constant function g over $[a, b]$ and

$$^{R\text{-}S}\!\int_a^b f \, dg = 0.$$

We can interpret the preceding examples in terms of Stieltjes' original motivating problem. When f is constant, it acts as a change of units factor and $R\text{-}S\int_a^b f\,dg$ reflects the mass over $(a, b]$ expressed in the selected units. If g is constant, then there is no mass in the interval $[a, b]$. Hence the mass, force, and moment of inertia will all be zero.

Example 53 (Dirichlet). The Dirichlet function d is Riemann-Stieltjes integrable with respect to g over $[a, b]$ only if g is constant.

To see why, take any $\alpha < \beta \in [a, b]$ and let $\mathcal{P} = \{[x_{i-1}, x_i]\}$ be a partition of $[a, b]$ that includes α and β as division points, say $x_j = \alpha$ and $x_k = \beta$. Create two tagged partitions from $\mathcal{P}$. Create $\mathcal{P}_1$ by choosing rational numbers for the tags of the subintervals between α and β and irrational tags for the remaining intervals. Construct $\mathcal{P}_2$ using all irrational tags. Then, irrespective of the mesh of $\mathcal{P}$,

$$S_{R\text{-}S}(d, g, \mathcal{P}_1) = \sum_{i=j}^{k} 1 \, [g(x_i) - g(x_{i-1})] = g(\alpha) - g(\beta)$$

and

$$S_{R\text{-}S}(d, g, \mathcal{P}_2) = \sum_{i=1}^{n} 0 \, [g(x_i) - g(x_{i-1})] = 0.$$

If $d(x)$ is integrable, then we can make $S_{R\text{-}S}(d, g, \mathcal{P}_1)$ be arbitrarily close to $S_{R\text{-}S}(d, g, \mathcal{P}_2)$ by choosing $\mathcal{P}$ of sufficiently small mesh. But this means that $g(\alpha) - g(\beta) = 0$. Thus d is Riemann-Stieltjes integrable with respect to g over $[a, b]$ only if g is constant.

Thus far, the examples have not exhibited any behaviors that are contrary to what we would expect from a Riemann integral. The next three examples break this pattern.

Example 54 (Step functions). Let

$$f(x) = \begin{cases} 1, & 0 \le x \le 1 \\ 0, & 1 < x \le 2 \end{cases}$$

and

$$g(x) = \begin{cases} 0, & 0 \le x < 1 \\ 1, & 1 \le x \le 2. \end{cases}$$

By examples 51 and 52, $R\text{-}S\int_0^1 f\,dg = 1$ and $R\text{-}S\int_1^2 f\,dg = 0$. However, f is not Riemann-Stieltjes integrable with respect to g over $[0, 2]$.

To see why, let $\mathcal{P} = \{[x_{i-1}, x_i]\}$ be a partition of $[0, 2]$ that does not include 1 as a division point. Then there is an integer k such that $x_{k-1} < 1 < x_k$. Let $\mathcal{P}_1$ be a tagged partition based on $\mathcal{P}$ where $t_k = x_{k-1}$ and let $\mathcal{P}_2$ be a tagged partition based on $\mathcal{P}$ with $t_k = x_k$. Since $\Delta g = 0$ for all but the kth interval, $S_{R\text{-}S}(f, g, \mathcal{P}_1) = 1$ and $S_{R\text{-}S}(f, g, \mathcal{P}_2) = 0$. As the mesh of $\mathcal{P}$ can be arbitrarily small, f cannot be Riemann-Stieltjes integrable with respect to g over $[0, 2]$.

Contrary to the situation with the Riemann integral, Riemann-Stieltjes integrability over $[a, b]$ and $[b, c]$ is not sufficient to conclude Riemann-Stieltjes integrability over $[a, c]$. Moreover, while bounded functions are Riemann integrable as long as the set of discontinuities has measure zero, example 54 shows that discontinuity at even a single point can render a function non-Riemann-Stieltjes integrable. Even monotonicity can fail for Riemann-Stieltjes integrals.

Example 55 (Not monotone). Let $f_1(x) = 1$, $f_2(x) = 2$, and

$$g(x) = \begin{cases} 1, & 0 \leq x \leq 1 \\ 0, & 1 < x \leq 2. \end{cases}$$

Then $f_1 < f_2$, but $^{R\text{-}S}\!\int_0^2 f_1 \, dg = -1$ while $^{R\text{-}S}\!\int_0^2 f_2 \, dg = -2$.

One would not necessarily expect the fundamental theorem of calculus to apply to Riemann-Stieltjes integrals, but even more basic properties fail. The function $F(x) = {}^{R\text{-}S}\!\int_a^x f \, dg$ need not be continuous.

Example 56 (F not continuous). Let $f(x) = 1$ and

$$g(x) = \begin{cases} 0, & 0 \leq x \leq 1 \\ 1, & 1 < x \leq 2. \end{cases}$$

Then by example 51,

$$F(x) = {}^{R\text{-}S}\!\int_0^x f \, dg = g(x)$$

for $x \in [0, 2]$. Thus F is not continuous at $x = 1$.

8.2   Basic integrability theorems

In light of the previous examples, it seems that we are more than ready for some general positive results. We begin with a pair of theorems that should seem very familiar.

Theorem 84 (Cauchy criterion). *Given functions f and g on $[a, b]$, f is Riemann-Stieltjes integrable with respect to g over $[a, b]$ if and only if given any $\varepsilon > 0$ there is a $\delta > 0$ such that*

$$|S_{R\text{-}S}(f, g, \mathcal{P}_1) - S_{R\text{-}S}(f, g, \mathcal{P}_2)| < \varepsilon$$

whenever $\mathcal{P}_1$ and $\mathcal{P}_2$ are tagged partitions of $[a, b]$ whose meshes are both less than δ.

Proof. The proof of this theorem follows the same lines as previous Cauchy criteria and is left as an exercise. □

The Cauchy criterion allows us to prove integrability over subintervals even though example 54 (page 229) shows that the converse is false.

Theorem 85 (Subintervals). *If f and g are functions on $[a, b]$ with f Riemann-Stieltjes integrable with respect to g on $[a, b]$, then f is Riemann-Stieltjes integrable with respect to g over $[c, d]$ for any $[c, d] \subset [a, b]$.*

Proof. This is a straightforward exercise in the application of the Cauchy criterion. □

We can also prove a fairly general existence result for the Riemann-Stieltjes integral.

Theorem 86 (Integrability conditions). *Let f be a continuous function on $[a, b]$ and suppose that g is increasing on $[a, b]$. Then f is Riemann-Stieltjes integrable with respect to g over $[a, b]$.*

Proof. Fix $\varepsilon > 0$ and set $B = g(b) - g(a) + 1$. Since f is uniformly continuous on $[a, b]$ we can find a $\delta > 0$ so that

$$|f(x) - f(y)| < \frac{\varepsilon}{2B}$$

for all $x, y \in [a, b]$ satisfying $|x - y| < \delta$. Suppose that $\mathcal{P}_1 = \{(t_i, [x_{i-1}, x_i])\}$ and $\mathcal{P}_2$ are two tagged partitions of $[a, b]$ with mesh less than δ. Then $\mathcal{P}_3 = \mathcal{P}_1 \cup \mathcal{P}_2$ is a common refinement to which we assign midpoint tags.[1] Define the step functions $\phi_{\underline{\mathcal{P}_1}} = f(a) \cdot 1_{\{a\}} + \sum_{\mathcal{P}} \inf_{(x_{i-1}, x_i]} f \cdot 1_{(x_{i-1}, x_i]}$ and $\phi_{\overline{\mathcal{P}_1}} = f(a) \cdot 1_{\{a\}} + \sum_{\mathcal{P}} \sup_{(x_{i-1}, x_i]} f \cdot 1_{(x_{i-1}, x_i]}$.
Because g is increasing,

$$S_{R\text{-}S}(\phi_{\underline{\mathcal{P}_1}}, g, \mathcal{P}_1) \leq S_{R\text{-}S}(f, g, \mathcal{P}_1) \leq S_{R\text{-}S}(\phi_{\overline{\mathcal{P}_1}}, g, \mathcal{P}_1)$$

---

[1] The particular choice of tags does not really matter.

and

$$S_{R\text{-}S}\left(\phi_{\underline{\mathcal{P}_1}}, g, \mathcal{P}_1\right) = S_{R\text{-}S}\left(\phi_{\underline{\mathcal{P}_1}}, g, \mathcal{P}_3\right)$$
$$\leq S_{R\text{-}S}\left(f, g, \mathcal{P}_3\right)$$
$$\leq S_{R\text{-}S}\left(\phi_{\overline{\mathcal{P}_1}}, g, \mathcal{P}_1\right).$$

Moreover,

$$S_{R\text{-}S}\left(\phi_{\overline{\mathcal{P}_1}}, g, \mathcal{P}_1\right) - S_{R\text{-}S}\left(\phi_{\underline{\mathcal{P}_1}}, g, \mathcal{P}_1\right)$$
$$= \sum_{\mathcal{P}_1}\left(\phi_{\overline{\mathcal{P}_1}}(t_i) - \phi_{\underline{\mathcal{P}_1}}(t_i)\right)\left(g(x_i) - g(x_{i-1})\right)$$
$$\leq \sum_{\mathcal{P}_1}\frac{\varepsilon}{2B}\left(g(x_i) - g(x_{i-1})\right)$$
$$= \frac{\varepsilon}{2B}\left(g(b) - g(a)\right) < \frac{\varepsilon}{2}.$$

Hence

$$\left|S_{R\text{-}S}\left(f, g, \mathcal{P}_1\right) - S_{R\text{-}S}\left(f, g, \mathcal{P}_3\right)\right| < \frac{\varepsilon}{2}.$$

Similarly,

$$\left|S_{R\text{-}S}\left(f, g, \mathcal{P}_2\right) - S_{R\text{-}S}\left(f, g, \mathcal{P}_3\right)\right| < \frac{\varepsilon}{2}.$$

Thus

$$\left|S_{R\text{-}S}\left(f, g, \mathcal{P}_1\right) - S_{R\text{-}S}\left(f, g, \mathcal{P}_2\right)\right| < \varepsilon.$$

Integrability follows from the Cauchy criterion.   □

By linearity, theorem 86 can immediately be extended to the case where g is of bounded variation. This extension is quite useful in the context of theorem 89 (page 236) below. On the other hand, example 54 (page 229) serves as a warning that the restriction that f be continuous cannot be removed easily. The restriction can be loosened, but to do so requires additional theory that we have not yet developed. By the time we have developed the theory, the effort to extend theorem 86 will be superseded by Lebesgue-like results.

8.3   Evaluation theorems

While the theorems of the previous section allow us to determine that a Riemann-Stieltjes integral exists, they provide little help in evaluating an integral. For the Riemann integral, FTC-1 is the basic evaluation tool. There is no such theorem for the Riemann-Stieltjes integral.

One typically encounters two types of Riemann-Stieltjes integrals corresponding to the two basic types of probability distributions: continuous and discrete. The following two theorems provide useful tools for their evaluation.

Theorem 87 (Continuous distributions). *Suppose that f and g are functions on $[a, b]$ where f is continuous and g is differentiable with g' being Riemann integrable over $[a, b]$. Then f is Riemann-Stieltjes integrable with respect to g over $[a, b]$ and*

$$R\text{-}S\int_a^b f \, dg = {}^R\!\int_a^b fg'.$$

Proof. Since g' is Riemann integrable, g' and so fg' are bounded and continuous a.e. Thus fg' is Riemann integrable. Since g' is bounded we can select a real value B so that $|g'| < B$.

Fix $\varepsilon > 0$ and choose $\delta > 0$ so that $|f(x) - f(y)| < \frac{\varepsilon}{B(b-a)}$ whenever $x, y \in [a, b]$ with $|x - y| < \delta$. Let $\mathcal{P} = \{(t_i, I_i)\}$ be a tagged partition of $[a, b]$ with mesh less than δ. By FTC-1 for the Riemann integral, $\Delta g_i = {}^R\!\int_{I_i} g'$ so that

$$\left| S_{R\text{-}S}(f, g, \mathcal{P}) - {}^R\!\int_a^b fg' \right| = \left| \sum_{\mathcal{P}} f \, \Delta g - \sum_{\mathcal{P}} {}^R\!\int_{I_i} fg' \right|$$

$$= \left| \sum_{\mathcal{P}} {}^R\!\int_{I_i} [f(t_i) - f] g' \right|$$

$$< \sum_{\mathcal{P}} {}^R\!\int_{I_i} \frac{\varepsilon}{B(b-a)} B$$

$$= \frac{\varepsilon}{b-a} \sum_{\mathcal{P}} \Delta x_i = \varepsilon.$$

Hence f is Riemann-Stieltjes integrable with respect to g over $[a, b]$ and

$$R\text{-}S\int_a^b f \, dg = {}^R\!\int_a^b fg'. \qquad \square$$

Theorem 87 can be strengthened by allowing g to have points of nondifferentiability. Assuming g' is differentiable a.e. and Riemann integrable when set to zero at the points of nondifferentiability, the theorem remains valid. (See exercise 25.)

Before stating the evaluation theorem for the discrete case, we introduce some additional notation. Let f be defined on $[a, b]$ and let $c \in [a, b]$. Then

$$f\left(c^+\right) = \lim_{x \to c^+} f(x) \text{ and } f\left(c^-\right) = \lim_{x \to c^-} f(x)$$

where it is understood that

$$f\left(a^-\right) = f(a) \text{ and } f\left(b^+\right) = f(b).$$

These values are always defined for monotone functions.

Theorem 88 (Discrete distributions). *Let g be a step function on $[a, b]$ with discontinuities at $\{c_i\}_{k=1}^{n}$ and suppose that f is a continuous function on $[a, b]$. Then f is Riemann-Stieltjes integrable with respect to g over $[a, b]$ and*

$$\text{R-S} \int_a^b f \, dg = \sum_{i=1}^{n} f(c_i)\left[g\left(c_i^+\right) - g\left(c_i^-\right)\right].$$

Proof. Fix $\varepsilon > 0$ and set $M = \max_{x \in [a,b]} |g(x)|$. By the uniform continuity of f, there is a $\delta_0 > 0$ so that $|f(x) - f(y)| < \frac{\varepsilon}{4nM}$ whenever $x, y \in [a, b]$ with $|x - y| < \delta_0$. Take $\delta_1 = \min_{1 \le i, j \le n}\left\{|c_i - c_j|\right\}$ and set $\delta = \min\{\delta_0, \delta_1/2\}$. Note that $g(x) = g\left(c_i^-\right)$ for $x \in [a, b] \cap (c_i - \delta, c_i)$ and $g(x) = g\left(c_i^+\right)$ for $x \in [a, b] \cap (c_i, c_i + \delta)$.

Now suppose that $\mathcal{P} = \{(t_i, [x_{i-1}, x_i])\}$ is a tagged partition of $[a, b]$ with mesh less than δ. Since the mesh of $\mathcal{P}$ is less than δ_1, any interval $\left[x_{j-1}, x_j\right]$ can contain at most one point where g is discontinuous and any point of discontinuity is contained in at most two intervals from $\mathcal{P}$. If $\left[x_{j-1}, x_j\right]$ contains no point of discontinuity, then $f(t_j)\left[g(x_j) - g(x_{j-1})\right] = 0$. If $x_{j-1} < c_k < x_j$ then

$$\left| f(t_j)\left[g(x_j) - g(x_{j-1})\right] - f(c_k)\left[g\left(c_k^+\right) - g\left(c_k^-\right)\right] \right|$$

$$= \left| f(t_j)\left[g\left(c_k^+\right) - g\left(c_k^-\right)\right] - f(c_k)\left[g\left(c_k^+\right) - g\left(c_k^-\right)\right] \right|$$

$$= \left| f(t_j) - f(c_k) \right| \left| g\left(c_k^+\right) - g\left(c_k^-\right) \right|$$

$$< \frac{\varepsilon}{4nM} 2M = \frac{\varepsilon}{2n}.$$

If c_k is a division point with $x_{j-1} < c_k = x_j < x_{j+1}$, then

$$\left| f(t_j)[g(x_j) - g(x_{j-1})] + f(t_{j+1})[g(x_{j+1}) - g(x_j)] \right.$$
$$\left. - f(c_k)[g(c_k^+) - g(c_k^-)] \right|$$
$$= \left| f(t_j)[g(c_k) - g(c_k^-)] \right.$$
$$+ f(t_{j+1})[g(c_k^+) - g(c_k)]$$
$$\left. - f(c_k)[g(c_k^+) - g(c_k^-)] \right|$$
$$\leq \left| f(t_j) - f(c_k) \right| \left| g(c_k) - g(c_k^-) \right|$$
$$+ \left| f(t_{j+1}) - f(c_k) \right| \left| g(c_k^+) - g(c_k) \right|$$
$$< \frac{\varepsilon}{4nM} 2M + \frac{\varepsilon}{4nM} 2M = \frac{\varepsilon}{n}.$$

The cases where $c_k = x_0$ or $c_k = x_n$ are similar but simpler. Hence

$$\left| S_{R\text{-}S}(f, g, \mathcal{P}) - \sum_{i=1}^{n} f(c_i) \left[g(c_i^+) - g(c_i^-) \right] \right| < \varepsilon$$

so that f is Riemann-Stieltjes integrable with respect to g over $[a, b]$ with

$$^{\text{R-S}}\!\!\int_a^b f \, dg = \sum_{i=1}^{n} f(c_i) \left[g(c_i^+) - g(c_i^-) \right]. \qquad \square$$

Theorem 92 (page 239) in the next section can be used to extend theorem 88 to conclude that

$$^{\text{R-S}}\!\!\int_a^b f \, dg = \sum_k f(c_k) \left[g(c_k^+) - g(c_k^-) \right]$$

when g has a countable number of discontinuities $\{c_k\}$ satisfying $\sum_k \left| g(c_k^+) - g(c_k^-) \right| < \infty$. This situation is quite common when working with discrete probability distributions.

Linearity allows us to apply the previous theorems in mixed situations.

Example 57. Compute $^{\text{R-S}}\!\!\int_0^2 x^2 \, dg$ where

$$g(x) = \begin{cases} 0, & x = 0 \\ 3x + 1, & 0 < x < 1 \\ 1, & x = 1 \\ x^3, & 1 < x < 2 \\ 9, & x = 2. \end{cases}$$

First note that we can write $g = g_1 + g_2$ where

$$g_1(x) = \begin{cases} 3x, & 0 \le x \le 1 \\ x^3 + 2, & 1 < x \le 2 \end{cases}$$

and

$$g_2(x) = \begin{cases} 0, & x = 0 \\ 1, & 0 < x < 1 \\ -2, & 1 \le x < 2 \\ -1, & x = 2. \end{cases}$$

Then, by theorems 86 and 87 (pages 231 and 233),

$$^{R\text{-}S}\!\!\int_0^2 x^2 \, dg_1 = {}^{R}\!\!\int_0^1 x^2 \, 3 \, dx + {}^{R}\!\!\int_1^2 x^2 \, 3x^2 \, dx = 1 + \frac{93}{5} = \frac{98}{5}$$

and, by theorem 88,

$$^{R\text{-}S}\!\!\int_0^2 x^2 \, dg_2 = 0 \cdot 1 + 1 \cdot (-3) + 4 \cdot 1 = 1.$$

So by linearity, $^{R\text{-}S}\!\!\int_0^2 x^2 \, dg = \frac{103}{5}$.

On occasion, it is simpler to consider the integral $^{R\text{-}S}\!\!\int_a^b g \, df$ instead of the integral $^{R\text{-}S}\!\!\int_a^b f \, dg$. The following theorem (which should have a very familiar appearance) allows us to interchange the roles of the two functions in a Riemann-Stieltjes integral.

Theorem 89 (Integration by parts). *Suppose that f and g are bounded functions on $[a, b]$ with no common discontinuities. If f is Riemann-Stieltjes integrable with respect to g over $[a, b]$ then g is Riemann-Stieltjes integrable with respect to f over $[a, b]$ and*

$$^{R\text{-}S}\!\!\int_a^b g \, df = f(b) g(b) - f(a) g(a) - {}^{R\text{-}S}\!\!\int_a^b f \, dg.$$

Proof. Let $\varepsilon > 0$ be given. By assumption, there is a $\delta > 0$ so that

$$\left| S_{R\text{-}S}(f, g, \mathcal{Q}) - {}^{R\text{-}S}\!\!\int_a^b f \, dg \right| < \varepsilon$$

for any tagged partition $\mathcal{Q}$ of $[a, b]$ with mesh less than δ. Now suppose that $\mathcal{P} = \{(t_i, [x_{i-1}, x_i])\}_{i=1}^n$ is a tagged partition of $[a, b]$ with mesh less than

$\delta/2$. Define $t_0 = a$ and $t_{n+1} = b$. Then $\mathcal{P}^* = \{(x_{k-1}, [t_{k-1}, t_k])\}_{k=1}^{n+1}$ is a tagged partition of $[a, b]$ with mesh less than δ. Hence

$$\left| S_{R\text{-}S}(g, f, \mathcal{P}) - \left(f(b) g(b) - f(a) g(a) - {}^{R\text{-}S}\!\!\int_a^b f \, dg \right) \right|$$

$$= \left| \sum_{i=1}^n g(t_i) f(x_i) - \sum_{i=1}^n g(t_i) f(x_{i-1}) \right.$$

$$\left. - f(x_n) g(t_{n+1}) + f(x_0) g(t_0) + {}^{R\text{-}S}\!\!\int_a^b f \, dg \right|$$

$$= \left| {}^{R\text{-}S}\!\!\int_a^b f \, dg - \sum_{i=1}^{n+1} f(x_{i-1}) [g(t_i) - g(t_{i-1})] \right|$$

$$= \left| {}^{R\text{-}S}\!\!\int_a^b f \, dg - S_{R\text{-}S}(f, g, \mathcal{P}^*) \right| < \varepsilon.$$

Hence g is Riemann-Stieltjes integrable with respect to f over $[a, b]$ with

$$^{R\text{-}S}\!\!\int_a^b g \, df = f(b) g(b) - f(a) g(a) - {}^{R\text{-}S}\!\!\int_a^b f \, dg. \qquad \square$$

Example 58. By theorem 87 (page 233),

$$^{R\text{-}S}\!\!\int_0^3 x^2 \, dx^2 = {}^{R}\!\!\int_0^3 x^2 2x \, dx = \frac{81}{2}.$$

By theorem 89,

$$^{R\text{-}S}\!\!\int_0^3 x^2 \, dx^2 = 9 \cdot 9 - 0 \cdot 0 - {}^{R\text{-}S}\!\!\int_0^3 x^2 \, dx^2.$$

Solving, we again find that

$$^{R\text{-}S}\!\!\int_0^3 x^2 \, dx^2 = \frac{81}{2}.$$

8.4 Convergence theorems

We will prove one convergence theorem for the Riemann-Stieltjes integral. In preparation, we establish two technical lemmas. Notice the alignment of the hypotheses for these theorems with the integrability conditions from Section 2.

Lemma 90. *Suppose that f is a continuous function on $[a, b]$ and that $g \in$*
$BV([a,b])$.[2] *Then*

$$\left| {}^{R\text{-}S}\!\!\int_a^b f\,dg \right| \le \max_{[a,b]} f \cdot V_a^b g.$$

Proof. Exercise 31.   □

Lemma 91. *Suppose that f is a continuous function on $[a, b]$ and $\varepsilon, \delta > 0$*
with $|f(x) - f(y)| < \varepsilon$ for all $x, y \in [a, b]$ satisfying $|x - y| < \delta$. Further
suppose that $g \in BV([a,b])$. If $\mathcal{P} = \{(t_i, [x_{i-1}, x_i])\}$ is a tagged partition
of $[a, b]$ with mesh less than δ, then

$$\left| S_{R\text{-}S}(f, g, \mathcal{P}) - {}^{R\text{-}S}\!\!\int_a^b f\,dg \right| < \varepsilon V_a^b g.$$

Proof. First note that $\max_{x \in [x_{i-1}, x_i]} |f(t_i) - f(x)| < \varepsilon$ and, by example
51 (page 228),

$$f(t_i)\,[g(x_i) - g(x_{i-1})] = {}^{R\text{-}S}\!\!\int_{x_{i-1}}^{x_i} f(t_i)\,dg.$$

Thus, by lemma 90 (page 238) and property (2) of theorem 60 (page 181),

$$\left| S_{R\text{-}S}(f, g, \mathcal{P}) - {}^{R\text{-}S}\!\!\int_a^b f\,dg \right| = \left| \sum_i {}^{R\text{-}S}\!\!\int_{x_{i-1}}^{x_i} (f(t_i) - f)\,dg \right|$$

$$< \sum_i \varepsilon V_{x_{i-1}}^{x_i} g$$

$$= \varepsilon V_a^b g. \qquad \square$$

Since the Riemann-Stieltjes integral involves two functions, it is natural
that convergence theorems consider two sequences of functions. Understanding the consequences of the convergence of $\{g_n\}$ is critical in the study of
probability theory. In particular, the central limit theorem, arguably the most
important theorem in probability and statistics, concerns just the sort of convergence of $\{g_n\}$ identified in the following theorem.[3]

---

[2] Recall that $V(f, \mathcal{P}) = \sum_k |f(x_k) - f(x_{k-1})|$ where $\mathcal{P} = \{[x_{k-1}, x_k]\}$ is a partition
of $[a, b]$, $V_a^b f = \sup_{\mathcal{P}} V(f, \mathcal{P})$, and $BV([a,b])$ is the set of all functions on $[a, b]$ for
which $V_a^b f$ is finite.

[3] The relevant type of convergence is convergence in distribution. The sequence $\{g_n\}$ **converges in distribution** to g if $\{g_n\}$ converges pointwise to g at all points where g is continuous. In a probability theory context, g is monotone increasing so g is continuous except at
a countable number of points and all the functions $\{g_n\}$ and g have a variation of 1.

Theorem 92. *Suppose that $\{f_n\}$ is a sequence of continuous functions that converges uniformly to f on $[a, b]$. Suppose further that $\{g_n\}$ is a sequence of functions converging to g on a dense subset S of $[a, b]$ that includes a and b. If there is a uniform bound M on the variations of $\{g_n\}$ and g then*

$$\lim_{n \to \infty} {}^{R\text{-}S}\!\!\int_a^b f_n \, dg_n = {}^{R\text{-}S}\!\!\int_a^b f \, dg.$$

Proof. As f is continuous on $[a, b]$, it is bounded by some value $|f| < B$.

Let $\varepsilon > 0$ and choose $\delta > 0$ so that $|f(x) - f(y)| < \frac{\varepsilon}{4M}$ for all $x, y \in [a, b]$ satisfying $|x - y| < \delta$. Let $\mathcal{P} = \{(t_i, [x_{i-1}, x_i])\}_{i=1}^m$ be a tagged partition of $[a, b]$ with mesh less than δ whose division points are selected from S. Because $\{g_n\}$ converges to g on S, we can find a natural number N_g so that $|g_n(x_i) - g(x_i)| < \frac{\varepsilon}{4mB}$ for all $n \geq N_g$ and $1 \leq i \leq m$. Since $\{f_n\}$ converges uniformly to f we can find a natural number N_f so that $|f_n - f| < \frac{\varepsilon}{4M}$ for all $n \geq N_f$. Set $N = \max\{N_f, N_g\}$ and suppose that $n \geq N$.

By the triangle inequality and lemma 90 (page 238),

$$\left| {}^{R\text{-}S}\!\!\int_a^b f_n \, dg_n - {}^{R\text{-}S}\!\!\int_a^b f \, dg \right|$$

$$\leq \left| {}^{R\text{-}S}\!\!\int_a^b (f_n - f) \, dg_n \right| + \left| {}^{R\text{-}S}\!\!\int_a^b f \, d(g_n - g) \right|$$

$$\leq \frac{\varepsilon}{4M} M + \left| {}^{R\text{-}S}\!\!\int_a^b f \, d(g_n - g) \right|.$$

For the second term, we use the triangle inequality and lemma 91 (page 238) to conclude that

$$\left| {}^{R\text{-}S}\!\!\int_a^b f \, d(g_n - g) \right| \leq \left| {}^{R\text{-}S}\!\!\int_a^b f \, dg_n - S_{R\text{-}S}(f, g_n, \mathcal{P}) \right|$$

$$+ |S_{R\text{-}S}(f, g_n - g, \mathcal{P})|$$

$$+ \left| {}^{R\text{-}S}\!\!\int_a^b f \, dg - S_{R\text{-}S}(f, g, \mathcal{P}) \right|$$

$$< \frac{\varepsilon}{4M} V_a^b g_n + \sum_{i=1}^m B \frac{\varepsilon}{4mB} + \frac{\varepsilon}{4M} V_a^b g < \frac{3}{4} \varepsilon.$$

Hence

$$\lim_{n \to \infty} {}^{R\text{-}S}\!\!\int_a^b f_n \, dg_n = {}^{R\text{-}S}\!\!\int_a^b f \, dg. \qquad \square$$

Example 59. Compute $\lim_{n\to\infty} {}^{R\text{-}S}\!\int_0^1 \arctan nx \, dx^n$.

Note that $\{x^n\}$ converges pointwise to

$$g(x) = \begin{cases} 0, & 0 \le x < 1 \\ 1, & x = 1 \end{cases}$$

on $[0, 1]$ and $V_0^1 x^n = V_0^1 g = 1$. Unfortunately, $\{\arctan nx\}_n$ does not converge uniformly on $[0, 1]$. However, it does converge uniformly to $\frac{\pi}{2}$ on $\left[\frac{1}{2}, 1\right]$ and is uniformly bounded by $\frac{\pi}{2}$ on $\left[0, \frac{1}{2}\right]$. Thus

$$\lim_{n\to\infty} {}^{R\text{-}S}\!\int_{\frac{1}{2}}^1 \arctan nx \, dx^n = {}^{R\text{-}S}\!\int_{\frac{1}{2}}^1 \frac{\pi}{2} \, dg = \frac{\pi}{2}$$

and

$$0 \le \lim_{n\to\infty} {}^{R\text{-}S}\!\int_0^{\frac{1}{2}} \arctan nx \, dx^n \le \lim_{n\to\infty} {}^{R\text{-}S}\!\int_0^{\frac{1}{2}} \frac{\pi}{2} \, dx^n = {}^{R\text{-}S}\!\int_0^{\frac{1}{2}} \frac{\pi}{2} \, dg = 0.$$

Hence

$$\lim_{n\to\infty} {}^{R\text{-}S}\!\int_0^1 \arctan nx \, dx^n = \frac{\pi}{2}.$$

Example 60. Set

$$f_n(x) = \begin{cases} 0, & x < 0 \\ \frac{1}{2}, & 0 \le x < 1 + \frac{1}{n} \\ 1, & 1 + \frac{1}{n} \le x \end{cases}$$

and evaluate $\lim_{n\to\infty} {}^{R\text{-}S}\!\int_{-1}^2 f_n(x) \, dx^5$.

In this case, the functions $\{f_n\}$ are not continuous and $\{f_n\}$ converges pointwise but not uniformly on $[0, 1]$ to

$$f(x) = \begin{cases} 0, & x < 0 \\ \frac{1}{2}, & 0 \le x \le 1 \\ 1, & 1 < x. \end{cases}$$

We can still use theorem 92 by applying integration by parts.

$$\lim_{n\to\infty} {}^{R\text{-}S}\!\int_{-1}^2 f_n(x) \, dx^5 = \lim_{n\to\infty} \left[1 \cdot 2^5 - 0 \cdot (-1)^5 - {}^{R\text{-}S}\!\int_{-1}^2 x^5 \, df_n \right]$$

$$= 2^5 - {}^{R\text{-}S}\!\int_{-1}^2 x^5 \, df = 2^5 - \left\{ 0 \cdot \frac{1}{2} + 1 \cdot \frac{1}{2} \right\} = \frac{63}{2}$$

by theorem 88 (page 234).

Observe that if one thinks of the f in the previous example as a weighting function, then it places a weight of $\frac{1}{2}$ at $x = 0$ and at $x = 1$. In this case, it is more common to work with the function

$$f^*(x) = \begin{cases} 0, & x < 0, \\ \frac{1}{2}, & 0 \le x < 1, \\ 1, & 1 \le x. \end{cases}$$

While $\{f_n\}$ does not converge to f^* at $x = \frac{1}{2}$, $\{f_n\}$ does converge to f^* on a dense set that includes -1 and 2. Thus we still have

$$\lim_{n \to \infty} {}^{R\text{-}S}\!\!\int_{-1}^{2} f_n(x)\, dx^5 = 2^5 - {}^{R\text{-}S}\!\!\int_{1}^{2} x^5\, df^* = 2^5 - \frac{1}{2}$$

even though $\{f_n\}$ fails to converge to f^* on a set of weight $\frac{1}{2}$. While this situation initially seems rather odd, it makes a bit more sense if we think of a sequence of random variables $\{X_n\}$ where X_n takes on the values 0 or $1 + \frac{1}{n}$ with probability $\frac{1}{2}$ each. The random variables converge in a natural sense[4] to a random variable that takes on the values 0 and 1 with probability $\frac{1}{2}$ each.

8.5 Connecting to measure theory

The fact that f fails to be integrable with respect to g over $[a, b]$ when f and g have a common point of discontinuity is rather awkward. This runs counter to all our previous experience where a few discontinuities could be ignored. One way to remedy this difficulty would be to develop a theory of the gauge-Stieltjes integral (see exercise 10). We will take a different path.

The fundamental idea underlying the Riemann-Stieltjes integral is that the interval $(s, t]$ is treated according to its weight, $w\,((s, t]) = g(t) - g(s)$, rather than its length, $l\,((s, t]) = t - s$. This statement should remind you of the Lebesgue integral. In fact, we can develop a general theory of measure very similar to that of Lebesgue measure using w instead of l. For this development, we assume that g is monotone increasing and right-continuous ($g\,(x^+) = g(x)$ for x in the domain of g). For the rest of this section, we will assume that g satisfies these conditions without reiteration in subsequent theorems and lemmas.

If we are to mimic the development of Lebesgue measure, we need to know that the weight function $w\,((s, t]) = g(t) - g(s)$ shares some fundamental properties with the length function $l\,((s, t]) = t - s$.

---

[4] They converge in distribution.

Theorem 93 (Properties of w). *Under the assumptions of this section,*

1. $w\left((a,b]\right) = w\left((a,c]\right) + w\left((c,b]\right)$ *whenever* $a < c < b$.

2. *If* $(a,b] \subseteq \cup_{k=1}^{n} (a_k, b_k]$, *then* $w\left((a,b]\right) \leq \sum_k w\left((a_k, b_k]\right)$.

3. *If the intervals in* $\{(a_k, b_k]\}_{k=1}^{n}$ *are disjoint and* $\cup_k (a_k, b_k] \subseteq (a,b]$, *then* $\sum_k w\left((a_k, b_k]\right) \leq w\left((a,b]\right)$.

4. *If* $(a,b] \subseteq \cup_{k=1}^{\infty} (a_k, b_k]$, *then* $w\left((a,b]\right) \leq \sum_k w\left((a_k, b_k]\right)$.

5. *If the intervals in* $\{(a_k, b_k]\}_{k=1}^{\infty}$ *are disjoint and* $\cup_k (a_k, b_k] \subseteq (a,b]$, *then* $\sum_k w\left((a_k, b_k]\right) \leq w\left((a,b]\right)$.

Proof. We will prove only part (4). The other properties are more straightforward and are left as exercises.

Fix $\varepsilon > 0$. Since g is increasing and right-continuous, we can find an $a^* > a$ so that $g(a^*) < g(a) + \varepsilon$ and for each $k \in \mathbb{N}$ we can choose a $b_k^* > b_k$ so that $w\left((a_k, b_k^*]\right) \leq w\left((a_k, b_k]\right) + \frac{\varepsilon}{2^k}$. Then $\{(a_k, b_k^*)\}$, being an open cover of the compact interval $[a^*, b]$, has a finite subcover $\{(a_j, b_j^*)\}_{j=1}^{n}$. Then from (2),

$$w\left((a,b]\right) \leq w\left((a, a^*]\right) + \sum_{j=1}^{n} w\left((a_j, b_j^*]\right)$$

$$\leq \varepsilon + \sum_{k=1}^{\infty} w\left((a_k, b_k]\right) + \varepsilon.$$

Since $\varepsilon > 0$ is arbitrary, $w\left((a,b]\right) \leq \sum_k w\left((a_k, b_k]\right)$. $\square$

Since w behaves sufficiently like l, we can use the weight function based on g to define the outer measure of any subset of $\mathbb{R}$.

Definition 32 (Outer measure). *Let* $A \subset \mathbb{R}$. *The* **Lebesgue-Stieltjes outer measure** *of* A *is*

$$\mu_g^*(A) = \inf$$

$$\left\{ \sum_k w\left((a_k, b_k]\right) : \{(a_k, b_k]\} \text{ is a countable cover of } A \text{ by finite intervals} \right\}.$$

Example 61 $((a,b])$. For any outer measure μ_g^*, we have $\mu_g^*\left((a,b]\right) = w\left((a,b]\right) = g(b) - g(a)$.

To see why, note that $(a,b]$ covers itself so that $\mu_g^*\left((a,b]\right) \leq w\left((a,b]\right)$. The reverse inequality follows from (4) of theorem 93.

Example 62 ($[a, b]$). For any outer measure μ_g^*, we have $\mu_g^*([a, b]) = g(b) - g(a^-)$.

To see why, note that for any $\delta > 0$, $[a, b] \subset (a - \delta, b]$ so that $\mu_g^*([a, b]) \leq g(b) - g(a - \delta)$. Since the inequality holds for all $\delta > 0$, we can conclude that $\mu_g^*([a, b]) \leq g(b) - g(a^-)$. For the other inequality, observe that if $\{(a_k, b_k]\}$ is a countable cover of $[a, b]$, then there is an index k^* so that $a \in (a_{k^*}, b_{k^*}]$. Then $[a, b] \subset (a_{k^*}, b] \subseteq \cup_k (a_k, b_k]$. Hence $g(b) - g(a^-) \leq g(b) - g(a_{k^*}) \leq w((a_{k^*}, b]) \leq \sum_k w((a_k, b_k])$ by (4) of theorem 93.

Example 63 (Point mass). Define

$$g(x) = \begin{cases} 0, & x < 0 \\ 1, & 0 \leq x \end{cases}$$

and let A be a subset of $\mathbb{R}$. Then

$$\mu_g^*(A) = \begin{cases} 1, & 0 \in A \\ 0, & 0 \notin A. \end{cases}$$

To see why, note that

1. $w((a, b]) = g(b) - g(a)$ is 1 if $a < 0 \leq b$ and is zero otherwise.

2. $A \subseteq \cup_{k \in \mathbb{Z}} (k - 1, k]$ and $\sum_k w((k - 1, k]) = 1$ so $\mu_g^*(A) \leq 1$.

3. If $0 \in A$ and $A \subseteq \cup_k (a_k, b_k]$, then $0 \in (a_{k^*}, b_{k^*}]$ for some k^*. Thus $\sum_k w((a_k, b_k]) \geq 1$ so $\mu_g^*(A) \geq 1$.

4. If $0 \notin A$, then $A \subseteq \cup_n \{(-n, -\frac{1}{n}] \cup (0, n]\}$ so $\mu_g^*(A) = 0$.

Example 64 (Bernoulli distribution). Let $0 < p < 1$ and define

$$g(x) = \begin{cases} 0, & x < 0 \\ 1 - p, & 0 \leq x < 1 \\ 1, & 1 \leq x. \end{cases}$$

Let A be a subset of $\mathbb{R}$. An analysis similar to that of the previous example shows that

$$\mu_g^*(A) = \begin{cases} 0, & 0, 1 \notin A \\ 1 - p, & 0 \in A, 1 \notin A \\ p, & 0 \notin A, 1 \in A \\ 1 & 0, 1 \in A. \end{cases}$$

Thus μ_g^* describes the probabilities associated with a random variable that takes on the value 1 with probability p and the value 0 with probability $1 - p$.[5]

Of course, the outer measure generated by g need not be translation invariant. However, it will satisfy the remaining properties of Lebesgue outer measure.

Theorem 94 (Properties of outer measure). *Any Lebesgue-Stieltjes outer measure μ_g^* satisfies*

1. $\mu_g^* (\emptyset) = 0$.

2. If $A \subseteq B$ then $\mu_g^* (A) \leq \mu_g^* (B)$.

3. $\mu_g^* (\cup_k A_k) \leq \sum_k \mu_g^* (A_k)$ for any countable collection of sets $\{A_k\}$.

Proof. The verification of these properties follows very closely the proof of the corresponding properties for Lebesgue outer measure and is left as an exercise. $\square$

As with Lebesgue measure, one of the critical problems is to identify the collection of sets on which μ_g is countably additive. We can address this question the same way—with the Carathéodory criterion.

Definition 33 (Carathéodory's measurability condition). *A set $E \subseteq \mathbb{R}$ is (μ_g) measurable if*

$$\mu_g^* (A) = \mu_g^* (A \cap E) + \mu_g^* (A \cap E^c)$$

for every subset $A \subseteq \mathbb{R}$. When E is measurable, we write $\mu_g (E) = \mu_g^* (E)$.

Depending on the function g, the set of μ_g-measurable sets can be quite different from the Lebesgue-measurable sets.

Example 65 (Point mass). Define

$$g (x) = \begin{cases} 0, & x < 0 \\ 1, & 0 \leq x \end{cases}$$

as in example 63 (page 243). Then every subset E of $\mathbb{R}$ is measurable. To see why, let A be an arbitrary subset of $\mathbb{R}$. If $0 \notin A$, then

$$\mu_g^* (A) = \mu_g^* (A \cap E) = \mu_g^* (A \cap E^c) = 0.$$

---

[5] In other words, μ_g^* describes the probabilities associated with the binomial distribution $B(1, p)$.

If $0 \in A$, then 0 is in exactly one of $A \cap E$ or $A \cap E^c$. In either case,

$$\mu_g^* (A) = \mu_g^* (A \cap E) + \mu_g^* (A \cap E^c)$$

so that E is measurable.

Review the development of Lebesgue measure on pages 98 through 104. Except for the proof that $(-\infty, a)$ is a measurable set and the proof of theorem 22 (page 104), none of the proofs uses properties or lengths of intervals. The proofs depend only on the Carathéodory measurability condition, monotonicity, and subadditivity. We have already verified that μ_g^* is monotone and subadditive so we can assert that

1. μ_g is countably additive on μ_g-measurable sets and

2. the collection of μ_g-measurable sets is a sigma algebra.

Once we verify that $(-\infty, a)$ is μ_g-measurable, we can conclude that, in addition,

3. The collection of μ_g-measurable sets contains the Borel sets.

Theorem 95. *Intervals of the form $(-\infty, a)$ are μ_g-measurable.*

Proof. Since we know that the collection of μ_g-measurable sets is a sigma algebra, it will suffice to show that intervals of the form $(-\infty, a]$ are μ_g-measurable. Let A be a subset of $\mathbb{R}$ and choose a countable cover $\{I_k\}$ of A by intervals of the form $I_k = (a_k, b_k]$ such that

$$\sum_k w(I_k) \le \mu_g^* (A) + \varepsilon.$$

Since $A \subseteq \cup_k I_k$, monotonicity and subadditivity imply that

$$\mu_g^* (A \cap (-\infty, a]) + \mu_g^* (A \cap (a, +\infty))$$
$$\le \mu_g^* ((\cup_k I_k) \cap (-\infty, a]) + \mu_g^* ((\cup_k I_k) \cap (a, +\infty))$$
$$\le \sum_k \mu_g^* (I_k \cap (-\infty, a]) + \sum_k \mu_g^* (I_k \cap (a, +\infty)).$$

Now

$$(a_k, b_k] \cap (-\infty, a] = \begin{cases} (a_k, b_k], & b_k \le a \\ (a_k, a], & a_k \le a < b_k \\ \emptyset, & a < a_k \end{cases}$$

and

$$(a_k, b_k] \cap (a, +\infty) = \begin{cases} \emptyset, & b_k \leq a \\ (a, b_k], & a_k \leq a < b_k \\ (a_k, b_k], & a \leq a_k. \end{cases}$$

Hence μ_g^* and w agree for all the sets in the summation and, by part (1) of theorem 93 (page 242), $w(I_k) = w(I_k \cap (-\infty, a]) + w(I_k \cap (a, +\infty))$. Moreover, all the terms in the summations are nonnegative so they can be reordered. Therefore we can continue the chain of inequalities with

$$= \sum_k w(I_k \cap (-\infty, a]) + \sum_k w(I_k \cap (a, +\infty))$$

$$= \sum_k [w(I_k \cap (-\infty, a]) + w(I_k \cap (a, +\infty))]$$

$$= \sum_k w(I_k) \leq \mu^*(A) + \varepsilon.$$

Since $\varepsilon > 0$ was arbitrary, we conclude that

$$\mu^*(A \cap (-\infty, a]) + \mu^*(A \cap (a, +\infty)) \leq \mu^*(A).$$

The reverse inequality is a consequence of subadditivity. The interval $(-\infty, a]$ is measurable. □

Since the collection of μ_g-measurable sets is a sigma algebra containing intervals of the form $(-\infty, a]$, the collection of μ_g-measurable sets contains all the Borel sets. As we have seen in earlier examples, there may be additional μ_g-measurable sets. It is relatively straightforward to prove that any set of μ_g^*-measure zero is μ_g-measurable. These sets fill out the collection of μ_g-measurable sets as they do for Lebesgue measure.

Theorem 96. *A subset E of $\mathbb{R}$ is μ_g-measurable if and only if E can be written as $E = B \cup Z$ where B is a Borel set and Z is a set of μ_g-measure zero.*

Proof. Let E be a μ_g-measurable set and set $F = E^c$. First suppose that $\mu_g(F) < \infty$. Then for any natural number n there is a countable cover $\{I_k\}$ of F by intervals of the form $I_k = (a_k, b_k]$ such that

$$\mu_g(F) \leq \mu_g(\cup_k I_k) \leq \sum_k w(I_k) < \mu_g(F) + \frac{1}{n}.$$

If we set $G_n = \cup_k I_k$, then G_n is a Borel set satisfying $F \subseteq G_n$ and $\mu_g(G_n) < \mu_g(F) + \frac{1}{n}$. When $\mu_g(F) = \infty$, apply the same argument to $F_k = F \cap (k, k+1]$ for $k \in \mathbb{Z}$ to find Borel sets $\{G_{n,k}\}_{k=-\infty}^{\infty}$ containing F_k such that $\mu^*(G_{n,k} \setminus F_k) < \frac{1}{n2^{k+2}}$. Then $G_n = \cup_k G_{n,k}$ is a Borel set containing F with $\mu^*(G_n \setminus F) < \frac{1}{n}$.

The measurability of F now implies that

$$\mu_g(F) + \frac{1}{n} > \mu_g(G_n) = \mu_g(G_n \cap F) + \mu_g(G_n \cap F^c)$$
$$= \mu_g(F) + \mu_g(G_n \setminus F).$$

Subtracting, we find that $\mu_g(G_n \setminus F) < \frac{1}{n}$. Set $G = \cap_n G_n$ and $Z = G \setminus F = G \cap E$. Since $G \setminus F \subseteq G_n \setminus F$, we have $\mu_g(Z) = \mu_g(G \setminus F) = 0$. Taking complements, $G^c \subseteq F^c = E$ so that $G^c \cup Z = G^c \cup (G \cap E) = E$. Thus E is the union of a Borel set and a set of μ_g-measure zero.

The converse is a consequence of the facts that sets of μ_g-measure zero are measurable and that the collection of μ_g-measurable sets is a sigma algebra.   □

8.6 Integration with measures

The motivation for considering a more general theory of measure is the observation that the properties of the length function l and the weight function w are quite similar. This suggests the possibility of constructing a theory of integration based on μ_g to extend the Riemann-Stieltjes integral like the Lebesgue integral extends the Riemann integral. As with the Lebesgue integral, the critical condition that must be satisfied before considering the integral of a function f is that f be measurable.

Definition 34 (Measurable function). *Let X be a μ_g-measurable subset of real numbers and let $f : X \to \mathbb{R}$. Then f is μ_g-**measurable** if the preimage under f of any interval is μ_g-measurable.*

Review the section on Lebesgue measurable functions on page 105 and following. You will see that all the proofs depend solely on

1. the properties of inverse images of functions,

2. the fact that all Borel sets are measurable, and

3. the fact that the collection of measurable sets is a sigma algebra.

Since these facts are also true for μ_g-measurable sets, all the proofs from Section 6 of Chapter 6 transfer without modification into the context of μ_g-measurable functions. In particular,

1. All increasing functions are μ_g-measurable.

2. All continuous functions are μ_g-measurable.

3. Linear combinations of μ_g-measurable functions are μ_g-measurable.

4. If f_1 is a μ_g-measurable function and $f_1 = f_2$ except on a set of μ_g-measure zero,[6] then g is μ_g-measurable.

5. Supremums and infimums of sets of μ_g-measurable functions are again μ_g-measurable.

6. Limits, limsups, and liminfs of sequences of μ_g-measurable functions are again μ_g-measurable.

Because of the close parallel between Lebesgue and μ_g-measurability, one commonly drops the modifier and speaks only of measurability. The particular sigma algebra of measurable sets is understood from the context.

At this point, we have essentially replicated the infrastructure for the Lebesgue integral found in Sections 2 through 6 of Chapter 6 for the more general measure μ_g. Given that we have the same supporting structure, we can define the Lebesgue-Stieltjes integral by making slight modifications to the definition of the Lebesgue integral.

Definition 35 (Lebesgue-Stieltjes integral). *Let f be a nonnegative, μ_g-measurable function defined on $[a, b]$. Let $\mathcal{P}$ be the partition of $[0, +\infty)$ defined by the division points $\{y_k\}_{k=1}^n$ and set $E_k = f^{-1}([y_{k-1}, y_k))$, $k = 1, 2, \ldots, n$, and $E_{n+1} = f^{-1}([y_n, \infty))$. The (lower)* **Lebesgue-Stieltjes sum** *of f with respect to $\mathcal{P}$ is $S_{L\text{-}S}(f, \mu_g, \mathcal{P}) = \sum_{k=1}^{n+1} y_{k-1} \cdot \mu_g(E_k)$. The ($\mu_g$-)* **integral of** *$f$ over $[a, b]$ is*

$$^M\!\!\int_a^b f \, \mathrm{d}\mu_g = \sup_{\mathcal{P}} S_{L\text{-}S}(f, \mu_g, \mathcal{P})$$

where the supremum is taken over all partitions $\mathcal{P}$ of $[0, +\infty)$. The function f is said to be (μ_g) **integrable** *if $^M\!\!\int_a^b f \, \mathrm{d}\mu_g < +\infty$.*[7]

---

[6] Alternatively, $f_1 = f_2$ μ_g^*-a.e. or $f_1 = f_2$ a.e. μ_g^*.

[7] We use $^M\!\!\int_a^b f \, \mathrm{d}\mu_g$ since the integral is based on a measure. Serendipitously, M also corresponds to the Greek letter μ.

When $f : [a, b] \to \mathbb{R}$ and the integral of at least one of $f^+ = \frac{1}{2}(|f| + f)$ or $f^- = \frac{1}{2}(|f| - f)$ is finite, we set

$$^M\!\!\int_a^b f \, d\mu_g = {}^M\!\!\int_a^b f^+ \, d\mu_g - {}^M\!\!\int_a^b f^- \, d\mu_g$$

*and say that f is (μ_g-) **integrable** if $^M\!\!\int_a^b |f| \, d\mu_g < +\infty$ or, equivalently, if the integrals of both f^+ and f^- are finite.*

The simple and Young versions of the Lebesgue integral can be similarly transformed and the resulting integrals are again equivalent to the Lebesgue-Stieltjes integral. The Lebesgue-Stieltjes integral is a generalization of the Lebesgue integral since $\mu_g = \lambda$ when $g(x) = x$.

It would appear that when g is an increasing, right-continuous function and f is Riemann-Stieltjes integrable with respect to g over $[a, b]$ then $^M\!\!\int_a^b f \, d\mu_g = {}^{R\text{-}S}\!\!\int_a^b f \, dg$. This is not quite correct since the Riemann-Stieltjes integral will not capture any weight the measure μ_g places at a. Instead,

$$^M\!\!\int_a^b f \, d\mu_g = {}^{R\text{-}S}\!\!\int_a^b f \, dg + f(a) \, \mu_g(\{a\}).$$

Thus the Lebesgue-Stieltjes integral agrees with the Riemann-Stieltjes integral over $[a, b]$ exactly when g is continuous at a. As shown by the following example, the Lebesgue-Stieltjes integral is a strict extension of the Riemann-Stieltjes integral.

Example 66 (Step functions). Let

$$f(x) = \begin{cases} 1, & 0 \le x \le 1 \\ 0, & 1 < x \le 2 \end{cases}$$

and let

$$g(x) = \begin{cases} 0, & 0 \le x < 1 \\ 1, & 1 \le x \le 2. \end{cases}$$

In example 54 (page 229), we showed that f is not Riemann-Stieltjes integrable with respect to g over $[0, 2]$. However $^M\!\!\int_0^2 f \, d\mu_g$ does exist and is 1.

Let $\mathcal{P}$ be a partition of $[0, +\infty)$ defined by the division points $\{y_k\}_{k=1}^n$ with at least one division point in $(0, 1)$. Set $E_k = f^{-1}([y_{k-1}, y_k))$, $k = 1, 2, \ldots, n$, and $E_{n+1} = f^{-1}([y_n, \infty))$. Now $y_0 = 0 < y_1$ and there is exactly one index k_1 satisfying $y_{k_1-1} \le 1 < y_{k_1}$. Thus $\mu_g(E_1) = \mu_g((1, 2]) = 0$, $\mu_g(E_{k_1}) = \mu_g([0, 1]) = 1$, and $\mu_g(E_k) = \mu_g(\emptyset) = 0$ for all other values of k. Hence $S_{L\text{-}S}(f, \mu_g, \mathcal{P}) = y_{k_1-1}$ so $^M\!\!\int_a^b f \, d\mu_g = \sup_{\mathcal{P}} S_{\underline{L\text{-}S}}(f, \mu_g, \mathcal{P}) = 1$.

Replacing λ by μ_g and ${}^L\!\int_a^b f$ by ${}^M\!\int_a^b f \, d\mu_g$ in the proofs of theorems 33 and 34 (page 135), we see that, as with the Lebesgue integral, we can ignore the behavior of f on a set of μ_g-measure zero. In other words, if f_1 is μ_g-integrable and $f_1 = f_2$ a.e. μ_g, then f_2 is μ_g-integrable and ${}^M\!\int_a^b f_2 \, d\mu_g = {}^M\!\int_a^b f_1 \, d\mu_g$. Note, however, that a single point can have positive measure. Changing the value of the function there will produce a different value for the integral.

What about the convergence properties of the Lebesgue-Stieltjes integral? Since the infrastructure is essentially the same, the convergence theorems for the Lebesgue integral extend to the Lebesgue-Stieltjes integral. The statements and proofs require only minor changes. The most significant change is that $(b - a)$ must be replaced by $\mu_g\,([a,b])$ in the proof of theorem 36 (page 136).[8]

Theorem 97 (Bounded convergence). *Let $\{f_k\}$ be a uniformly bounded sequence of μ_g-measurable functions that converges pointwise to f μ_g-a.e. on $[a,b]$. Then*

$$\lim_k {}^M\!\int_a^b f_k \, d\mu_g = {}^M\!\int_a^b \lim_k f_k \, d\mu_g = {}^M\!\int_a^b f \, d\mu_g.$$

Theorem 98 (Monotone convergence). *Let $\{f_k\}$ be a monotone increasing sequence of nonnegative μ_g-measurable functions that converges a.e. μ_g to f on $[a,b]$. Then*

$$ {}^M\!\int_a^b f \, d\mu_g = \lim_k {}^M\!\int_a^b f_k \, d\mu_g$$

where a value of $+\infty$ is permissible.

Corollary 99. *Let $\{f_k\}$ be a sequence of nonnegative, μ_g-measurable functions. Then*

$$ {}^M\!\int_a^b \underline{\lim}_k f_k \, d\mu_g \le \underline{\lim}_k {}^M\!\int_a^b f_k \, d\mu_g.$$

Theorem 100 (Dominated convergence). *Suppose that $\{f_k\}$ is a sequence of μ_g-measurable functions that converges to f μ_g-a.e. on $[a,b]$. If there is a measurable function h with ${}^M\!\int_a^b h \, d\mu_g < \infty$ and for which $|f_k| \le h$ a.e.*

---

[8] This change implies that in any subsequent generalization we need to guarantee that $\mu\,([a,b])$ is finite. This is not a problem for measures built from increasing functions on $[a,b]$ but can be an issue for general measures.

μ_g *on* $[a, b]$ *for all* k, *then*

$$M\int_a^b f \, d\mu_g = \lim_k {}^M\!\int_a^b f_k \, d\mu_g.$$

The ease with which one can move from integration with respect to λ to integration with respect to μ_g goes a long way toward explaining the preference for the Lebesgue integral over the gauge integral. But we are not done.

8.7 Extensions to other types of measures

While we have restricted our attention to measures of the form μ_g where g is an increasing, right continuous function, this is not necessary. We can dispense with the function g and develop a theory of integration starting with any triple $(\mathbb{R}, \mathcal{A}, \mu)$ called a **measure space** where $\mathcal{A}$ is a sigma algebra of subsets of $\mathbb{R}$ and μ is a measure on $\mathcal{A}$. In other words, μ is a function from $\mathcal{A}$ to the extended real numbers satisfying

1. $\mu(\emptyset) = 0$,

2. $\mu(E) \geq 0$ for all $E \in \mathcal{A}$,[9] and

3. $\mu(\cup_i E_i) = \sum_i \mu(E_i)$ for any countable, disjoint collection of sets $\{E_i\}$ from $\mathcal{A}$.

As long as the measure of every bounded interval is finite, everything that we have developed so far readily extends in a natural way to this new context. We simply replace the sum $S_{L\text{-}S}(f, \mu_g, \mathcal{P}) = \sum_{k=1}^{n+1} y_{k-1} \cdot \mu_g(E_k)$ with the almost imperceptible modification $S_M(f, \mu, \mathcal{P}) = \sum_{k=1}^{n+1} y_{k-1} \cdot \mu(E_k)$.

Example 67. Let $\mathcal{A}$ be the sigma algebra of all subsets of $\mathbb{R}$ and define μ by

$$\mu(E) = \begin{cases} 0, & 0, 1 \notin E \\ \frac{1}{3}, & 0 \in E, \ 1 \notin E \\ \frac{2}{3}, & 0 \notin E, \ 1 \in E \\ 1, & 0, 1 \in E, \end{cases}$$

---

[9] The condition that $\mu(E) \geq 0$ for all $E \in \mathcal{A}$ is equivalent to $\mu(A) \leq \mu(B)$ for all $A, B \in \mathcal{A}$ satisfying $A \subseteq B$. See Exercise 65.

for $E \in \mathcal{A}$. Then μ is a measure on $\mathcal{A}$. As long as $a \leq 0$ and $b \geq 1$,

$$M\int_a^b 1 \, d\mu = 1 \cdot \frac{1}{3} + 1 \cdot \frac{2}{3} = 1,$$

$$M\int_a^b x \, d\mu = 0 \cdot \frac{1}{3} + 1 \cdot \frac{2}{3} = \frac{2}{3}, \text{ and}$$

$$M\int_a^b \left(x - \frac{2}{3}\right)^2 \, d\mu = \frac{4}{9} \cdot \frac{1}{3} + \frac{1}{9} \cdot \frac{2}{3} = \frac{2}{9}.$$

To completely justify the value of the second integral from the definition, set $f(x) = x$ and $f_1 = f^+$. Let $\mathcal{P}$ be the partition of $[0, +\infty)$
defined by the division points $\{y_k\}_{k=1}^n$ and set $E_k = f_1^{-1}([y_{k-1}, y_k))$,
$k = 1, 2, \ldots, n$, and $E_{n+1} = f_1^{-1}([y_n, \infty))$. Then there are integers k_0
and k_1 such that $y_{k_0-1} \leq 0 < y_{k_0}$ and $y_{k_1-1} \leq 1 < y_{k_1}$.[10] Assuming
$k_0 < k_1$,

$$\mu(E_{k_0}) = \mu([a, y_{k_0})) = \frac{1}{3},$$

$$\mu(E_{k_1}) = \mu([y_{k_1-1}, y_{k_1})) = \frac{2}{3},$$

and $\mu(E_k) = 0$ for all other values of k. Thus

$$S_M(f^+, \mu, \mathcal{P}) = 0 \cdot \frac{1}{3} + y_{k_1-1} \cdot \frac{2}{3} = y_{k_1-1} \cdot \frac{2}{3}.$$

Hence $M\int_a^b f^+ \, d\mu = \sup_{\mathcal{P}} S_M(f^+, \mu, \mathcal{P}) = \frac{2}{3}$. Taking $f_2 = f^-$
and $F_k = f_2^{-1}([y_{k-1}, y_k)) = (-y_k, -y_{k-1}]$ for $k > 1$, $\mu(F_k) = \mu((-y_k, -y_{k-1}]) = 0$ so that

$$S_M(f^-, \mu, \mathcal{P}) = 0 \cdot \mu(F_1) = 0.$$

Hence $M\int_a^b f^- \, d\mu = 0$ and thus $M\int_a^b x \, d\mu = \frac{2}{3}$.

In practice, for discrete measures such as the measure in this example it
is easier to take an approach similar to that in theorem 88 (page 234) for
the Riemann-Stieltjes integral. Note those points in $[a, b]$ that have positive
measure and compute the sum over those points of the value of the function
times the point's measure.

If you are familiar with probability distributions, you will note that μ is
the probability measure associated with the binomial distribution $B\left(1, \frac{1}{3}\right)$.

---

[10] In fact, $k_0 = 1$.

The three integrals reflect the facts that the total probability is 1, the mean of the distribution is $\frac{2}{3}$, and the variance is $\frac{2}{9}$.

Example 68. Let $\mathcal{A}$ be the sigma algebra $\mathcal{M}$ of Lebesgue measurable sets and define μ as the measure on $\mathcal{M}$ generated by the weight function

$$w\left((a,b]\right) = \arctan(b) - \arctan(a) + v\left((a,b]\right)$$

where v places weights of $\frac{1}{2}, 1$, and $\frac{1}{2}$ at $-1, 0$, and 1 respectively. Splitting μ into μ_1 and μ_2 defined respectively by the weight functions $w_1\left((a,b]\right) = \arctan(b) - \arctan(a)$ and $w_1\left((a,b]\right) = v\left((a,b]\right)$ we can sum the two contributions to find that

$$^M\!\!\int_{-\sqrt{3}}^{\sqrt{3}} 1 \, d\mu = {}^M\!\!\int_{-\sqrt{3}}^{\sqrt{3}} 1 \, d\mu_1 + {}^M\!\!\int_{-\sqrt{3}}^{\sqrt{3}} 1 \, d\mu_2 = \frac{2}{3}\pi + 2,$$

$$^M\!\!\int_{-\sqrt{3}}^{\sqrt{3}} x \, d\mu = {}^M\!\!\int_{-\sqrt{3}}^{\sqrt{3}} x \, d\mu_1 + {}^M\!\!\int_{-\sqrt{3}}^{\sqrt{3}} x \, d\mu_2$$

$$= {}^M\!\!\int_{-\sqrt{3}}^{\sqrt{3}} \frac{x}{1+x^2} \, dx + 0 = 0, \text{ and}$$

$$^M\!\!\int_{-\sqrt{3}}^{\sqrt{3}} x^2 \, d\mu = {}^M\!\!\int_{-\sqrt{3}}^{\sqrt{3}} \frac{x^2}{1+x^2} \, dx + {}^M\!\!\int_{-\sqrt{3}}^{\sqrt{3}} x^2 \, d\mu_2$$

$$= 2\sqrt{3} - \frac{2}{3}\pi + 1.$$

On the one hand, any generalization gained by starting from $(\mathbb{R}, \mathcal{A}, \mu)$ instead of $(\mathbb{R}, \mathcal{A}, \mu_g)$ is largely illusory since, as long as $\mathcal{A}$ contains the Borel sets and bounded intervals have finite measure, $\mu = \mu_g$ where

$$g(x) = \begin{cases} \mu\left((0,x]\right), & 0 < x \\ 0, & x = 0 \\ -\mu\left((x,0]\right), & x < 0. \end{cases}$$

On the other hand, the pure measure-theoretic point of view invites other types of generalizations. The most simple generalization is to replace $\mathbb{R}$ by an alternate set such as $\mathbb{R}^n$ or $\mathbb{C}$. In this case, $\mathcal{A}$ must be a sigma algebra of subsets of the underlying space and μ must be a measure on $\mathcal{A}$.

Example 69. Let $\mathcal{A}$ be the sigma algebra of all subsets of $\mathbb{C}$ and define $\mu(A)$ to be the number of elements in $A \cap \{1, -1, i, -i\}$.[11] Then μ is a measure

---

[11] Alternatively, μ places a weight of 1 on each of the points $1, -1, i$, and $-i$.

on $\mathcal{A}$ and

$$^M\!\!\int_{\mathbb{C}} 1 \, d\mu = 4,$$

$$^M\!\!\int_{\mathbb{C}} z \, d\mu = 0,$$

$$^M\!\!\int_{\mathbb{C}} z^2 \, d\mu = 0, \text{ and}$$

$$^M\!\!\int_{\mathbb{C}} |z| \, d\mu = 4.$$

Even greater generalizations can be obtained by allowing μ to take on other types of values. We have seen hints of this idea when working with the Riemann-Stieltjes integral when g is nonmonotone but of bounded variation. This generalization produces a theory of signed measures. Similarly, it is relatively easy to construct a theory of complex-valued measures by considering measures of the form $\mu = \mu_1 - \mu_2 + i\mu_3 - i\mu_4$ where the μ_i are nonnegative measures on a common sigma algebra $\mathcal{A}$.

We will not pursue this path. Instead, we conclude with a brief introduction of the idea of projection-valued measures. For simplicity, we will restrict our attention to finite-dimensional vector spaces. For such spaces, the projection-valued measure will always be discrete. The true power of these ideas starts to become evident when working with continuous linear mappings between complete, infinite-dimensional vector spaces (Hilbert spaces), but the development of this theory is beyond the scope of this text.[12]

A projection P on a vector space is a linear mapping satisfying $P^2 = P$. The idea behind the algebraic expression is that, once a vector has been projected into a subspace, repeating the projection will have no additional effect.

Definition 36 (Projection-valued measure). *Given a set X and a sigma algebra $\mathcal{A}$ of subsets of X, a **projection-valued measure** on $\mathcal{A}$ is a function μ from $\mathcal{A}$ to the set of projections onto subspaces of a fixed vector space V that satisfies*

1. $\mu(\emptyset) = 0,$[13]

---

[12] In infinite-dimensional (or finite dimensional) contexts, the inner product can be used to ground the theory of projection-valued measures in the theory of complex-valued measures. For any pair of vectors v and w from the vector space V, the complex-valued measure $\mu_{v,w}$ is defined by $\mu_{v,w}(E) = \langle \mu(E)v, w \rangle$. The operator $A = \int f \, d\mu$ is defined by the way that it acts on vectors: $\langle Av, w \rangle = \int f \, d\mu_{v,w}$.

[13] The zero in this case is the zero mapping, not the real number 0.

2. $\mu(E) \geq 0$ for all $E \in \mathcal{A}$ (i.e. the inner product $\langle \mu(E) v, v \rangle$ is nonnegative for all $v \in V$),[14] and

3. $\mu(\cup_i E_i) = \sum_i \mu(E_i)$ for any countable, disjoint collection of sets $\{E_i\}$ from $\mathcal{A}$.

Example 70. Let

$$P_2 = \frac{1}{2} \begin{bmatrix} 1 & 1 \\ 1 & 1 \end{bmatrix}$$

and

$$P_4 = \frac{1}{2} \begin{bmatrix} 1 & -1 \\ -1 & 1 \end{bmatrix}.$$

Then P_2 and P_4 are projections onto the spaces spanned by $\{\begin{bmatrix} 1 \\ 1 \end{bmatrix}\}$ and $\{\begin{bmatrix} 1 \\ -1 \end{bmatrix}\}$ respectively. For $E \subseteq \mathbb{R}$, define

$$\mu(E) = \begin{cases} 0, & 2, 4 \notin E \\ P_2, & 2 \in E, 4 \notin E \\ P_4, & 2 \notin E, 4 \in E \\ P_2 + P_4, & 2, 4 \in E. \end{cases}$$

Then μ is a projection-valued measure on $\mathcal{A}$, the sigma algebra of all subsets of $\mathbb{R}$. In this case,

$$^M\!\!\int_0^5 1 \, d\mu = P_2 + P_4 = I,$$

$$^M\!\!\int_0^5 x \, d\mu = 2P_2 + 4P_4 = \begin{bmatrix} 3 & -1 \\ -1 & 3 \end{bmatrix}, \text{ and}$$

$$^M\!\!\int_0^5 x^2 \, d\mu = 4P_2 + 16P_4 = \begin{bmatrix} 10 & -6 \\ -6 & 10 \end{bmatrix}.$$

To justify the value of the second integral, set $f(x) = x$. Let $\mathcal{P}$ be the partition of $[0, +\infty)$ defined by the division points $\{y_k\}_{k=1}^n$ and set $E_k = f^{-1}([y_{k-1}, y_k))$, $k = 1, 2, \ldots, n$, and $E_{n+1} = f^{-1}([y_n, \infty))$. Then, assuming $\mathcal{P}$ has division points between 2 and 4, there are two values k_2 and k_4 such that $y_{k_2-1} \leq 2 < y_{k_2}$ and $y_{k_4-1} \leq 4 < y_{k_4}$. Then $\mu(E_{k_2}) = P_2$, $P(E_{k_4}) = P_4$, and $\mu(E_k) = 0$ for all other values of k. Thus the analog of the Lebesgue-Stieltjes sum is $S_M(f, \mu, \mathcal{P}) = y_{k_2-1} \cdot P_2 + y_{k_4-1} \cdot P_4$ which will converge to $2P_2 + 4P_4$ as the as the mesh of $\mathcal{P}$ gets finer. Thus $^M\!\!\int_0^5 x \, d\mu = 2P_2 + 4P_4$.

---

[14] Alternatively, $(\mu(E))^* = \mu(E)$ where A^* is the adjoint of the operator A. Or (2) can be omitted with the stipulation that $\mu(E)$ is an orthogonal projection.

If in the preceding example we set $A = {}^M\!\int_0^5 x \, d\mu$, then $A^2 = {}^M\!\int_0^5 x^2 \, d\mu$. This observation suggests that we can compute $f(A)$ as ${}^M\!\int_0^5 f \, d\mu$. This is indeed the case. We shall not prove this result as it would involve the introduction of theorems from operator theory. Instead we give a verifiable example.

Example 71. Under the same conditions as example 70,

$$
{}^M\!\int_0^5 \sqrt{x} \, d\mu = \sqrt{2}P_2 + 2P_4 =
\begin{bmatrix}
\frac{1}{2}\sqrt{2}+1 & \frac{1}{2}\sqrt{2}-1 \\
\frac{1}{2}\sqrt{2}-1 & \frac{1}{2}\sqrt{2}+1
\end{bmatrix}.
$$

A quick calculation shows that

$$
\begin{bmatrix}
\frac{1}{2}\sqrt{2}+1 & \frac{1}{2}\sqrt{2}-1 \\
\frac{1}{2}\sqrt{2}-1 & \frac{1}{2}\sqrt{2}+1
\end{bmatrix}^2
=
\begin{bmatrix}
3 & -1 \\
-1 & 3
\end{bmatrix}
= A.
$$

While a more complete development of these ideas lies outside the scope of this text, we note that the notion of a projection-valued measure together with convergence theorems like those of the previous section provide a significant set of tools for the study of the structure of linear operators. In particular, one way to search for invariant subspaces of a linear operator A (a subspace W such that $AW \subseteq W$) is to search for a projection P that commutes with A ($AP = PA$).

If $A = {}^M\!\int_X z \, d\mu$, then A will commute with any operator of the form $p(A) = {}^M\!\int_X p \, d\mu$ where p is a polynomial. Given any characteristic function 1_E, the operator $P = {}^M\!\int_X 1_E \, d\mu$ is a projection since $P^2 = {}^M\!\int_X (1_E)^2 \, d\mu = P$. Thus if one can find a sequence $\{p_k\}$ of polynomials that converges μ-a.e. to 1_E, then P will commute with A. In this case we have found an invariant subspace for the linear operator A.

Beyond purely mathematical applications, projection-valued measures are used to model measurement in the study of quantum theory in physics.

8.8 Exercises

8.0 Stieltjes-Type integrals: filling the gaps

1. Verify the standard integral properties for the Riemann-Stieltjes integral.

   (a) **Uniqueness.** The value of the Riemann-Stieltjes integral is unique (if it exists).

(b) **Linearity I.** If f and g are Riemann-Stieltjes integrable with respect to h over the interval $[a, b]$ and $c \in \mathbb{R}$, then so are $f + g$ and cf. Moreover,

$$R\text{-}S\int_a^b (f + g) \, dh = R\text{-}S\int_a^b f \, dh + R\text{-}S\int_a^b g \, dh$$

and

$$R\text{-}S\int_a^b cf \, dh = c \ R\text{-}S\int_a^b f \, dh.$$

(c) **Linearity II.** If f is Riemann-Stieltjes integrable with respect to g and h over the interval $[a, b]$ and $c \in \mathbb{R}$, then f is Riemann-Stieltjes integrable with respect to $g + h$ and cg. Moreover,

$$R\text{-}S\int_a^b f \, d(g + h) = R\text{-}S\int_a^b f \, dg + R\text{-}S\int_a^b f \, dh$$

and

$$R\text{-}S\int_a^b f \, d(cg) = c \ R\text{-}S\int_a^b f \, dg.$$

(d) **Monotonicity.** If f and g are Riemann-Stieltjes integrable with respect to h over the interval $[a, b]$ with $f(x) \leq g(x)$ for $x \in [a, b]$, then it is not necessarily true that $R\text{-}S\int_a^b f \, dh \leq R\text{-}S\int_a^b g \, dh$. Find and justify sufficient conditions for monotonicity.

(e) **Triangle inequality.** Find and justify sufficient conditions such that if f and $|f|$ are Riemann-Stieltjes integrable with respect to h over $[a, b]$, then $\left| R\text{-}S\int_a^b f \, dh \right| \leq R\text{-}S\int_a^b |f| \, dh$.

8.0 Stieltjes-Type integrals: deeper reflections

2. Suppose that $g(x)$ represents the mass in the interval $(-\infty, x)$. What would $g(x_i) - g(x_{i-1})$ represent?

3. Find a pair of functions f and g on $[0, 1]$ so that f is not Riemann-Stieltjes integrable with respect to g over $[0, 1]$ but $|f|$ is.

8.1 Examples: filling the gaps

4. In example 51 (page 228), why is $\sum_{i=1}^n \Delta g = (g(b) - g(a))$?

5. Fill in the details of example 54 (page 229).

(a) Explain why $S_{R\text{-}S}(f, g, \mathcal{P}_1) = 1$ and $S_{R\text{-}S}(f, g, \mathcal{P}_2) = 0$.

(b) Use part (a) with partitions of arbitrarily small mesh to explain why f cannot be Riemann-Stieltjes integrable with respect to g over $[0, 2]$.

6. Fill in the details of example 55 (page 230) to show that $^{R\text{-}S}\!\int_0^2 f_1 \, dg = -1$ and $^{R\text{-}S}\!\int_0^2 f_2 \, dg = -2$.

8.1 Examples: deeper reflections

When one encounters unexpected results such as those in example 54 (page 229), it is good practice to try to place the phenomena in a larger context. The next four exercises ask you to do this.

7. Example 54 (page 229) extends to the case where f and g are functions on $[a, b]$ with a common point of discontinuity. Give a proof of the fact that when f and g have a common point of discontinuity then f is not Riemann-Stieltjes integrable over $[a, b]$ in the special case where g is increasing. (Work with a partition of $[a, b]$ that straddles the discontinuity.)

8. Explain example 54 (page 229) in light of theorem 20 (page 63). (What is the weight/measure of $\{1\}$?)

9. Example 54 (page 229) shows that Riemann-Stieltjes integrability over $[a, b]$ and $[b, c]$ is not sufficient to conclude Riemann-Stieltjes integrability over $[a, c]$. Prove that if f is Riemann-Stieltjes integrable with respect to g over all three intervals, then $^{R\text{-}S}\!\int_a^c f \, dg = {}^{R\text{-}S}\!\int_a^b f \, dg + {}^{R\text{-}S}\!\int_b^c f \, dg$.

10. Define f and g as in example 54 (page 229). Show that f is gauge-Stieltjes integrable with respect to g over $[0, 1]$ where gauge-Stieltjes integrable has the obvious interpretation. What is $^{g\text{-}S}\!\int_a^b f \, dg$? (Use tag forcing.)

11. Find a pair of functions f and g on $[0, 1]$ such that f is Riemann-Stieltjes integrable with respect to g over $[0, 1]$ and $F(x) = {}^{R\text{-}S}\!\int_0^x f \, dg$ is discontinuous at every point in $[0, 1]$.

12. Prove that if f and g are functions on $[a, b]$ where f is bounded, g is absolutely continuous, and f is Riemann-Stieltjes integrable with respect to g over $[a, b]$, then $F(x) = {}^{R\text{-}S}\!\int_0^x f \, dg$ is continuous on $[a, b]$. (Bound and telescope.)

8.2 Integrability theorems: filling the gaps

13. Prove theorem 84 (page 231).

14. Prove theorem 85 (page 231). (Extend partitions of $[c, d]$ to partitions of $[a, b]$ using a common extension.)

15. In the proof of theorem 86 (page 231)

   (a) Why can't we just define $\phi_{\mathcal{P}_1} = \sum_{\mathcal{P}} \inf_{[x_{i-1}, x_i]} f \cdot 1_{[x_{i-1}, x_i]}$?
   (b) Explain why $S_{R\text{-}S}(\phi_{\mathcal{P}_1}, g, \mathcal{P}_1) \leq S_{R\text{-}S}(f, g, \mathcal{P}_1) \leq S_{R\text{-}S}(\phi_{\overline{\mathcal{P}_1}}, g, \mathcal{P}_1)$.
   (c) Explain why $S_{R\text{-}S}(\phi_{\mathcal{P}_1}, g, \mathcal{P}_1) = S_{R\text{-}S}(\phi_{\mathcal{P}_1}, g, \mathcal{P}_3)$.
   (d) Explain why $S_{R\text{-}S}(\phi_{\overline{\mathcal{P}_1}}, g, \mathcal{P}_3) \leq S_{R\text{-}S}(f, g, \mathcal{P}_3)$.

8.2 Integrability theorems: deeper reflections

16. Prove that if f is continuous and g is monotone increasing on $[a, b]$, then there is a $c \in [a, b]$ that satisfies

$$R\text{-}S\int_a^b f \, dg = f(c) \; R\text{-}S\int_a^b dg.$$

17. Modify the proof of theorem 20 (page 63) to prove that f is Riemann-Stieltjes integrable with respect to g over $[a, b]$ if and only if f is continuous μ_g-a.e. on $[a, b]$. (See Section 5.)

8.3 Evaluation theorems: filling the gaps

18. In the proof of theorem 87 (page 233)

   (a) Why can we choose $\delta > 0$ so that $|f(x) - f(y)| < \frac{\varepsilon}{B(b-a)}$ whenever $x, y \in [a, b]$ with $|x - y| < \delta$?
   (b) Why is

$$\left| \sum_P \int_{I_i} [f(t_i) - f] \, g' \right| < \sum_P \int_{I_i} \frac{\varepsilon}{B(b-a)} B = \varepsilon?$$

19. Why are $f(c^+)$ and $f(c^-)$ always defined when f is a monotone function?

20. In the proof of theorem 88 (page 233)

   (a) Why can we assume that $M > 0$?
   (b) Why is $f(x) = f(c_i^-)$ for $x \in [a, b]$ with $0 < c_i - x < \delta$?

(c) Verify that if $c_k = x_0$, then
$$\left| f(t_1)\left[g(x_1) - g(x_0)\right] - f(c_k)\left[g\left(c_k^+\right) - g\left(c_k^-\right)\right] \right| < \frac{\varepsilon}{2n}.$$

(d) Explain why $\left| S_{R\text{-}S}(f, g, \mathcal{P}) - \sum_{i=1}^{n} f(c_i)\left[g\left(c_i^+\right) - g\left(c_i^-\right)\right] \right| < \varepsilon.$

21. In g_1 of example 57 (page 235), why were $3x$ and $x^2 + 2$ used instead of $3x + 1$ and x^2?

22. In the proof of theorem 89 (page 236)

    (a) Why does $\mathcal{P}^*$ have mesh less than δ?
    (b) Why is x_{k-1} a tag for $[t_{k-1}, t_k]$? Pay particular attention to the cases $k = 1$ and $k = n + 1$.
    (c) It is possible that some of the intervals in $\mathcal{P}^*$ are degenerate. How? Why is this not a problem?

8.3 Evaluation theorems: deeper reflections

23. Compute the following Riemann-Stieltjes integrals. ($\lceil x \rceil$ is the ceiling function that returns the least integer greater than or equal to x.)

    (a) $R\text{-}S\!\int_{-1}^{2} x^3 \, dx^2.$
    (b) $R\text{-}S\!\int_{-1}^{1} |x| \, d\lceil x \rceil.$
    (c) $R\text{-}S\!\int_{-1}^{1} \lceil x \rceil \, d|x|.$
    (d) $R\text{-}S\!\int_{0}^{2} \sin \pi x \, d\lceil x \rceil.$
    (e) $R\text{-}S\!\int_{0}^{2} x \, d\lceil x^2 \rceil.$
    (f) $R\text{-}S\!\int_{0}^{\pi} \sin x \, d\sin x.$
    (g) $R\text{-}S\!\int_{0}^{1} c \, dc$ where c is the Cantor function.
    (h) $R\text{-}S\!\int_{0}^{1} x \, dc$ where c is the Cantor function.

24. Prove that theorem 87 (page 233) can be extended by allowing g to have a finite number of points of nondifferentiability as long as g is continuous on all of $[a, b]$. (Split the partition $\mathcal{P}$ into those subintervals that contain a point where g is not differentiable and those that do not.)

25. Prove that theorem 87 (page 233) can be extended to allow g to be differentiable a.e. as long as g is continuous on all of $[a, b]$ and g' is Riemann integrable when defined to be zero where g is not differentiable. (Make use of the Lebesgue integral.)

26. Extend exercise 25 by proving the conclusion still holds when f has a finite number of discontinuities but remains bounded. (Divide and conquer.)

27. The hypotheses of theorem 88 (page 234) can be relaxed. Keeping the same hypothesis on g, prove that it is necessary and sufficient that f be continuous at $\{c_i\}_{k=1}^n$.

28. Use theorem 92 (page 239) to extend theorem 88 (page 234) to the case where g has a countable number of discontinuities $\{c_k\}$ and $\sum_k \left| g\left(c_k^+\right) - g\left(c_k^-\right) \right| < \infty$. (For each point c_k of discontinuity define

$$
h_k(x) = \begin{cases} 0, & x < c_k \\ g(c_k) - g\left(c_k^-\right), & x = c_k \\ g\left(c_k^+\right) - g\left(c_k^-\right), & c_k < x \end{cases}
$$

and define $g_n = \sum_{k \le n} h_k$.)

29. Suppose that f is a bounded, continuous function on $[a, b]$. Justify a simple expression for evaluating $^{R\text{-}S}\!\int_a^b f \, df$. Can you extend your result to allow f to have a finite or countable set of discontinuities?

8.4 Convergence theorems: filling the gaps

30. Why is $^{R\text{-}S}\!\int_a^b f \, dg$ of lemma 90 (page 238) defined? In other words, why is f Riemann-Stieltjes integrable with respect to g over $[a, b]$?

31. Prove lemma 90 (page 238).

32. Justify the inequality

$$
\left| S_{R\text{-}S}(f, g, \mathcal{P}) - {}^{R\text{-}S}\!\int_a^b f \, dg \right| = \left| \sum_i {}^{R\text{-}S}\!\int_{x_{i-1}}^{x_i} (f(t_i) - f) \, dg \right|
$$
$$
< \sum_i \varepsilon V_{x_{i-1}}^{x_i} g
$$

from the proof of lemma 91 (page 238).

33. Why do the integrals in theorem 92 (page 239) exist?

34. In the proof of theorem 92 (page 239), why is it possible to select a tagged partition of $[a, b]$ with mesh less than δ and whose division points are from S?

8.4 Convergence theorems: deeper reflections

35. Why didn't we prove theorem 92 (page 239) by first proving the lemmas 90 and 91 (page 238) and theorem 92 (page 239) for monotone increasing functions $\{g_n\}$ and g? The extension to functions of bounded variation would follow immediately.

36. Let $\{r_k\}$ be an enumeration of the rational numbers in $[0, 1]$ and set $g(x) = \sum_{r_k \leq x} \frac{1}{2^k}$. Find an integral-free expression for $^{R\text{-}S}\!\int_0^1 x^2 \, dg$. (Work with $g_n(x) = \sum_{r_k \leq x, \, k \leq n} \frac{1}{2^k}$.)

37. Show how any infinite series $\sum_k a_k$ can be represented by a Riemann-Stieltjes integral. (The weighting function

$$
g(x) = \begin{cases}
0, & x = 0 \\
1 - \frac{1}{2^k}, & x \in \left(1 - \frac{1}{2^{k-1}}, 1 - \frac{1}{2^k}\right], \ k \in \mathbb{N} \\
1, & x = 1
\end{cases}
$$

provides one possible approach.)

8.5 Measure theory: filling the gaps

38. Prove parts (1)-(3) and (5) of theorem 93 (page 242).

39. In the proof of theorem 93 (page 242)

   (a) Why can we find an $a^* > a$ so that $g(a^*) < g(a) + \varepsilon$?
   (b) Why can we choose a $b_k^* > b_k$ so that $w\left((a_k, b_k^*]\right) \leq w\left((a_k, b_k]\right) + \frac{\varepsilon}{2^k}$?
   (c) Why is $\{(a_k, b_k^*)\}$ an open cover of $[a^*, b]$?
   (d) Why is $w\left((a, b]\right) \leq w\left((a, a^*]\right) + \sum_{j=1}^n w\left((a_j, b_j^*]\right) \leq \sum_{k=1}^\infty w\left((a_k, b_k]\right) + 2\varepsilon$?

40. Prove theorem 94 (page 244).

41. Why is it sufficient to prove that intervals of the form $(-\infty, a]$ are μ_g-measurable in the proof of theorem 95 (page 245)?

42. Prove that any set of μ_g^*-outer measure zero is μ_g-measurable. In other words, prove that if $\mu_g^*(E) = 0$ then E is μ_g-measurable.

43. If $F_k = F \cap (k, k+1]$ and $\mu^*\left(G_{n,k} \setminus F_k\right) < \frac{1}{n2^{k+2}}$ for all $k \in \mathbb{Z}$, explain why $\mu^*\left(G_n \setminus F\right) < \frac{1}{n}$ where $G_n = \cup_k G_{n,k}$.

44. In the proof of theorem 96 (page 246)

   (a) Why is F measurable?
   (b) Why is G^c a Borel set?

8.5 Measure theory: deeper reflections

45. Modify the proof of theorem 22 (page 104) so that it applies to μ_g-measurability. (Use the right-continuity of g to move from half-open to open intervals.)

46. Express the following in terms of g.

    (a) $\mu_g^* ((a,b))$.
    (b) $\mu_g^* (\{a\})$.
    (c) $\mu_g^* ([a,b))$.

47. Define

$$g(x) = \begin{cases} 2x, & x < 0 \\ x + 1, & 0 \le x. \end{cases}$$

    Calculate

    (a) $\mu_g^* ((-1,1))$.
    (b) $\mu_g^* (\{0\})$.
    (c) $\mu_g^* (\{1\})$.
    (d) $\mu_g^* (\{\text{rational numbers in } [0,1]\})$.

48. Define

$$g(x) = \begin{cases} 0, & x < 0 \\ x + 1, & 0 \le x < 1 \\ 3, & 1 \le x. \end{cases}$$

    Calculate

    (a) $\mu_g^* ((-1,1))$.
    (b) $\mu_g^* (\{0\})$.
    (c) $\mu_g^* ((-\infty,\infty))$.
    (d) $\mu_g^* (C)$ where C is the Cantor set.

49. Let $g(x) = \lceil x \rceil$. Calculate

    (a) $\mu_g^* ([-1,1])$.
    (b) $\mu_g^* (\{0\})$.
    (c) $\mu_g^* (C)$ where C is the Cantor set.
    (d) Give a general description of the value of $\mu_g^* (A)$.

50. Let c be the Cantor function. Calculate

    (a) $\mu_c^* ([0,1])$.

(b) $\mu_c^*(C)$ where C is the Cantor set.

(c) $\mu_c^*(\{\text{rational numbers in } [0, 1]\})$.

51. Let $g(x) = \lceil x \rceil$. Prove that every subset of $\mathbb{R}$ is μ_g-measurable.

8.6 Integration with measures: filling the gaps

52. Prove that any increasing function f is μ_g-measurable.

53. Prove that any continuous function f is μ_g-measurable.

54. Prove that if a function f is continuous μ_g-a.e., then f is μ_g-measurable.

55. Prove that if g is an increasing, right-continuous function and f is Riemann-Stieltjes integrable with respect to g over $[a, b]$, then f is μ_g-measurable and $^M\!\int_a^b f \, d\mu_g = {}^{R\text{-}S}\!\int_a^b f \, dg + f(a)\mu_g(\{a\})$. (See exercises 17 and 54.)

56. Prove that if f_1 is a μ_g-integrable function on $[a, b]$ and $f_1 = f_2$ a.e. μ_g, then f_2 is μ_g-integrable and $^M\!\int_a^b f_1 \, d\mu_g = {}^M\!\int_a^b f_2 \, d\mu_g$.

57. Prove that $^M\!\int_{E\cup F} f \, d\mu_g = {}^M\!\int_E f \, d\mu_g + {}^M\!\int_F f \, d\mu_g$ whenever E and F are disjoint μ_g-measurable sets and f is μ_g-integrable over $E \cup F$.

8.6 Integration with measures: deeper reflections

58. Suppose that if f_1 is μ_g-measurable and $f_2 = f_1$ except on a countable set S. Prove that f_2 is measurable even if $\mu_g(S) > 0$. (Why is $\{a\}$ a measurable set?)

59. Modify the definition of the Lebesgue-Young integral to apply to the measure μ_g and prove that it is equivalent to the Lebesgue-Stieltjes integral for μ_g-measurable functions.

60. Modify the definition of the simple-Lebesgue integral to apply to the measure μ_g and prove that it is equivalent to the Lebesgue-Stieltjes integral for μ_g-measurable functions.

61. Without using exercise 55 for the evaluation, give an example of a function f and an increasing, right-continuous function g on $\mathbb{R}$ such that $^M\!\int_a^b f \, d\mu_g \neq {}^{R\text{-}S}\!\int_a^b f \, dg$.

62. Prove that if f is nonnegative and μ_g-integrable over $[a, b]$, then for any $v > 0$

$$\mu_g(\{x \in [a, b] : f(x) > v\}) \leq \frac{1}{v} {}^M\!\int_a^b f \, d\mu_g.$$

63. Give an example of two functions f_1 and f_2 and an increasing, right-continuous function g for which $f_1 = f_2$ except at a single point, but for which $^M\!\int_a^b f_1 \, d\mu_g \neq {}^M\!\int_a^b f_2 \, d\mu_g$.

64. Example 54 (page 229) shows that Riemann-Stieltjes integrability over $[a, b]$ and $[b, c]$ is not sufficient to conclude Riemann-Stieltjes integrability over $[a, c]$.

    (a) Prove that if f is Lebesgue-Stieltjes integrable over $[a, b]$ and $[b, c]$, then f is Lebesgue-Stieltjes integrable over $[a, c]$.
    (b) Give an example where f is Lebesgue-Stieltjes integrable over $[a, b]$ and $[b, c]$, but $^M\!\int_a^c f \, d\mu_g \neq {}^M\!\int_a^b f \, d\mu_g + {}^M\!\int_b^c f \, d\mu_g$.
    (c) Explain how (b) can be true in spite of exercises 9 and 57.
    (d) Suggest a way to resolve this seeming contradiction.

8.7 Extensions: filling the gaps

65. Prove that $\mu(E) \geq 0$ for all $E \in \mathcal{A}$ is equivalent to $\mu(A) \leq \mu(B)$ for all $A, B \in \mathcal{A}$ satisfying $A \subseteq B$.

66. Prove that μ in example 67 (page 251) is a measure on $\mathcal{A}$. In other words, show that μ satisfies the three properties on page 251.

67. How do we know that μ in example 68 (page 253) is a measure on $\mathcal{M}$?

68. Justify the integral calculations in example 68 (page 253).

69. Let $(\mathbb{R}, \mathcal{A}, \mu)$ be a measure space such that $\mathcal{A}$ contains the Borel sets and every bounded interval has finite measure. Let

$$g(x) = \begin{cases} \mu((0, x]), & 0 < x \\ 0, & x = 0 \\ -\mu((x, 0]), & x < 0. \end{cases}$$

    Prove that

    (a) g is increasing and right-continuous. (Use monotonicity and that fact that $\sum_n \mu\left(\left(a + \frac{1}{n+1}, a + \frac{1}{n}\right]\right) = \mu((a, a + 1]) < \infty$.)
    (b) $\mu = \mu_g$.

70. Prove that $(\mathbb{C}, \mathcal{A}, \mu)$ of example 69 (page 253) is a measure space.

71. Verify that P_1 and P_2 of example 70 (page 255) are projections.

72. Verify that μ of example 70 (page 255) is a projection-valued measure on $\mathcal{A}$.

8.7 Extensions: deeper reflections

73. For the measure in example 69 (page 253)

   (a) Find the smallest positive integer n so that ${}^M\!\int_C z^n \, d\mu \neq 0$.

   (b) Compute ${}^M\!\int_C \frac{1}{z^2} \, d\mu$.

74. Prove that if $(\mathbb{R}, \mathcal{A}_1, \mu_1)$ and $(\mathbb{R}, \mathcal{A}_2, \mu_2)$ are two measure spaces then

   (a) $(\mathbb{R}, \mathcal{A}_1 \cap \mathcal{A}_2, \mu_1 + \mu_2)$ is a measure space.

   (b) For any μ-measurable function f, ${}^M\!\int_E f \, d\mu = {}^M\!\int_E f \, d\mu_1 + {}^M\!\int_E f \, d\mu_2$ where $\mu = \mu_1 + \mu_2$.

75. Suppose that μ is a discrete measure that assigns weights to the points $\{c_i\}_{i=1}^n$ and for which $\mu \left(R \setminus \{c_i\}_{i=1}^n \right) = 0$. Prove that for any function f defined on $\mathbb{R}$, ${}^M\!\int_{\mathbb{R}} f \, d\mu = \sum_i f(c_i) \mu(\{c_i\})$.

76. Suppose that

$$\mu(A) = \begin{cases} 0, & 1, 2 \notin A \\ 1, & 1 \in A, \, 2 \notin A \\ 2, & 1 \notin A, \, 2 \in A \\ 3, & 1, 2 \in A. \end{cases}$$

Compute

   (a) ${}^M\!\int_0^2 1 \, d\mu$.

   (b) ${}^M\!\int_0^1 x \, d\mu$.

   (c) ${}^M\!\int_0^2 x \, d\mu$.

   (d) ${}^M\!\int_0^2 x^2 \, d\mu$.

77. Let $\mathcal{A}$ be the sigma algebra of all subsets of $\mathbb{C}$ and define

$$\mu(E) = \sum_{\substack{n+im \in E, \\ n,m \in \mathbb{Z} \setminus \{0\}}} \frac{1}{nm}.$$

   (a) Prove that μ is a measure on $\mathcal{A}$.

      Let $D_2 = \{z \in \mathbb{C} : |z| \leq 2\}$. Compute

   (b) ${}^M\!\int_{D_2} 1 \, d\mu$.

   (c) ${}^M\!\int_{D_2} z \, d\mu$.

   (d) ${}^M\!\int_{D_2} z^2 \, d\mu$.

78. Consider the measure space $(\mathbb{R}, \mathcal{A}, \mu)$ where $\mathcal{A} = \{\emptyset, \mathbb{R}\}$ and

$$\mu(A) = \begin{cases} 1, & A = \mathbb{R} \\ 0, & A = \emptyset. \end{cases}$$

(a) Characterize the μ-measurable functions.

(b) Give a simple formula for evaluating $^M\!\int_{\mathbb{R}} f \, d\mu$ when f is μ-measurable.

79. Consider the measure space $(\mathbb{R}, \mathcal{A}, \mu)$ where $\mathcal{A}$ is the collection of all subsets of $\mathbb{R}$ and $\mu(A)$ is the number of elements in A when A is a finite set and $\mu(A) = \infty$ otherwise.

(a) Characterize the μ-measurable functions.

(b) Provide an alternate expression for $^M\!\int_{\mathbb{R}} f \, d\mu$ when f is μ-integrable.

80. Give an example of a measure space on which the bounded convergence theorem (page 250) fails.

81. Let $(\mathbb{R}, \mathcal{A}, \mu)$ be a measure space where $\mathcal{A}$ contains all intervals and $\mu([0, 1]) < \infty$. Prove that if μ is translation invariant, then μ is a multiple of the Lebesgue measure.

82. Let $X = [0, 1] \times [0, 1]$ and let w be the weight function that assigns a weight of $x^2 y$ to the rectangle $[0, x] \times [0, y]$. (Note that the degenerate rectangles $[0, 0] \times [0, y]$ and $[0, x] \times [0, 0]$ have weights of zero.) What are the measures of $[0, b] \times (c, d]$, $(a, b] \times [0, d]$, and $(a, b] \times (c, d]$?

83. Let

$$P_{-2} = \frac{1}{4} \begin{bmatrix} 1 & 1 & 1 & 1 \\ 1 & 1 & 1 & 1 \\ 1 & 1 & 1 & 1 \\ 1 & 1 & 1 & 1 \end{bmatrix},$$

$$P_{2} = \frac{1}{4} \begin{bmatrix} 1 & -1 & 1 & -1 \\ -1 & 1 & -1 & 1 \\ 1 & -1 & 1 & -1 \\ -1 & 1 & -1 & 1 \end{bmatrix}, \text{ and}$$

$$P_{4} = \frac{1}{2} \begin{bmatrix} -1 & 0 & 1 & 0 \\ 0 & -1 & 0 & 1 \\ 1 & 0 & -1 & 0 \\ 0 & 1 & 0 & -1 \end{bmatrix}.$$

Define μ by taking $\mu(\{z\}) = P_z, z = -2, 2, 4$, setting $\mu(\mathbb{R}\backslash\{-2, 2, 4\}) = 0$, and extending μ in the natural way to all subsets of $\mathbb{R}$.

(a) Compute

    i. $^{M}\!\int_{-5}^{5} 1 \, d\mu$,

    ii. $A = {}^{M}\!\int_{-5}^{5} x \, d\mu$, and

    iii. $B = {}^{M}\!\int_{-5}^{5} \sqrt{x} \, d\mu$.

(b) Show that $B^2 = A$.

84. Let

$$P_3 = \frac{1}{3} \begin{bmatrix} 2 & 1 & -1 \\ 1 & 2 & 1 \\ -1 & 1 & 2 \end{bmatrix}, \text{ and}$$

$$P_6 = \frac{1}{3} \begin{bmatrix} 1 & -1 & 1 \\ -1 & 1 & -1 \\ 1 & -1 & 1 \end{bmatrix}.$$

Define μ by setting $\mu(\{z\}) = P_z$, $z = 3, 9$, $\mu(\mathbb{R} \setminus \{3, 9\}) = 0$. Extend μ in the natural way to all subsets of $\mathbb{R}$.

(a) Compute

    i. $^{M}\!\int_{0}^{10} 1 \, d\mu$,

    ii. $A = {}^{M}\!\int_{0}^{10} x \, d\mu$, and

    iii. $B = {}^{M}\!\int_{0}^{10} \sqrt{x} \, d\mu$.

(b) Show that $B^2 = A$. (You may want to use a computer algebra system.)

(c) Find $\log_3 A$ using an integral.

(d) Another approach to finding $\log_2 A$ would be to approximate it using $\log_2 A = \frac{1}{\ln 2} \ln A$ and the power series

$$\ln x = \sum_{k=1}^{\infty} \frac{(-1)^{k+1}}{k} (x - 1)^k.$$

Why does this not work?

85. Prove that if $(X, \mathcal{A}, \mu)$ is a measure space where μ is a projection-valued measure, then $\mu(A)$ and $\mu(B)$ are orthogonal projections when A and B are disjoint elements of $\mathcal{A}$. (Two projections P_1 and P_2 are orthogonal if $P_1 P_2 = 0$.)

8.8 Other directions: deeper reflections

86. Show that the Dirichlet function is gauge-Stieltjes[15] integrable with respect to g over $[a, b]$ if and only if the rational numbers in $[a, b]$ have μ_g-measure zero. What must be true about g in order to have $\mu_g (\mathbb{Q} \cap [a, b]) = 0$?

87. Modify the definition of the Darboux integral to a definition of the Darboux-Stieltjes integral.

    (a) Prove that the Darboux-Stieltjes and Riemann-Stieltjes integrals are the same.
    (b) State and prove a Cauchy criterion for the Darboux-Stieltjes integral.
    (c) Prove that a function f is Darboux-Stieltjes integrable with respect to an increasing, right-continuous function g over $[a, b]$ if and only if f is continuous μ_g a.e.

88. The weight functions in this chapter have a specific form. Namely, they are constructed using $w((a, b]) = g(b) - g(a)$ where g is an increasing, right-continuous function. When one moves away from this form of weight function, odd things can happen. Let $w((a, b]) = \sqrt{b - a}$ and define

$$\mu_w^* (A) = \inf \left\{ \sum_k w((a_k, b_k]) : \begin{array}{l} \{w((a_k, b_k])\} \text{ is a countable} \\ \text{cover of } A \text{ by finite intervals} \end{array} \right\}$$

    for any $A \subseteq \mathbb{R}$.[16]

    (a) Prove that $\mu_w^* ((a, a + 1]) = 1$ for any $a \in \mathbb{R}$. ($\sqrt{x} \geq x$ for $0 \leq x \leq 1$.)
    (b) Show that $(0, 1]$ is not measurable. (Apply the Carathéodory criterion to $A = (0, 2]$.)

8.9 References

Burk, F.E. (2007). *A Garden of Integrals*. Mathematical Association of America.

Brandt, H.E. (1999). Positive operator valued measure in quantum information processing. *Am. J. Phys.* **67**, 434–440.

---

[15] Use the most obvious modification of the Riemann-Stieltjes integral.
[16] This exercise is based on an example from *Measure, Topology, and Fractal Geometry* by Gerald Edgar (page 139).

DePree, J. and C. Swartz (1988). *Introduction to Real Analysis*. John Wiley & Sons.

Halmos, P.R. (1976). *Measure Theory*. Graduate Texts in Mathematics **18**, Springer.

Hildebrandt, T.H. (1938). Definitions of Stieltjes integrals of the Riemann type. *The American Mathematical Monthly* **45** (5): 265–278. JSTOR 2302540.

Protter, Jr., M.H. and C.B. Morrey (1991). *A First Course in Real Analysis*. Springer.

Taylor, A.E. (1965). *General Theory of Functions and Integration*. Dover.

Teschl, G. (2009). *Mathematical Methods in Quantum Mechanics with Applications to Schrödinger Operators*. Graduate Studies in Mathematics **157**, American Mathematical Society. www.mat.univie.ac.at/~gerald/ftp/book-schroe/.

CHAPTER 9

A Look Back

Historically, the Riemann and Darboux integrals (and the equivalent Cauchy integral that preceded them) were introduced to solve a different set of problems than were the Lebesgue and gauge integrals. The Riemann and Darboux integrals were developed in response to foundational questions such as those identified by Bishop Berkeley (page 7). The Lebesgue and gauge integrals address issues of convergence arising from Fourier series. It should then be no great surprise that the integrals have somewhat different properties. This concluding chapter presents a comparative overview of the integrals covered in this text.

9.1 Basic approaches

Given a function f on $[a, b]$, how do we define $\int_a^b f$?

Riemann integral
The Riemann integral partitions $[a, b]$ into subintervals $\{I_k\}$. Then, using tags $\{t_k\}$ with $t_k \in I_k$, the sum $\sum_k f(t_k) \Delta x_k$ is computed. If all such sums approximate some fixed real number A when the subintervals in the partition are suitably controlled, the function f is integrable over $[a, b]$ and the value of the integral is A. Specifically, given an $\varepsilon > 0$ there must be a $\delta > 0$ such that, when all the subintervals in a partition have width less than δ, the sum must be within ε of A independent of the the choice of the tags. When proving results about the Riemann integral it is common to spend a significant amount of energy identifying a possible value for A. Cauchy sequences and cluster points figure prominently in this process.

Darboux Integral
Like the Riemann integral, the Darboux integral partitions $[a, b]$ into subintervals $\{I_k\}$. Instead of selecting tags, we compute the sums

271

$\sum_k \inf_{I_k} f \ \Delta x_k$ and $\sum_k \sup_{I_k} f \ \Delta x_k$. A function is Darboux integrable on $[a, b]$ if

$$\sup \sum_k \inf_{I_k} f \ \Delta x_k = \inf \sum_k \sup_{I_k} f \ \Delta x_k$$

where the outer supremum and infimum are taken over all possible partitions of $[a, b]$ by subintervals. In this case, the Darboux integral is the common value.

The Riemann and Darboux integrals are equivalent. However, proofs for the two integrals can follow quite different paths. For example, when proving theoretical results about the Darboux integral, there is no need to construct a value using Cauchy sequences or some related technique since the supremum and infimum already identify a value. Even apparent analogs have significant differences. For the Darboux integral, a partition can play the role played by the δ in the Riemann integral. A function is Darboux integrable on $[a, b]$ if and only if given any $\varepsilon > 0$, there is a partition $\{I_k\}$ such that $\left| \sum_k \inf_{I_k} f \ \Delta x_k - \sum_k \sup_{I_k} f \ \Delta x_k \right| < \varepsilon$ (theorem 12, page 55). Unlike the case for the Riemann integral, however, the inequality involves only the single partition $\{I_k\}$ and not other associated partitions.

Lebesgue integral

The Lebesgue integral takes what one might call an orthogonal approach. Given a measurable function satisfying $\alpha < f < \beta$, the interval $(\alpha, \beta]$ is partitioned using disjoint intervals of the form $(y_k, y_{k+1}]$. From such a partition, the sum $\sum_k y_{k-1} \lambda (E_k)$ is computed where E_k is the preimage of the kth interval and λ is Lebesgue measure. The sums will always converge in the sense that there is a real value A that the sums approach as the maximum length of the subintervals of the partition decreases to zero. By allowing A to be an extended real number, this approach can be generalized to handle functions that are not bounded above. (See page 132.)

Alternatively, one can partition $[a, b]$ with measurable sets $\{E_k\}$ and compute $\sup \sum_k \left(\inf_{E_k} f \right) \lambda (E_k)$ where the supremum is taken over all measurable partitions of $[a, b]$. For non-negative (possibly unbounded) functions, the supremum always exists in the extended real numbers. An arbitrary measurable function f is Lebesgue integrable if both f^+ and f^- are (finitely) Lebesgue integrable. (See page 133.)

The major catch with the Lebesgue integral is that the significant framework of measurability and Lebesgue measure must be constructed in order to make sense of $\lambda (E_i)$. The work done to develop measure theory is roughly equivalent to the effort required to develop the Darboux integral. Once the measure-theoretic foundation has been laid, one finds that the defining sums only make sense for measurable functions. Then one can begin the task of investigating the properties of the Lebesgue integral applied to measurable functions.

Gauge integral

The gauge integral returns to a Riemann-style approach where $[a, b]$ is partitioned with tagged intervals. Unlike the Riemann integral, the gauge integral does not permit a free choice of tags once the intervals have been selected. Instead, the intervals and tags are selected together subject to a gauge restriction. A gauge is a function that associates a positive value $\delta(t)$ with each point $t \in [a, b]$. A tagged partition $\{(t_k, I_k)\}$ is δ-fine if for all k the length of I_k is less than $\delta(t_k)$. A function is gauge integrable if there is a real value A such that given any $\varepsilon > 0$, it is possible to construct a gauge δ such that $\left| \sum_k f(t_k) \Delta x_k - A \right| < \varepsilon$ for any δ-fine partition. Once again, when proving general results, one must invoke Cauchy sequences (or some analog) to construct the value A. However, there is no need to develop the somewhat complicated (and for many students initially mystifying) theory of measure. Instead, one needs to develop facility in constructing gauges.

9.2 Integrable functions

In 1823, before Riemann introduced his integral, Cauchy gave a related definition of an integral that essentially uses right endpoints as tags. (See page 8.) While the Cauchy integral is equivalent to the Riemann and Darboux integrals (see exercise 38 on page 49 and exercise 32 on page 70), Cauchy assumed that all functions to be integrated were continuous. Only later, when the need to deal with functions arising from sources like Fourier series came into focus, was an effort made to determine the extent of the class of functions for which an integral makes sense. The Dirichlet function gave an early and relatively simple example of a function that is not Riemann integrable.

Riemann-Darboux integral

It is fairly easy to see that monotone and continuous functions are Riemann-Darboux integrable, but a complete characterization of the Riemann-Darboux integrable functions had to wait over 30 years. In 1902, Lebesgue proved that a function is Riemann-Darboux integrable over $[a, b]$ if and only if it is bounded on $[a, b]$ and continuous except on a set of measure zero[1] (theorem 20, page 63).

Lebesgue integral

The Lebesgue integral greatly extends the class of integrable functions. Any bounded, measurable function is Lebesgue integrable. No non-measurable function is Lebesgue integrable since the sum $\sum_i y_i \lambda(E_i)$ is not defined if one of the E_i is not measurable. For non-negative functions, the restriction that the integrand be bounded can be removed if integrals are permitted to take on values in the extended real numbers. In general, a function f is Lebesgue integrable if and only if f is measurable and both ${}^L\!\int_a^b f^+$ and ${}^L\!\int_a^b f^-$ are finite.

Gauge Integral

The gauge integral modestly extends the class of integrable functions. Every Lebesgue integrable function is gauge integrable (theorem 77, page 200) and all gauge integrable functions are measurable (theorem 78, page 200), but it is possible for f to be gauge integrable while both ${}^g\!\int_a^b f^+$ and ${}^g\!\int_a^b f^-$ are infinite. In particular, if f is a differentiable function, then f' is gauge integrable. Example 47 (page 192) illustrates that not all derivatives are Lebesgue integrable.

9.3 Convergence theorems

Given a sequence $\{f_n\}$ of integrable functions that converges to a function f, when is f integrable and when is

$$\lim_n \int_a^b f_n = \int_a^b \lim_n f_n?$$

Riemann-Darboux integral

For the Riemann-Darboux integral, a sequence of integrable functions must converge uniformly to guarantee that the limit function is integrable and that the order of the limit and integral may be interchanged. The biggest obstacle to convergence results for the Riemann-Darboux integral is that the limit function may not be integrable. In the absence of uniform convergence, introducing additional restrictions such as requiring the sequence to be uniformly

[1] As the Dirichlet function shows, being continuous except on a set of measure zero is not the same as being equal almost everywhere to a continuous function.

bounded and increasing will not ensure that the limit function is integrable. (See example 8 on page 41.) Even when the limit function is integrable, example 9 (page 42) shows that the limit and integral cannot always be interchanged. This second mode of failure becomes predominant as the class of integrable functions grows.

Lebesgue integral

The issue of integrability is substantially reduced for the Lebesgue integral. If a sequence of measurable functions converges almost everywhere, then the limit function is again measurable (theorem 25, page 107). The only way for the limiting function f to fail to be Lebesgue integrable is for ${}^{\text{L}}\!\int_a^b f^+$ or ${}^{\text{L}}\!\int_a^b f^-$ to be infinite. But even when the limit function is integrable, unbounded values of the functions $\{f_n\}$ are potential barriers to the interchange of limits and integrals. (See example 9, page 42.) Therefore, the convergence results for the Lebesgue integral tend to focus on controlling the size of the functions $\{f_n\}$. If the functions in $\{|f_n|\}$ are uniformly bounded by a constant, then the limit function f is Lebesgue integrable and the order of the integral and limit can be swapped (theorem 36, page 136). The bound need not be uniform. An integrable function bounding all the functions in $\{|f_n|\}$ will suffice (theorem 39, page 140). Boundedness can be dropped altogether if we restrict our attention to an increasing sequence of non-negative functions and allow integrals to assume values in the extended real numbers (theorem 37, page 138). Even if the sequence $\{f_k\}$ of non-negative functions is not increasing and does not converge, we can still say (theorem 38, page 139) that

$$
{}^{\text{L}}\!\int_a^b \lim_k f_k \le \lim_k {}^{\text{L}}\!\int_a^b f_k.
$$

These are far more powerful results than those that are available for the Riemann-Darboux integral.

Gauge integral

Similar results hold for the gauge integral. However, the proof techniques are quite different. The gauge integral has no analog to the fact that the limit of measurable functions is always measurable. Nevertheless, the existence of gauge integrable functions g_1 and g_2 on $[a, b]$ such that $g_1 \le f_k \le g_2$ for all k is sufficient to ensure that the limit function is gauge integrable and that the order of the integral and the limit can be reversed (theorem 69, page 189). This result is stronger than the corresponding theorem for the

Lebesgue integral since g_2 need not be non-negative as must be the case if $|f_k|$ is to be bounded. In the case of a monotone sequence of functions $\{f_k\}$, the boundedness of $\left\{ \sqrt[g]{\int_a^b} f_k \right\}$ is both necessary and sufficient for the limit function to be integrable and for the order of the integral and limit to be reversible (theorem 67, page 186). Note that, unlike the case for the Lebesgue integral, these results do not require $\left\{ \sqrt[g]{\int_a^b} f_k^+ \right\}$ or $\left\{ \sqrt[g]{\int_a^b} f_k^- \right\}$ to be bounded.

9.4   The fundamental theorems

The first fundamental theorem of calculus states that, under appropriate conditions, $\int_a^x F' = F(x) - F(a)$ for all $x \in [a, b]$. At a minimum, F must be continuous everywhere as illustrated by example 7 (page 40). Beyond this basic requirement, the conditions vary by the type of integral.

Riemann-Darboux integral

For the Riemann-Darboux integral, the required condition is that F be differentiable on $[a, b]$ with F' bounded on $[a, b]$ and continuous a.e. These are the conditions that ensure that F' is integrable. As illustrated by Volterra's function (page 81), F' need not be Riemann-Darboux integrable even if F has a bounded derivative at every point in $[a, b]$.

Lebesgue integral

For the Lebesgue integral, the appropriate condition is that F be absolutely continuous on $[a, b]$. Then F will be differentiable a.e. on $[a, b]$ and F' will be Lebesgue integrable with $\sqrt[L]{\int_a^x} F' = F(x) - F(a)$ (theorem 47, page 148). Conversely, the second fundamental theorem of calculus for the Lebesgue integral implies that F must be absolutely continuous if $\sqrt[L]{\int_a^x} F' = F(x) - F(a)$.

Gauge integral

The gauge integral has the simplest version of the first fundamental theorem of calculus. If F is differentiable on $[a, b]$, then F' is gauge integrable on $[a, b]$ and $\sqrt[g]{\int_a^x} F' = F(x) - F(a)$ (theorem 71, page 191). The discussion following the proof of theorem 75 (page 197) shows the existence of differentiable functions that are not absolutely continuous and so for which the fundamental theorem will fail when using the Lebesgue integral. As long as F remains continuous, the first fundamental theorem of calculus for the gauge integral can be generalized to admit a countable number of points at which F is not differentiable (theorem 72, page 193). Since Lebesgue integrable functions are also gauge integrable, absolutely continuous functions must also satisfy ${}^g\!\!\int_a^x F' = F(x) - F(a)$.

The second fundamental theorem
Given an integrable function f on $[a, b]$, the second fundamental theorem of calculus says that, under appropriate conditions, the function $F(x) = \int_a^x f$ will be differentiable with $F' = f$. In the cases of both the Riemann-Darboux and the Lebesgue integrals, F is absolutely continuous and so must be differentiable a.e. (theorem 42, page 143). Moreover, $F'(x_0) = f(x_0)$ whenever f is continuous at x_0. Since Riemann-Darboux integrable functions are continuous a.e., $F' = f$ a.e. Even though Lebesgue integrable functions need not be continuous anywhere (consider the Dirichlet function, for example), any Lebesgue integrable function f satisfies $F' = f$ a.e. (theorem 42, page 146). For gauge integrable functions f, the functions $F(x) = {}^g\!\!\int_a^x f$ may not be absolutely continuous. Nevertheless, except for a set of measure zero, F is differentiable with $F' = f$ (theorem 75, page 197).

9.5 Conclusion

In this text, we have investigated four integrals: Riemann, Darboux, Lebesgue, and gauge. In each case, the succeeding integral definition offers a distinct improvement over the previous integral. The Darboux integral supports more efficient proof techniques than the Riemann integral. The Lebesgue integral allows for a much larger class of integrable functions and has much stronger convergence properties than the Riemann-Darboux integral. The gauge integral modestly enlarges the class of integrable functions and has somewhat more flexible convergence properties while, at the same time, avoiding the need to develop measure theory. While the option of avoiding the early introduction of measure theory is appealing, the desirability of doing so is somewhat diminished by the fact that, as seen at the end of Chapter 8, measure theory itself has powerful and useful generalizations.

CHAPTER 10

Afterword: L_2 Spaces and Fourier Series

As is evident in the previous chapters, the task of resolving issues related to the interactions between Fourier series and integration was a significant motivator in the development of definitions and theories of integration. In some sense, we have completed the task. We have two integrals, Lebesgue and gauge, that interact well with sequences and series in general and so with Fourier series in particular.

But these integrals do more than resolve the issues related to the interchange of integrals and limits. They also provide a context, L_2 spaces, within which we can develop an alternative notion of convergence that is particularly suited to the study of Fourier series. In this new setting, functions have Fourier series whose coefficients can be computed using integration and which converge to the original function. The properties of the Lebesgue and gauge integrals are critical in developing L_2 spaces.

The goal of this epilogue is to introduce L_2 spaces and to examine how they resolve fundamental issues raised by Fourier series that, initially, seem to be unrelated to questions of integration.

Given a Lebesgue-integrable function f on the interval $[0, 1]$, the integrals $a_k = \int_0^1 f(x) \sin(2\pi k x) \, dx$ and $b_k = \int_0^1 f(x) \cos(2\pi k x) \, dx$ are defined. Moreover, as long as $a_0 + 2 \sum_{k=1}^{\infty} (a_k \sin(2\pi k x) + b_k \cos(2\pi k x))$ converges appropriately to a function g, then the convergence theorems developed in prior chapters along with a few basic facts about the integrals of trigonometric functions[1] show that, for any $n \in \mathbb{N}$,

$$\int_0^1 g(x) \sin(2\pi n x)\, dx$$

$$= \int_0^1 \left(a_0 + 2\sum_{k=1}^{\infty} (a_k \sin(2\pi k x) + b_k \cos(2\pi k x)) \right) \sin(2\pi n x)\, dx$$

$$= a_0 \int_0^1 \sin(2\pi n x)\, dx$$

$$+ 2\sum_{k=1}^{\infty} a_k \int_0^1 \sin(2\pi k x) \sin(2\pi n x)\, dx$$

$$+ 2\sum_{k=1}^{\infty} b_k \int_0^1 \cos(2\pi k x) \sin(2\pi n x)\, dx$$

$$= a_n.$$

A similar derivation shows that $\int_0^1 g(x) \cos(2\pi n x)\, dx = b_n$. Thus f and g produce the same coefficients. By translating and scaling, we can draw analogous conclusions for functions defined on any compact interval.

While the issues related to integration have been resolved, other questions remain. The relevant integrals now make sense and, with a bit of care, we are justified in interchanging the integrals and the summations. However, we have not verified what we really want to know: namely, does f equal g? In fact we cannot know that $f = g$. If the function f is modified on a set of measure zero, then all the associated integrals, and thus the coefficients a_k and b_k, remain unchanged. Consequently, it is clear that the best we could hope for is that $f = g$ a.e.

The words "as long as" that preceded "converges appropriately" at the bottom of page 279 are even more problematic. How do we know that the sum converges at all?

The question of the convergence of trigonometric series received a great deal of attention in the early 1800s. For example, Neils Henrik Abel published a result that, when combined with trigonometric identities like

$$\sum_{k=1}^{n} (-1)^{k-1} \cos\left(\frac{(2k-1)\pi x}{2} \right) = \frac{1 - (-1)^n \cos(\pi n x)}{2\cos(\pi x / 2)}$$

---

[1] For integers k and n, $\int_0^1 \cos(2\pi k x) \sin(2\pi n x)\, dx = 0$, $\int_0^1 \sin^2(2\pi n x)\, dx = \frac{1}{2}$, and, when $k \neq n$, $\int_0^1 \sin(2\pi k x) \sin(2\pi n x)\, dx = \int_0^1 \cos(2\pi k x) \cos(2\pi n x)\, dx = 0$.

provided a useful convergence test (Dirichlet's test) for the convergence of trigonometric series.[2] But knowing that $a_0 + 2 \sum_{k=1}^{\infty} (a_k \sin(2\pi k x) + b_k \cos(2\pi k x))$ converges is not the same as knowing the series converges to f.

We will not develop results like Dirichlet's test here as the Lebesgue and gauge integrals provide another path to resolving the questions raised by Fourier series. Moreover, since this chapter is not concerned with comparing the properties of various integrals, we will work exclusively with the Lebesgue integral and use the notation $\int_a^b f \, d\lambda$.

While we worked exclusively with real-valued functions in the main body of this text, L_2 spaces, the context within which we will be working in this chapter, are typically defined for complex-valued functions. Since this generalization requires little extra effort, going forward we will assume that we are working with complex-valued functions unless otherwise noted. A complex-valued function $f = u + iv$ is **measurable** if the real-valued functions u and v are both measurable. When u and v are integrable, $\int_a^b f \, d\lambda = \int_a^b u \, d\lambda + i \int_a^b v \, d\lambda$. A review of some of the basic properties of complex numbers is provided in exercises 1 through 3. Also, consult Appendix A.6, which contains a quick overview of the basic facts about complex numbers used in this chapter.

10.1 L_2 spaces

Fix a compact interval $[a, b]$ and let λ denote the Lebesgue measure.[3]

Definition 37. *A measurable function $f : [a, b] \rightarrow \mathbb{C}$ is called **square-integrable** if $\int_a^b |f|^2 \, d\lambda < \infty$. The set of all square-integrable functions on $[a, b]$ is denoted by $L_2([a, b])$.[4]*

More generally, for any measurable set X, $L_p(X)$ is defined as the set of measurable functions f satisfying $\int_X |f|^p \, d\lambda < \infty$. We will not develop the theory of L_p spaces in the main text, but some of the central ideas are introduced in the exercises.

---

[2] David Bressoud provides a well-written account of this work in pages 165–173 of *A Radical Approach to Real Analysis*, The Mathematical Association of America, Washington, D.C., 1994.

[3] While we will develop the theory of L_2 spaces using the Lebesgue integral, one could use the gauge integral. The Lebesgue integral is selected here since, with little additional effort, what follows could be extended to any finite measure μ on a compact set X.

[4] It is easy to see that if $f = u + iv$ where u and v are real-valued functions, then f is square-integrable if and only if u and v are both square integrable. (See Exercise 40.)

The set $L_2([a, b])$ has much more structure than its set description indicates. First, $L_2([a, b])$ is a vector space. In other words, using the usual operations of scalar multiplication and addition of functions, $L_2([a, b])$ satisfies the algebraic properties of closure, commutativity, associativity, and distributivity and has an additive identity and inverses.

In what follows, we assume the reader is familiar with the basic properties of vector spaces and with (orthogonal) projections in inner product spaces. Exercises 4 through 19 provide a review of the salient properties. Also, consult Appendix A.7.

Theorem 101. *The set of square-integrable functions forms a vector space under the usual operations of scalar multiplication and addition of functions.*

Proof. Suppose that f and g are square-integrable functions over $[a, b]$ and that c is a scalar. Note that cf and $f + g$ are measurable functions. The proof that the set of square-integrable functions is closed under scalar multiplication is left as a straightforward exercise (exercise 20).

To prove that $L_2([a, b])$ is closed under addition, note that

$$\int_a^b |f + g|^2 \, d\lambda \leq \int_a^b (|f| + |g|)^2 \, d\lambda$$

$$\leq \int_a^b (2 \max \{|f|, |g|\})^2 \, d\lambda$$

$$= \int_a^b (4 \max\{|f|^2, |g|^2\}) \, d\lambda$$

$$\leq \int_a^b (4(|f|^2 + |g|^2)) \, d\lambda$$

$$\leq 4 \left(\int_a^b |f|^2 \, d\lambda + \int_a^b |g|^2 \, d\lambda \right) < \infty.$$

The set of square-integrable functions inherits the remainder of the vector space properties from the vector space of all functions on $[a, b]$. □

In fact, $L_2([a, b])$ is more than a vector space. We can use the Lebesgue integral to create an inner product on $L_2([a, b])$.

Definition 38. *Let* $f, g \in L_2([a, b])$. *Define* $\langle f, g \rangle = \int_a^b f \overline{g} \, d\lambda$ *where* $\overline{f}(x) = \overline{f(x)}$.

Theorem 102. *The function* $\langle \cdot, \cdot \rangle$ *is an inner product on* $L_2([a, b])$. *In other words, for any functions* $f, g,$ *and* $h \in L_2([a, b])$ *and any scalar* $c \in \mathbb{C}$

1. $\langle f, g \rangle$ *is finite,*

2. $\langle g, f \rangle = \overline{\langle f, g \rangle}$,

3. $\langle cf, g \rangle = c \langle f, g \rangle$,

4. $\langle f + g, h \rangle = \langle f, h \rangle + \langle g, h \rangle$, *and*

5. $\langle f, f \rangle \geq 0$ *with* $\langle f, f \rangle = 0$ *if and only if* $f = 0$ *a.e.*

Proof. To verify (1), note that, since $(|f| - |g|)^2 \geq 0$, $|fg| \leq 2|fg| \leq |f|^2 + |g|^2$. Also, $|f\overline{g}| = |fg|$. Hence

$$|\langle f, g \rangle| = \left| \int_a^b f\overline{g} \, d\lambda \right| \leq \int_a^b |fg| \, d\lambda \leq \int_a^b (|f|^2 + |g|^2) \, d\lambda < \infty.$$

To prove the second half of (5), suppose that $\langle f, f \rangle = 0$. Set $Z = \{x \in [a, b] : f(x) \neq 0\}$ and note that Z is the disjoint union of $Z_n = \{x \in [a, b] : \frac{1}{n} \leq |f(x)| < \frac{1}{n-1}\}$, $n \in \mathbb{N}$. (Here we interpret $\frac{1}{n-1}$ as ∞ when $n = 1$.) Since

$$0 = \langle f, f \rangle = \int_a^b |f|^2 \, d\lambda \geq \int_{Z_n} \frac{1}{n^2} \, d\lambda = \frac{1}{n^2} \lambda(Z_n) \geq 0,$$

we conclude that $\lambda(Z_n) = 0$ for $n \in \mathbb{N}$. Hence $\lambda(Z) = \sum_n \lambda(Z_n) = 0$ and $f = 0$ a.e.

The proofs of the remaining properties are left as an exercise (exercise 21) as they are straightforward consequences of properties of the Lebesgue integral. □

Since $L_2([a, b])$ has an inner product, it also has the standard norm induced by the inner product.

Definition 39. *Let* $f \in L_2([a, b])$. *Define* $\|f\|_2 = \sqrt{\langle f, f \rangle} = (\int_a^b |f|^2 \, d\lambda)^{1/2}$.

Theorem 103. $\|f\|_2$ *is a norm on* $L_2([a, b])$. *That is, for* $f, g \in L_2([a, b])$ *and any scalar* $c \in \mathbb{C}$,

1. $\|cf\|_2 = |c| \|f\|_2$ *and*

2. $\|f + g\|_2 \leq \|f\|_2 + \|g\|_2$.

Proof. Exercises 11 and 16. □

We refer to $\|f\|_2$ as the L_2 norm of f.

While we have referred to the elements of $L_2([a,b])$ as functions (and we will continue to do so), the elements of $L_2([a,b])$ are more accurately characterized as equivalence classes of functions that are equal a.e. In a vector space with a norm $\|\cdot\|$, one wants $\|\mathbf{v}\|$ to be zero if and only if the vector $\mathbf{v}$ is zero. By (5) of theorem 102, $\|f - g\|_2 = 0$ is equivalent to $f = g$ a.e. Hence, for example, the zero function and the Dirichlet function

$$d(x) = \begin{cases} 1, & x \in \mathbb{Q} \\ 0, & x \notin \mathbb{Q} \end{cases}$$

are considered to be the same function in $L_2([a,b])$.

At this point it is worthwhile pausing to consider how all this is related to Fourier series. The functions involved in creating a Fourier series (or, more generally, a trigonometric series) on the interval $[0,1]$ are $1, \sin(2\pi kx)$, and $\cos(2\pi kx)$ where $k = 1, 2, 3, \ldots$. As elements of the inner product space $L_2([0,1])$, these functions are mutually orthogonal as

$$\langle 1, \sin(2\pi kx) \rangle = \int_0^1 \sin(2\pi kx) \, d\lambda = 0,$$

$$\langle 1, \cos(2\pi kx) \rangle = \int_0^1 \cos(2\pi kx) \, d\lambda = 0, \text{ and}$$

$$\langle \sin(2\pi jx), \cos(2\pi kx) \rangle = \int_0^1 \sin(2\pi jx) \cos(2\pi kx) \, d\lambda = 0$$

for all $j, k \in \mathbb{N}$. Additionally, for $j \neq k \in \mathbb{N}$,

$$\langle \sin(2\pi jx), \sin(2\pi kx) \rangle = \int_0^1 \sin(2\pi jx) \sin(2\pi kx) \, d\lambda = 0 \text{ and}$$

$$\langle \cos(2\pi jx), \cos(2\pi kx) \rangle = \int_0^1 \cos(2\pi jx) \cos(2\pi kx) \, d\lambda = 0.$$

Similar calculations show that $\langle 1, 1 \rangle = 1$ and that

$$\langle \sin(2\pi kx), \sin(2\pi kx) \rangle = \langle \cos(2\pi kx), \cos(2\pi kx) \rangle = \frac{1}{2}$$

for $k \in \mathbb{N}$.

Returning to our original definition of Fourier series, we see that

$$\langle f, 1 \rangle = \int_0^1 f(x) \, d\lambda = a_0$$

and, for $k \in \mathbb{N}$,

$$\langle f, \sin(2\pi k x) \rangle = \int_0^1 f(x) \sin(2\pi k x) \, d\lambda = a_k, \text{ and}$$

$$\langle f, \cos(2\pi k x) \rangle = \int_0^1 f(x) \cos(2\pi k x) \, d\lambda = b_k.$$

Applying the theory of (orthogonal) projections[5] from linear algebra (see exercises 17 to 19), we see that $f_n(x) = a_0 + 2 \sum_{k=1}^n (a_k \sin(2\pi k x) + b_k \cos(2\pi k x))$ is the projection of f into the finite-dimensional subspace V_n spanned by

$$\mathcal{B}_n = \{\cos(2\pi n x), \ldots, \cos(2\pi 2 x), \cos(2\pi) x, 1, \sin(2\pi x),$$
$$\sin(2\pi 2 x), \ldots, \sin(2\pi n x)\}.$$

(See exercise 19.) In this reframed context, the principle question becomes: When does the sequence $\{f_n\}$ of projections into the nested sequence of subspaces $\{V_n\}$ converge to f?

We begin by computing the norm of f_n. Using the orthogonality of the elements of $\mathcal{B}_n$, we find that

$$\|f_n\|_2^2 = \langle f_n, f_n \rangle$$

$$= \left\langle a_0 + 2 \sum_{k=1}^n (a_k \sin(2\pi k x) + b_k \cos(2\pi k x)), \right.$$

$$\left. a_0 + 2 \sum_{k=1}^n (a_k \sin(2\pi k x) + b_k \cos(2\pi k x)) \right\rangle$$

$$= |a_0^2| + \sum_{k=1}^n (|a_k|^2 + |b_k|^2).$$

Thus the L_2 norm of f_n is $\|f_n\|_2 = \sqrt{|a_0^2| + \sum_{k=1}^n (|a_k|^2 + |b_k|^2)}$, the usual Euclidean norm of the vector $\widehat{f}_n = (b_n, \ldots, b_1, a_0, a_1, \ldots, a_n)$. The dimension of the vector $\widehat{f}_n$ changes with n, but it is fairly natural to add zeros on both ends of $\widehat{f}_n$ so that we can view $\widehat{f}_n = (\ldots, 0, 0, 0, b_n, \ldots, b_1, a_0, a_1, \ldots, a_n, 0, 0, 0, \ldots)$ as belonging to a single infinite dimensional space that does not depend on n. In this context, the

---

[5] Given a subspace W of an inner product space V and $\mathbf{v} \in V$ the (orthogonal) **projection** of $\mathbf{v}$ into W is the unique vector $\hat{\mathbf{v}} \in W$ such that $\mathbf{v} - \hat{\mathbf{v}}$ is orthogonal to every element in W.

original function f naturally corresponds to the doubly infinite sequence $\widehat{f} = (\ldots, b_2, b_1, a_0, a_1, a_2, \ldots)$. When we consider the norm of $\widehat{f}$ we are led to the definition of l_2 spaces.

Definition 40 (l_2 spaces). *Let $\mathcal{I}$ be a countable index set. A sequence $\{x_i\}_{i \in \mathcal{I}}$ of scalars is **square-summable** if $\sum_{i \in \mathcal{I}} |x_i|^2$ is finite. The set of all square-summable sequences is denoted by $l_2(\mathcal{I})$.*

The most common choices for the index set $\mathcal{I}$ are $\mathbb{N}$ and $\mathbb{Z}$. Using the natural addition and scalar multiplication of sequences, $l_2(\mathcal{I})$ becomes a vector space. The proof that $l_2(\mathcal{I})$ is closed under addition follows the same contours as the proof of theorem 101. The standard dot product and norm for vectors in $\mathbb{R}^n$ and $\mathbb{C}^n$ extend to $l_2(\mathcal{I})$.

Definition 41. *Let $\mathbf{x} = \{x_i\}_{i \in \mathcal{I}}$ and $\mathbf{y} = \{y_i\}_{i \in \mathcal{I}}$ be elements of $l_2(\mathcal{I})$. Define $\langle \mathbf{x}, \mathbf{y} \rangle = \sum_{i \in \mathcal{I}} x_i \overline{y}_i$ and $\|\mathbf{x}\|_2 = \sqrt{\sum_{i \in \mathcal{I}} |x_i|^2}$.*

Theorem 104. *$\langle \mathbf{x}, \mathbf{y} \rangle$ is an inner product on $l_2(\mathcal{I})$ with associated norm $\|\mathbf{x}\|_2 = \sqrt{\langle \mathbf{x}, \mathbf{x} \rangle}$. That is, for $\mathbf{x}, \mathbf{y}, \mathbf{z} \in l_2(\mathcal{I})$ and any scalar $c \in \mathbb{C}$,*

1. $\langle \mathbf{x}, \mathbf{y} \rangle$ is finite,

2. $\langle \mathbf{y}, \mathbf{x} \rangle = \overline{\langle \mathbf{x}, \mathbf{y} \rangle}$,

3. $\langle c\mathbf{x}, \mathbf{y} \rangle = c \langle \mathbf{x}, \mathbf{y} \rangle$,

4. $\langle \mathbf{x} + \mathbf{y}, \mathbf{z} \rangle = \langle \mathbf{x}, \mathbf{z} \rangle + \langle \mathbf{y}, \mathbf{z} \rangle$,

5. $\langle \mathbf{x}, \mathbf{x} \rangle \geq 0$ with $\langle \mathbf{x}, \mathbf{x} \rangle = 0$ if and only if $\mathbf{x} = 0$,

6. $\|c\mathbf{x}\|_2 = |c| \, \|\mathbf{x}\|_2$, and

7. $\|\mathbf{x} + \mathbf{y}\|_2 \leq \|\mathbf{x}\|_2 + \|\mathbf{y}\|_2$ (triangle inequality).

Proof. The only properties whose proofs require significant effort are (1) and (7). The proofs of the remaining properties involve straightforward algebraic manipulation of series and are left as exercises. (exercise 24.)

To prove (1), note that $(|x_i| - |y_i|)^2 \geq 0$ so that $2 |x_i| \, |y_i| \leq |x_i|^2 + |y_i|^2$. Hence

$$|\langle \mathbf{x}, \mathbf{y} \rangle| = \left| \sum_{i \in \mathcal{I}} x_i \overline{y}_i \right| \leq \sum_{i \in \mathcal{I}} |x_i \overline{y}_i| \leq \sum_{i \in \mathcal{I}} 2 |x_i| \, |y_i| \leq \sum_{i \in \mathcal{I}} (|x_i|^2 + |y_i|^2) < \infty.$$

For (7), we may assume that $\mathbf{y} \neq \mathbf{0}$ and set $\mathbf{z} = \mathbf{x} - \frac{\langle \mathbf{x}, \mathbf{y} \rangle}{\langle \mathbf{y}, \mathbf{y} \rangle} \mathbf{y}$. Observe that

$$\langle \mathbf{z}, \mathbf{y} \rangle = \left\langle \mathbf{x} - \frac{\langle \mathbf{x}, \mathbf{y} \rangle}{\langle \mathbf{y}, \mathbf{y} \rangle} \mathbf{y}, \mathbf{y} \right\rangle = \langle \mathbf{x}, \mathbf{y} \rangle - \frac{\langle \mathbf{x}, \mathbf{y} \rangle}{\langle \mathbf{y}, \mathbf{y} \rangle} \langle \mathbf{y}, \mathbf{y} \rangle = 0.$$

Hence $\mathbf{z}$ is orthogonal to $\mathbf{y}$ so that we can apply the Pythagorean theorem to $\mathbf{x} = \frac{\langle \mathbf{x}, \mathbf{y} \rangle}{\langle \mathbf{y}, \mathbf{y} \rangle} \mathbf{y} + \mathbf{z}$ to conclude that

$$\|\mathbf{x}\|_2^2 = \left| \frac{\langle \mathbf{x}, \mathbf{y} \rangle}{\langle \mathbf{y}, \mathbf{y} \rangle} \right|^2 \|\mathbf{y}\|_2^2 + \|\mathbf{z}\|_2^2 \geq \left| \frac{\langle \mathbf{x}, \mathbf{y} \rangle}{\langle \mathbf{y}, \mathbf{y} \rangle} \right|^2 \|\mathbf{y}\|_2^2 = \frac{|\langle \mathbf{x}, \mathbf{y} \rangle|^2}{\|\mathbf{y}\|_2^2}.$$

Multiplying by $\|\mathbf{y}\|_2^2$ and taking square roots, we derive the *Cauchy-Schwarz inequality*

$$|\langle \mathbf{x}, \mathbf{y} \rangle| \leq \|\mathbf{x}\|_2 \|\mathbf{y}\|_2.$$

Now use the Cauchy-Schwarz inequality to determine that

$$\begin{aligned} \|\mathbf{x} + \mathbf{y}\|_2^2 &= \langle \mathbf{x} + \mathbf{y}, \mathbf{x} + \mathbf{y} \rangle \\ &= \langle \mathbf{x}, \mathbf{x} \rangle + \langle \mathbf{x}, \mathbf{y} \rangle + \langle \mathbf{y}, \mathbf{x} \rangle + \langle \mathbf{y}, \mathbf{y} \rangle \\ &\leq \|\mathbf{x}\|_2^2 + 2 \|\mathbf{x}\|_2 \|\mathbf{y}\|_2 + \|\mathbf{y}\|_2^2 \\ &= (\|\mathbf{x}\|_2 + \|\mathbf{y}\|_2)^2. \end{aligned}$$

Property (7) follows by taking square roots.   □

In this context, there is a natural analog to the question of the convergence of the projections $\{f_n\}$ of f. Namely, if $\mathbf{x} = (\ldots, b_2, b_1, a_0, a_1, a_2, \ldots) \in l_2(\mathbb{Z})$ and $\mathbf{x}_n = (\ldots, 0, 0, b_n, \ldots, b_2, b_1, a_0, a_1, a_2, \ldots, a_n, 0, 0, \ldots)$, does the sequence $\{\mathbf{x}_n\}_{n=1}^{\infty}$ converge to $\mathbf{x}$? In this case, it is clear that the answer is yes since, by the definition of convergence for infinite sums,

$$\begin{aligned} \lim_n \|\mathbf{x} - \mathbf{x}_n\|_2^2 &= \lim_n \sum_{k > n} (|a_k|^2 + |b_k|^2) \\ &= \lim_n \left(\sum_{k=1}^{\infty} (|a_k|^2 + |b_k|^2) - \sum_{k=1}^{n} (|a_k|^2 + |b_k|^2) \right) = 0. \end{aligned}$$

This fact leads naturally to the question of whether or not the vector of coefficients $\widehat{f} = (\ldots, b_2, b_1, a_0, a_1, a_2, \ldots)$ used in the Fourier series for the function f belongs to $l_2(\mathbb{Z})$. Shortly, we will answer this question in the affirmative and so conclude that $\{\widehat{f_n}\}$ converges to $\widehat{f}$. However, the convergence of $\{\widehat{f_n}\}$ to $\widehat{f}$ does not imply that $\{f_n\}$ converges to f.

To see why not, consider the problem of expressing $f(x) = \sin 2\pi x$ as an infinite linear combination of the functions in

$$\mathcal{B}^* = \{1, \cos(2\pi x), \cos(2\pi 2x), \cos(2\pi 3x), \ldots\}.$$

Since $\langle \sin(2\pi x), 1 \rangle = \langle \sin(2\pi x), \cos(2\pi nx) \rangle = 0$ for all $k \in \mathbb{N}$, we see that the (one-sided) sequence of coefficients relative to $\mathcal{B}^*$ is $\widehat{f} = (0, 0, 0, 0, \ldots)$. Consequently $\widehat{f}_n = (0, 0, 0, 0, \ldots)$ and $\{\widehat{f}_n\}$ not only converges to $\widehat{f}$ but is uniformly identical to $\widehat{f}$. However, f_n is always the zero function so that $\{f_n\}$ most certainly does not converge to f. At this point, we have no reason to believe that a similar, but perhaps more subtle, phenomenon does not occur when using the set

$$\mathcal{B} = \{\ldots, \cos(2\pi 2x), \cos(2\pi x), 1, \sin(2\pi x), \sin(2\pi 2x), \ldots\}.$$

10.2 Completeness

Continuing our analysis of

$$f_n(x) = a_0 + 2\sum_{k=1}^{n}(a_k \sin(2\pi nx) + b_k \cos(2\pi kx)q),$$

we note that

$$f_{k+1}(x) - f_k(x) = a_{k+1}\sin(2\pi(k+1)x) + b_{k+1}\cos(2\pi(k+1)x).$$

Thus $f_{k+1} - f_k$ and $f_{j+1} - f_j$ are mutually orthogonal when $j \neq k$. Additionally, f_n and $f - f_n$ are orthogonal since f_n is a projection of f onto the subspace V_n. By the Pythagorean theorem for inner product spaces (see exercise 13), we can conclude that

$$
\begin{aligned}
\|f\|_2^2 &= \|f_n\|_2^2 + \|f - f_n\|_2^2 \\
&\geq \|f_n\|_2^2 \\
&= \sum_{k=0}^{n-1}\|f_{k+1} - f_k\|_2^2 + \|f_0\|_2^2.
\end{aligned}
$$

Hence $\sum_{k=0}^{\infty}\|f_{k+1} - f_k\|_2^2$ converges. Since $\|f_{k+1} - f_k\|_2^2 = |a_{k+1}|^2 + |b_{k+1}|^2$, we see that $\widehat{f} \in l_2(\mathbb{Z})$. As noted previously, this fact implies that $\{\widehat{f}_n\}$ converges to $\widehat{f}$.

Since $\sum_{k=0}^{\infty}\|f_{k+1} - f_k\|_2^2$ converges, given $\varepsilon > 0$ we can find an N so that $\|f_m - f_n\|_2^2 = \sum_{k=n}^{m-1}\|f_{k+1} - f_k\|_2^2 < \varepsilon^2$ or $\|f_m - f_n\|_2 < \varepsilon$ for all $m > n \geq N$. In other words, $\{f_n\}$ is a Cauchy sequence.

This observation raises the question of whether or not all Cauchy sequences in $L_2([a,b])$ converge to a function in $L_2([a,b])$. They do, but the convergence need not be pointwise. Exercise 51 gives an example of a Cauchy sequence in $L_2 ([0, 1])$ that does not converge at any point in $[0, 1)$. However, Cauchy sequences will converge in norm to a function in $L_2 ([0, 1])$.

Definition 42. *A sequence of functions $\{g_n\}$ from $L_2 ([a, b])$ is said to con-verge in norm to the function $g \in L_2 ([a, b])$ provided that, for each $\varepsilon > 0$, it is possible to find a natural number N so that $\|g_n - g\|_2 < \varepsilon$ whenever $n > N$. In other words, $\{g_n\}$ converges in norm to g if $\|g_n - g\|_2$ converges to 0.*

Theorem 105 ($L_2([a, b])$ **is complete**). *Any Cauchy sequence of functions from $L_2 ([a, b])$ converges in norm to a function in $L_2 ([a, b])$.*

Proof. Let $\{f_n\}$ be a Cauchy sequence of functions from $L_2 ([a, b])$. We begin by creating a related, positive function to be used in the dominated convergence theorem (theorem 39, page 140).

Since $\{f_n\}$ is a Cauchy sequence, we can find a strictly increasing se-quence of indices $\{n_k\}$ such that $\| f_{n_{k+1}} - f_{n_k} \|_2 < 2^{-k}$ for $k \geq 1$. Set $f_{n_0} = 0$ and define $g (x) = \sum_{k=1}^{\infty} | f_{n_k} (x) - f_{n_{k-1}} (x)|$. It may be that $g (x)$ is unbounded for some values of x. However, g is defined in the ex-tended real numbers and, being the limit of measurable functions, is measur-able. Moreover, by Fatou's lemma (theorem 38, page 139),

$$\|g\|_2 = \left(\int_a^b \left(\sum_{k=1}^{\infty} | f_{n_k} - f_{n_{k-1}}| \right)^2 \, d\lambda \right)^{1/2}$$

$$\leq \left(\varliminf_n \int_a^b \left(\sum_{k=1}^{n} | f_{n_k} - f_{n_{k-1}}| \right)^2 \, d\lambda \right)^{1/2}$$

$$= \varliminf_n \left(\int_a^b \left(\sum_{k=1}^{n} | f_{n_k} - f_{n_{k-1}}| \right)^2 \, d\lambda \right)^{1/2}$$

$$= \varliminf_n \left\| \sum_{k=1}^{n} | f_{n_k} - f_{n_{k-1}}| \right\|_2$$

$$\leq \varliminf_n \sum_{k=1}^{n} \| f_{n_k} - f_{n_{k-1}} \|_2$$

$$\leq \| f_{n_1} \|_2 + 1.$$

Thus g^2 is integrable and so the set of values for which $g(x)$ is not finite has measure zero. In other words, g is finite a.e.

Define $f(x) = \sum_{k=1}^{\infty}(f_{n_k}(x) - f_{n_{k-1}}(x))$ if $g(x)$ is finite and $f(x) = 0$ otherwise. Then $\{f_{n_k}\}$ converges a.e. to f. Moreover, $|f_{n_k}| \le \sum_{j=1}^{k}|f_{n_j} - f_{n_{j-1}}| \le g$ for all $k \in \mathbb{N}$, so $|f| \le g$. Thus $f \in L_2([a,b])$ with

$$\|f\|_2 = \left(\int_a^b |f|^2 \, d\lambda\right)^{1/2} \le \left(\int_a^b |g|^2 \, d\lambda\right)^{1/2} \le \|g\|_2.$$

Similarly, $\|f_{n_k}\|_2 \le \|g\|_2$ for all $k \in \mathbb{N}$. Since $|f_{n_k} - f|^2 \le (g+g)^2 = 4g^2$ for all $k \in \mathbb{N}$, the dominated convergence theorem implies that

$$\lim_k \|f_{n_k} - f\|_2 = \left(\lim_k \int_a^b |f_{n_k} - f|^2 \, d\lambda\right)^{1/2} = 0.$$

Hence the subsequence $\{f_{n_k}\}$ converges in norm to f.

To show that the original Cauchy sequence $\{f_n\}$ converges in norm to f, suppose that $\varepsilon > 0$ and, noting that $n_n \ge n$, select an N such that $\|f_{n_n} - f_n\|_2 < \varepsilon/2$ and $\|f_{n_n} - f\| < \varepsilon/2$ whenever $n > N$. Then

$$\|f_n - f\|_2 \le \|f_{n_n} - f_n\|_2 + \|f_{n_n} - f\|_2 < \varepsilon$$

for all $n > N$. In other words, $\{f_n\}$ converges in norm to f. $\square$

A set S with a norm is called **complete** if every Cauchy sequence in S converges to an element of S. Since every Cauchy sequence in $L_2([a,b])$ converges to a function in $L_2([a,b])$, we see that $L_2([a,b])$ is complete.

It is worth noting that the proof of theorem 105 provides the tools to conclude that while a sequence of functions $\{f_n\}$ that converges in norm to a function f in $L_2([(a,b)])$ need not converge at any point in $[a,b]$, there is a subsequence of $\{f_n\}$ that will converge to f a.e.

Theorem 106. *Suppose that $\{f_n\}$ is a sequence of functions from $L_2([(a,b)])$ that converges in norm to $f \in L_2([(a,b)])$. Then there is a strictly increasing sequence of indices $\{n_k\}$ such that $\{f_{n_k}\}$ converges to f a.e.*

Proof. Since $\{f_n\}$ converges in norm in $L_2([a,b])$, $\{f_n\}$ is a Cauchy sequence. Thus we can apply the construction used in the proof of theorem 105 to find a strictly increasing sequence of indices $\{n_k\}$ such that

$\sum_{k=1}^{\infty} (f_{n_k}(x) - f_{n_{k-1}}(x))$ converges both almost everywhere and in norm to a function f^*. But

$$\sum_{k=1}^{m} \left(f_{n_k}(x) - f_{n_{k-1}}(x) \right) = f_{n_m},$$

so that $\{f_{n_k}\}$ converges almost everywhere and in norm to f^*. The proof will be complete when we show that $f = f^*$ a.e.

Since $\|f - f^*\|_2 \le \|f - f_{n_k}\|_2 + \|f^* - f_{n_k}\|_2$ and both $\|f - f_{n_k}\|_2$ and $\|f^* - f_{n_k}\|_2$ can be made arbitrarily small by selecting a sufficiently large value for k, we conclude that $\|f - f^*\|_2 = 0$ and hence $f = f^*$ a.e.   □

Note that the proof of theorem 106 also shows that if $\{f_n\}$ converges in norm to f then all a.e.-convergent subsequences of $\{f_n\}$ converge a.e. to f.

10.3   Density

We now know that, when $f \in L_2([a,b])$, the sequence $\{f_n\}$ defined by $f_n(x) = a_0 + 2\sum_{k=1}^{n} (a_k \sin(2\pi kx) + b_k \cos(2\pi kx))$ converges in norm to some function $g \in L_2([a,b])$.[6] The remaining question is whether or not $g = f$. Since f_n is the projection of f onto the space spanned by

$$\mathcal{B}_n = \{\cos(2\pi nx), \dots, \cos(2\pi 2x), \cos(2\pi x), 1,$$
$$\sin(2\pi x), \sin(2\pi 2x), \dots, \sin(2\pi nx)\},$$

the condition that $f = g$ for all $f \in L_2([a,b])$ is equivalent to the condition that linear combinations of the functions in $\{\dots, \cos(2\pi 2x), \cos(2\pi x), 1, \sin(2\pi x), \sin(2\pi 2x), \dots\}$ are dense in $L_2([a,b])$.

The first step in verifying this density is to note that the three trigonometric identities

$$2 \cos\alpha \cos\beta = \cos(\alpha - \beta) + \cos(\alpha + \beta),$$
$$2 \sin\alpha \sin\beta = \cos(\alpha - \beta) - \cos(\alpha + \beta), \text{ and}$$
$$2 \sin\alpha \cos\beta = \sin(\alpha + \beta) + \sin(\alpha - \beta)$$

can be used to prove that the vector space V spanned by

$$\mathcal{B} = \{\dots, \cos(2\pi 2x), \cos(2\pi x), 1, \sin(2\pi x), \sin(2\pi 2x), \dots\}$$

---

[6] Recall that the coefficients a_k and b_k are computed from the function f. See page 280.

is also an **algebra**. In other words, V is closed not only under scalar multiplication and addition of functions but also under the multiplication of functions. (See exercise 39.)

Instead of beginning with the trigonometric polynomials and asking which functions can be approximated by them, we will approach the problem from the other end by asking what sets of functions will generate algebras that are dense in $L_2([a, b])$. The algebra **generated** by a set S of functions is the smallest set of functions that contains S and is also closed under scalar multiplication, function addition, and function multiplication.

We will identify a finite sequence of sets each of which generates an algebra whose members can approximate elements of the previous set arbitrarily closely. Expressed alternatively, we will create a finite sequence of sets each of which generates an algebra whose closure in the L_2 norm contains the previous set. We will choose the initial set so that it generates $L_2([a, b])$. Hence the final set also generates an algebra whose closure contains $L_2([a, b])$.

Begin by noting that if $f = u + iv \in L_2([a, b])$, we can express $f = u^+ - u^- + iv^+ - iv^-$ with each of the component functions belonging to $L_2([a, b])$. (See exercise 40.) Hence the algebra generated by the non-negative, measurable functions in $L_2([a, b])$ is $L_2([a, b])$. Next, note that, given any non-negative function f, the sequence $\{|f - f_n|^2\}$ where f_n is defined by $f_n = \min\{f, n\}$ converges pointwise to 0. Since $|f - f_n|^2 < |f|^2$ and $|f|^2$ is integrable, we can use the dominated convergence theorem (theorem 39, page 140) to conclude that

$$\lim_n \|f - f_n\|_2 = \left(\lim_n \int_a^b |f - f_n|^2 \, d\lambda\right)^{1/2} = 0.$$

Thus the closure of the algebra generated by the set of bounded, non-negative, measurable functions on $[a, b]$ contains the set of non-negative, measurable functions in $L_2([a, b])$.

Now suppose f is a measurable function with $0 \le f \le B$. The measurable simple function $\varphi_n = \sum_{k=0}^{n-1} k\frac{B}{n} \cdot 1_{E_k}$ where $E_k = f^{-1}\left(\left[k\frac{B}{n}, (k+1)\frac{B}{n}\right]\right)$ satisfies

$$\|f - \varphi_n\|_2 = \left(\int_a^b |f - \varphi_n|^2 \, d\lambda\right)^{1/2} \le \left(\int_a^b \left(\frac{B}{n}\right)^2 \, d\lambda\right)^{1/2} = \frac{B\sqrt{b-a}}{n}.$$

Since we can make $\frac{B\sqrt{b-a}}{n}$ arbitrarily small by taking n sufficiently large, the closure of the algebra generated by the measurable characteristic functions contains the set of bounded, non-negative, measurable functions.

Theorem 107. *The closure of the algebra of continuous functions on* $[a,b]$ *contains the measurable characteristic functions.*

Proof. Given a non-empty subset S of $[a,b]$ and $t \in [a,b]$, define $d_S(t) = \inf\{|x - t| : x \in S\}$. We claim that d_S is a continuous function. To see why, fix $t_0 \in [a,b]$ and $\varepsilon > 0$. Suppose that $t \in [a,b]$ with $|t_0 - t| < \varepsilon/2$. Select $x_0 \in S$ such that $d_S(t_0) \le |x_0 - t_0| < d_S(t_0) + \varepsilon/2$. Then the triangle inequality implies that

$$d_S(t) \le |x_0 - t| \le |x_0 - t_0| + |t_0 - t| < d_S(t_0) + \varepsilon.$$

A symmetric argument shows that $d_S(t_0) < d_S(t) + \varepsilon$. Hence d_S is continuous.

If $d_F(x) = 0$, then there is a sequence $\{x_n\}$ of points in F such that $\lim_n |x - x_n| = 0$. Thus if F is a closed set, $d_F(x) = 0$ implies that $x \in F$.

Now let E be an arbitrary non-empty, measurable subset of $[a,b]$ and let $\varepsilon > 0$. By theorem 22 (page 104), we can find an open set G and a closed set F satisfying $F \subseteq E \subseteq G$, $\lambda(G \backslash E) < \varepsilon^2/2$, and $\lambda(E \backslash F) < \varepsilon^2/2$. Then $\lambda(G \backslash F) < \varepsilon^2$. Set $H = G^c$ and note that, since F and H are disjoint, closed sets, $d_F(x) + d_H(x) > 0$. Consequently,

$$f(x) = \frac{d_F(x)}{d_F(x) + d_H(x)}$$

is a continuous function on $[a,b]$ satisfying $0 \le f \le 1$ and $f = 1_E$ on $F \cup H$. Thus

$$\|1_E - f\|_2 = \left(\int_a^b |1_E - f|^2 \, d\lambda \right)^{1/2} \le \left(\int_{G \backslash F} 1 \, d\lambda \right)^{1/2} = \sqrt{\lambda(G \backslash F)} < \varepsilon.$$

Hence the closure of the continuous functions on $[a,b]$ contains the measurable characteristic functions on $[a,b]$. $\qquad\square$

At this point, we know that the closure of the algebra of continuous functions contains $L_2([a,b])$. To finish the task of verifying that the algebra of functions generated by

$$\mathcal{B} = \{\ldots, \cos(2\pi 2x), \cos(2\pi x), 1, \sin(2\pi x), \sin(2\pi 2x), \ldots\}$$

is dense in $L_2([0,1])$, we will prove three versions of the Stone-Weierstrass theorem. The first two versions of the theorem apply to real-valued functions and the third to complex-valued functions. The Stone-Weierstrass theorems draw the conclusion that an algebra of functions is uniformly dense in the set of continuous functions. In other words, any continuous function can be uniformly approximated by a function from the algebra. Since a sequence of functions converging uniformly on $[a, b]$ to a function f will also converge to f in the L_2 norm, the algebra of functions is also dense in the set of continuous functions when using the norm of $L_2([a, b])$.

The proof of the first version of the Stone-Weierstrass theorem proceeds by dividing the range of f into thirds, finding a function from the algebra that roughly approximates f on the three corresponding preimages, and then repeating the process on the error.

Theorem 108 (Stone-Weierstrass 1). *Let X be a nonempty, compact set and let C be an algebra of continuous real-valued functions that satisfies*

1. *$1 \in C$,*

2. *if $f, g \in C$, then $\max\{f, g\} \in C$, and*

3. *the functions of C separate the points of X in the sense that, if x_0 and y_0 are distinct elements of X, then there is a function $f \in C$ such that $f(x_0) \neq f(y_0)$.*

Then given any continuous function f on X and any $\varepsilon > 0$, we can find a $g \in C$ such that $|f - g| < \varepsilon$. In other words, C is uniformly dense in the continuous real-valued functions on X.[7]

Proof. The algebra C contains all constant functions. So suppose that f is a non-constant, real-valued, continuous function on X. Because f is a continuous function on the compact set X, f is bounded and takes on its supremum and infimum. Since the algebra C is closed under vertical scaling and translation, we may assume that f's maximum and minimum are 1 and -1 respectively.

Define $E = f^{-1}([\frac{1}{3}, 1])$ and $F = f^{-1}([-1, -\frac{1}{3}])$. Then E and F, being closed (non-empty) subsets of X, are compact. Now for every $x \in E$ and $y \in F$ there is a function $g_{x,y} \in C$ such that $g_{x,y}(x) \neq g_{x,y}(y)$. Scale and translate $g_{x,y}$ to define

$$h_{x,y} = \frac{2}{3} - \frac{4}{3} \frac{g_{x,y} - g_{x,y}(x)}{g_{x,y}(y) - g_{x,y}(x)}.$$

By construction, $h_{x,y}(x) = \frac{2}{3}$ and $h_{x,y}(y) = -\frac{2}{3}$ and, since C is an algebra, $h_{x,y} \in C$

Temporarily fix $y \in F$. For each $x \in E$, the continuity of $h_{x,y}$ implies the existence of a neighborhood U_x of x on which $h_{x,y} > \frac{1}{3}$. Since $\{U_x\}_{x \in E}$

---

[7] Note that in this theorem and the next we are taking the set of scalars to be $\mathbb{R}$ rather than $\mathbb{C}$.

is an open cover of the compact set E, there are points $x_1, x_2, \ldots, x_n$ such that $E \subset \cup_{i=1}^n U_{x_i}$. Let $\varphi_y = \max\{h_{x_1,y}, h_{x_2,y}, \ldots, h_{x_n,y}\}$. Then $\varphi_y \in \mathcal{C}$, $\varphi_y(y) = -\frac{2}{3}$, and $\varphi_y(x) > \frac{1}{3}$ for all $x \in E$.

Now for each $y \in F$, the continuity of φ_y implies that we can find a neighborhood U_y of y on which $\varphi_y < -\frac{1}{3}$. Again, the compactness of F means that we can find a set of points $y_1, y_2, \ldots, y_m \in F$ such that $F \subset \cup_{i=1}^m U_{y_i}$. Then

$$\phi = \min\{\varphi_{y_1}, \varphi_{y_2}, \ldots, \varphi_{y_m}\}$$
$$= -\max\{-\varphi_{y_1}, -\varphi_{y_2}, \ldots, -\varphi_{y_m}\}$$

satisfies $\phi \in \mathcal{C}$, $\phi(x) > \frac{1}{3}$ for all $x \in E$, and $\phi(y) < -\frac{1}{3}$ for all $y \in F$.

Finally, define $f_1 = \min\{\max\{\phi, -\frac{1}{3}\}, \frac{1}{3}\}$. Then f_1 satisfies $f_1 \in \mathcal{C}$, $-\frac{1}{3} \le f_1 \le \frac{1}{3}$, $f_1(x) = \frac{1}{3}$ for all $x \in E$, and $f_1(y) = -\frac{1}{3}$ for all $y \in F$. In addition, the maximum and minimum of $f - f_1$ are $\frac{2}{3}$ and $-\frac{2}{3}$ respectively.

Now note that $\frac{3}{2}(f - f_1)$ is a continuous function with a maximum and minimum of 1 and -1 respectively. Apply the previous procedure to create a new function $f_2 \in \mathcal{C}$ such that $\frac{3}{2}(f - f_1) - f_2$ has a maximum and minimum of $\frac{2}{3}$ and $-\frac{2}{3}$ respectively. Thus

$$-\left(\frac{2}{3}\right)^2 \le f - \left(f_1 + \frac{2}{3}f_2\right) \le \left(\frac{2}{3}\right)^2$$

and the upper and lower bounds are attained. Repeating the process on

$$\left(\frac{3}{2}\right)^2 \left(f - \left(f_1 + \frac{2}{3}f_2\right)\right)$$

will produce another function $f_3 \in \mathcal{C}$ such that

$$-\left(\frac{2}{3}\right)^3 \le f - \left(f_1 + \frac{2}{3}f_2 + \left(\frac{2}{3}\right)^2 f_3\right) \le \left(\frac{2}{3}\right)^3$$

where, again, $f - \left(f_1 + \frac{2}{3}f_2 + \left(\frac{2}{3}\right)^2 f_3\right)$ takes on the values of the bounds. The construction can be iterated to produce a sequence of functions $\{f_n\}$ in $\mathcal{C}$ with

$$-\left(\frac{2}{3}\right)^n \le f - \sum_{k=1}^n \left(\frac{2}{3}\right)^{k-1} f_k \le \left(\frac{2}{3}\right)^n.$$

Since $\lim_{n \to \infty} \left(\frac{2}{3}\right)^n = 0$, we see that any continuous function on X can be uniformly approximated by an element of $\mathcal{C}$. $\qquad\square$

An alternate way of expressing the conclusion of theorem 108 is to state that the uniform closure of C contains all the continuous real-valued functions on X. In the next version of the Stone-Weierstrass theorem, we use the fact that the closure of an algebra is again an algebra (see exercise 45) to drop the hypothesis that if $f, g \in C$, then max $\{f, g\} \in C$.

Theorem 109 (Stone-Weierstrass 2). *Let X be a nonempty, compact set and let C be an algebra of continuous real-valued functions that satisfies*

1. $1 \in C$, and

2. the functions of C separate the points of X in the sense that, if x_0 and y_0 are distinct elements of X, then there is a function $f \in C$ such that $f(x_0) \neq f(y_0)$.

Then the uniform closure of C contains all the continuous real-valued functions on X. In other words, given any continuous function f on X and any $\varepsilon > 0$ we can find a $g \in C$ such that $|f - g| < \varepsilon$.

Proof. We begin by noting that, since max $\{f, g\} = \frac{1}{2}(|f - g| + (f + g))$ and $|f| = $ max $\{f, 0\} + $ max $\{-f, 0\}$, the two statements $f, g \in C$ implies max $\{f, g\} \in C$ and $f \in C$ implies $|f| \in C$ are equivalent. We will show that if f is a continuous function on X, then $|f|$ can be uniformly approximated by a polynomial in f. In other words, given $\varepsilon > 0$, we can find a polynomial p_ε such that $|p_\varepsilon(f) - |f|| < \varepsilon$. Since $p_\varepsilon(f) \in C$, $|f|$ is in $\overline{C}$, the uniform closure of C. We can then conclude that $\overline{C}$ contains all the continuous functions on X because $\overline{C}$ is an algebra satisfying the conditions of theorem 108 and the closure of $\overline{C}$ is $\overline{C}$ (see exercises 45 and 46).

Let f be an arbitrary continuous function on X. Since f is bounded and $|f| = B|\frac{f}{B}|$ for any positive constant B, we may assume that $|f|$ is bounded by 1. We will construct a sequence of polynomials $\{p_n\}$ that converges uniformly to $\sqrt{t}$ on $[0, 1]$. Then $\{p_n \circ f^2\}$ is a sequence of functions in C that converges uniformly to $|f| = \sqrt{f^2}$.

Define $p_0 = 0$ and $p_{n+1}(t) = p_n(t) + \frac{1}{2}(t - p_n^2(t))$ for $n \geq 0$. Then $p_n(t) \leq \sqrt{t} \leq 1$ for $0 \leq t \leq 1$. To see why, observe that $p_0(t) = 0 \leq \sqrt{t}$ and, by induction,

$$\sqrt{t} - p_{n+1}(t) = \sqrt{t} - p_n(t) - \frac{1}{2}(t - p_n^2(t))$$

$$= (\sqrt{t} - p_n(t))\left(1 - \frac{1}{2}(\sqrt{t} + p_n(t))\right)$$

is nonnegative for $0 \leq t \leq 1$ and $n \geq 0$. Hence $\{p_n\}$, being an increasing and bounded sequence of functions on $[0, 1]$, must converge on $[0, 1]$ to some function p. Since $p \geq 0$ and must satisfy $p(t) = p(t) - \frac{1}{2}(t - p^2(t))$, we see that $p(t) = \sqrt{t}$ on $[0, 1]$.

To verify that the convergence is uniform, suppose for the purpose of contradiction that the convergence is not uniform. In other words, assume there is an $\varepsilon > 0$, a strictly increasing sequence of indices $\{n_k\}$, and a sequence of points $\{x_k\}$ such that $\sqrt{x_k} - p_{n_k}(x_k) > \varepsilon$ for all $k \in \mathbb{N}$. Since X is compact, $\{x_k\}$ must have a cluster point x^*.

Fix $n \in \mathbb{N}$. Then, for any $k > n$, the facts that $\{p_k\}$ is increasing and $n_k \geq k > n$ imply that

$$\sqrt{x_k} - p_n(x_k) \geq \sqrt{x_k} - p_{n_k}(x_k) > \varepsilon.$$

Because $\sqrt{t} - p_n(t)$ is continuous and x^* is a cluster point of $\{x_k\}$, we conclude that $\sqrt{x^*} - p_n(x^*) \geq \varepsilon$. But n was arbitrary. So we have contradicted the fact that $\{p_n\}$ converges pointwise to $\sqrt{t}$ on $[0, 1]$. This contradiction implies that $\{p_n\}$ must converge uniformly to $\sqrt{t}$.

The uniform closure of C contains all the continuous functions on X. $\square$

The final version of the Stone-Weierstrass theorem extends the result to complex-valued functions.

Theorem 110 (Stone-Weierstrass 3). *Let X be a nonempty, compact set and let C be an algebra of complex-valued continuous functions that satisfies*

1. $1 \in C$,

2. *if $f \in C$, then $\bar{f} \in C$, and*

3. *the functions of C separate the points of X in the sense that, if x_0 and y_0 are distinct elements of X, then there is a function $f \in C$ such that $f(x_0) \neq f(y_0)$.*

Then the uniform closure of C contains all the complex-valued continuous functions on X. In other words, given any continuous function f on X and any $\varepsilon > 0$ we can find a $g \in C$ such that $|f - g| < \varepsilon$.

Proof. Let $C_\mathbb{R}$ be the algebra of all real-valued functions in C (using $\mathbb{R}$ for the scalars). By hypothesis, given distinct points x_0 and y_0 from X there is a function $f = u + iv \in C$ such that $f(x_0) \neq f(y_0)$. This implies that either $u(x_0) \neq u(y_0)$ or $v(x_0) \neq v(y_0)$. Since $u = \frac{1}{2}(f + \bar{f})$ and $v = \frac{1}{2i}(f - \bar{f})$ belong to C, $u, v \in C_\mathbb{R}$. Thus $C_\mathbb{R}$ fulfills the hypotheses of theorem 109 and so the uniform closure of $\mathcal{C}_{\mathbb{R}}$ contains all real-valued, continuous functions on X. As $\mathcal{C}$ contains all functions of the form $u + iv$ with $u, v \in \mathcal{C}_{\mathbb{R}}$, the closure of $\mathcal{C}$ contains all complex-valued, continuous functions on X.   □

We now return to the original question of this section and prove that $\mathrm{Span}\,(\mathcal{B})$ is dense in $L_2\,([0, 1])$ where $\mathcal{B} = \{\ldots, \cos 2\pi 2x, \cos 2\pi x, 1, \sin 2\pi x, \sin 2\pi 2x, \ldots\}$. We cannot directly apply the Stone-Weierstrass theorem because the functions in $\mathcal{B}$ do not separate the points of $[0, 1]$. Every function $f \in \mathcal{B}$ satisfies $f\,(0) = f\,(1)$. Since there is only one problematic point and our interest is in approximating a function in the L_2 norm rather than uniformly, we can work around this difficulty.

One way to deal with this issue is to identify the points 0 and 1 turning the interval $[0, 1]$ into a circle. Since the circle is compact, we can apply theorem 110 to the algebra generated by $\mathcal{B}$ to conclude that the uniform closure of the algebra contains all continuous functions on the circle or, equivalently, all continuous functions g on $[0, 1]$ with $g\,(0) = g\,(1)$.

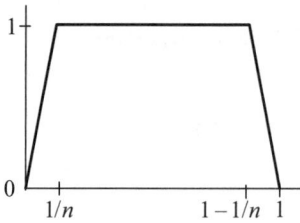

Figure 10.1. φ_n

Now define a sequence of continuous functions $\{\varphi_n\}$ that approximate $1_{(0,1)}$ by taking φ_n to be 1 on $[1/n, 1 - 1/n]$, 0 on $\{0, 1\}$, and linear on the intervals $[0, 1/n]$ and $[1 - 1/n, 1]$. The sequence $\{\varphi_n\}$ converges pointwise to $1_{(0,1)}$. Given a continuous function f on $[0, 1]$, define $f_n = f \cdot \varphi_n$. Then $\{f - f_n\} = \{f \cdot (1 - \varphi_n)\}$ converges a.e. to zero. By the dominated convergence theorem,

$$\lim_n \|f - f_n\|_2 = \left(\lim_n \int_0^1 (|f| \cdot (1 - \varphi_n))^2 \, d\lambda \right)^{1/2} = 0.$$

Hence given any $\varepsilon > 0$, we can find an n so that $\|f - f_n\|_2 < \varepsilon/2$. Moreover, f_n is continuous with $f_n\,(0) = 0 = f_n\,(1)$ so f_n is in the uniform closure of the algebra generated by B. Thus there is a function g in the algebra generated by B such that $|f_n - g| < \varepsilon/2$. But then

$$\|f_n - g\|_2 = \left(\int_0^1 |f_n - g|^2 \, d\lambda\right)^{1/2} < \left(\int_0^1 \frac{\varepsilon^2}{4} \, d\lambda\right)^{1/2} = \varepsilon/2$$

so that

$$\|f - g\|_2 \le \|f - f_n\|_2 + \|f_n - g\|_2 < \varepsilon.$$

Linear combinations of functions from B are dense in $L_2\left([0,1]\right)$.

Reflecting back on the example just before Section 10.2, we can see why linear combinations of the functions in $B^* = \{1, \cos 2\pi x, \cos 2\pi 2x, \cos 2\pi 3x, \ldots\}$ are not dense in $L_2\left([0,1]\right)$. The functions in B^* do not separate points. Any function $f \in B^*$ satisfies $f(x) = f(1-x)$ for $x \in [0,1]$. If an identification similar to that used above is performed, we see that functions in Span (B^*) can be used to uniformly approximate functions in $L_2([0,\frac{1}{2}])$. This approximation can be extended to approximate functions f satisfying $f(x) = f(1-x)$ for $x \in [0,1]$ but does not extend to all functions in $L_2\left([0,1]\right)$. (See exercises 52 and 53.)

Returning to our original function $f \in L_2\left([0,1]\right)$ and its corresponding Fourier series, we know that the sequence $\{f_n\}$ defined by

$$f_n(x) = a_0 + 2 \sum_{k=1}^{n} (a_k \sin(2\pi kx) + b_k \cos(2\pi kx))$$

converges in norm to some function $g \in L_2\left([0,1]\right)$. Moreover, since f_n is the projection into the subspace V_n spanned by

$$B_n = \{\cos(2\pi nx), \ldots, \cos 2\pi 2x, \cos 2\pi x, 1, \sin 2\pi x,$$
$$\sin 2\pi 2x, \ldots, \sin(2\pi nx)\},$$

we know that $\|f - f_n\|_2 \le \|f - h\|_2$ for all functions $h \in V_n$.

Let $\varepsilon > 0$. Then there is a function h in the span of B such that $\|f - h\|_2 < \varepsilon$. Now $h \in V_N$ for some N and $V_N \subset V_n$ for $n > N$. Thus for $n > N$ we have $\|f - f_n\|_2 \le \|f - h\|_2 < \varepsilon$. Hence $\{f_n\}$ converges in norm to f as desired. In $L_2\left([0,1]\right)$, Fourier series behave exactly as one would hope.

10.4 Conclusion

The problem of making sense of Fourier series motivated mathematical developments for over a century. In particular, a great deal of work in the area of theories of integration was motivated by the desire to place Fourier series on a firm theoretical foundation. The Lebesgue integral addresses the problems related to the interaction between limits of functions and integration. In addition, the Lebesgue integral points toward a more satisfying resolution of the problems related to the convergence of Fourier series. The convergence properties of the Lebesgue integral are instrumental in proving that L_2 spaces are complete. The completeness of $L_2\left([0, 1]\right)$, together with some basic vector space theory and the Stone-Weierstrass theorem, create a robust context in which Fourier series always converge. L_2 spaces provide a clean foundation for the theory of Fourier series. Specifically, any function belonging to $L_2\left([0, 1]\right)$ has a well-defined Fourier series that converges in the L_2 norm to the original function. Moreover, the coefficients in the Fourier series provide a norm-preserving linear correspondence between functions in $L_2\left([0, 1]\right)$ and sequences in $l_2\left(\mathbb{Z}\right)$. In the end, we have a very satisfying resolution of issues raised by Fourier's approach to solving differential equations.

10.5   Exercises

10.01 Complex numbers: filling the gaps

The set $\mathbb{C}$ of complex numbers is the set of all numbers of the form $x = a + ib$ where $a, b \in \mathbb{R}$ and $i^2 = -1$. Exercises 1 through 3 provide a review of the basic properties of $\mathbb{C}$ needed for this chapter.

The **complex conjugate** of a complex number $x = a + ib$ is $\overline{x} = a - ib$.

1. Let x and y be complex numbers. Prove that

    (a) $\overline{x + y} = \overline{x} + \overline{y}$,
    (b) $\overline{xy} = \overline{x}\,\overline{y}$, and
    (c) $x\overline{x} \geq 0$ with $x\overline{x} = 0$ only if $x = 0$.

We define the **modulus** of a complex number $x = a + ib$ to be $|x| = \sqrt{x\overline{x}} = \sqrt{a^2 + b^2}$.

2. Let x and y be complex numbers. Prove that

    (a) $|\overline{x}| = |x|$,
    (b) $|xy| = |x|\,|y|$, and
    (c) $|x + y| \leq |x| + |y|$.

(Work with the squares of the expressions.)

3. Let $x = a + ib$ be a complex number. The **real** and **imaginary parts** of x, denoted by $\operatorname{Re} x$ and $\operatorname{Im} x$, are $\operatorname{Re} x = a$ and $\operatorname{Im} x = b$. Prove that $\operatorname{Re} x = \frac{1}{2}(x + \overline{x})$ and $\operatorname{Im} x = \frac{1}{2i}(x - \overline{x})$.

10.02 Vector and inner product spaces: filling the gaps

A **vector space** is a set V of objects called vectors together with two operations, called addition and scalar multiplication, subject to the following ten axioms. In the axioms, **u**, **v**, and **w** are any elements of V and a and b are any scalars. (The set of scalars may be taken to be either $\mathbb{R}$ or $\mathbb{C}$.)

(a) $\mathbf{u} + \mathbf{v} \in V$.
(b) $\mathbf{u} + \mathbf{v} = \mathbf{v} + \mathbf{u}$.
(c) $(\mathbf{u} + \mathbf{v}) + \mathbf{w} = \mathbf{u} + (\mathbf{v} + \mathbf{w})$.
(d) There is a vector $\mathbf{0} \in V$ such that $\mathbf{u} + \mathbf{0} = \mathbf{u}$ for all $\mathbf{u} \in V$.
(e) For each $\mathbf{u} \in V$, there is a vector $-\mathbf{u} \in V$ such that $\mathbf{u} + (-\mathbf{u}) = 0$.
(f) $a\mathbf{u} \in V$.
(g) $a(\mathbf{u} + \mathbf{v}) = a\mathbf{u} + a\mathbf{v}$.
(h) $(a + b)\mathbf{v} = a\mathbf{v} + b\mathbf{v}$.
(i) $a(b\mathbf{v}) = (ab)\mathbf{v}$.
(j) $1\mathbf{v} = \mathbf{v}$.

Exercises 4 through 19 provide a review of the properties of vector spaces required in this chapter.

4. Prove that the set of scalar-valued functions on a fixed set X is a vector space.

A subset W of a vector space V is a **subspace** of V if W is a vector space using the operations of V.

5. Let V be a vector space. Prove that if W is a subset of V that contains the vector $\mathbf{0}$ and is closed under addition and scalar multiplication, then W is subspace of V.

Let S be a subset of a vector space V. Any vector **v** that can be expressed as $\mathbf{v} = c_1\mathbf{v}_1 + c_2\mathbf{v}_2 + \cdots + c_n\mathbf{v}_n$ where $c_1, c_2, \ldots, c_n$ are scalars and $\mathbf{v}_1, \mathbf{v}_2, \ldots, \mathbf{v}_n \in S$ is a **linear combination** of the elements of S. The **span** of S, written $\operatorname{Span}(S)$, is the set of all possible linear combinations of vectors from S.

6. Use exercise 5 to prove that Span (S) is a subspace of V.

   Let V be a vector space. An **inner product** on V is a mapping $\langle \cdot, \cdot \rangle$: $V \times V \to \mathbb{C}$ (or $\mathbb{R}$) that for vectors $\mathbf{u}, \mathbf{v}, \mathbf{w}$ and scalar c satisfies

   (a) $\langle \mathbf{u}, \mathbf{v} \rangle = \overline{\langle \mathbf{v}, \mathbf{u} \rangle}$,
   (b) $\langle \mathbf{u} + \mathbf{v}, \mathbf{w} \rangle = \langle \mathbf{u}, \mathbf{w} \rangle + \langle \mathbf{v}, \mathbf{w} \rangle$,
   (c) $\langle c\mathbf{u}, \mathbf{v} \rangle = c \langle \mathbf{u}, \mathbf{v} \rangle$, and
   (d) $\langle \mathbf{u}, \mathbf{u} \rangle \geq 0$, with $\langle \mathbf{u}, \mathbf{u} \rangle = 0$ if and only if $\mathbf{u} = 0$.

   An **inner product space** is a vector space with an inner product.

7. Use the properties of an inner product space to prove that for vectors $\mathbf{u}$ and $\mathbf{v}$ and scalar c

   (a) $\langle \mathbf{u}, \mathbf{v} + \mathbf{w} \rangle = \langle \mathbf{u}, \mathbf{v} \rangle + \langle \mathbf{u}, \mathbf{w} \rangle$, and
   (b) $\langle \mathbf{u}, c\mathbf{v} \rangle = \overline{c} \langle \mathbf{u}, \mathbf{v} \rangle$.

   The elements of a set S of vectors in an inner product space are mutually **orthogonal** if $\langle \mathbf{u}, \mathbf{v} \rangle = 0$ for all $\mathbf{u}, \mathbf{v} \in S$ with $\mathbf{u} \neq \mathbf{v}$.

8. Suppose that $\mathbf{u}$ and $\mathbf{v}$ are orthogonal vectors. Prove that $a\mathbf{u}$ and $b\mathbf{v}$ are also orthogonal for any choice of scalars a and b.

9. Suppose that $\mathbf{v}$ is orthogonal to all the vectors in a set S. Prove that $\mathbf{v}$ is also orthogonal to any vector in Span (S).

10. Let S be a set of vectors from an inner product space V. Prove that the set of vectors in V that are orthogonal to all the vectors in S is a subspace of V.

    Given a vector $\mathbf{v}$ in an inner product space V, define the **norm** of $\mathbf{v}$ to be $\|\mathbf{v}\| = \sqrt{\langle \mathbf{v}, \mathbf{v} \rangle}$.

11. Prove that if $\mathbf{v}$ is a vector in an inner product space and c is a scalar, then $\|c\mathbf{v}\| = |c| \, \|\mathbf{v}\|$. (Use properties (a) and (b) of an inner product space.)

12. **Pythagorean theorem for inner product spaces.** Let $\mathbf{u}$ and $\mathbf{v}$ be members of an inner product space V. Prove that if $\mathbf{u}$ and $\mathbf{v}$ are orthogonal, then $\|\mathbf{u} + \mathbf{v}\|^2 = \|\mathbf{u}\|^2 + \|\mathbf{v}\|^2$.

13. Let $\{\mathbf{v}_1, \mathbf{v}_2, \ldots, \mathbf{v}_n\}$ be a set of orthogonal vectors in an inner product space V. Prove that $\| \sum_{k=1}^{n} c_k \mathbf{v}_k \|^2 = \sum_{k=1}^{n} |c_k|^2 \, \|\mathbf{v}_k\|^2$ for any choice of scalars $c_1, c_2, \ldots, c_n$. (Use induction and exercise 9.)

14. Prove the converse of exercise 12 for inner product spaces over the real numbers.

15. Prove the **Cauchy-Schwarz inequality.** Given vectors $\mathbf{u}$ and $\mathbf{v}$ from an inner product space V, $|\langle \mathbf{u}, \mathbf{v} \rangle| \leq \|\mathbf{u}\| \|\mathbf{v}\|$. (Assuming $\mathbf{v} \neq 0$, set $\mathbf{z} = \mathbf{u} - \frac{\langle \mathbf{u}, \mathbf{v} \rangle}{\langle \mathbf{v}, \mathbf{v} \rangle} \mathbf{v}$. Show that $\mathbf{z}$ and $\mathbf{v}$ are orthogonal. Solve for $\mathbf{u}$ and use exercises 8 and 12 or exercise 13 to show that $\|\mathbf{u}\|^2 \|\mathbf{v}\|^2 \geq |\langle \mathbf{u}, \mathbf{v} \rangle|^2$.)

16. Prove the **triangle inequality for inner product spaces.** If $\mathbf{u}$ and $\mathbf{v}$ are elements of an inner product space, then $\|\mathbf{u} + \mathbf{v}\| \leq \|\mathbf{u}\| + \|\mathbf{v}\|$. (Begin by expanding $\|\mathbf{u} + \mathbf{v}\|^2$. Then use the Cauchy-Schwarz inequality from exercise 15.)

Let W be a subspace of an inner product space V and let $\mathbf{v} \in V$. Then the **projection** of $\mathbf{v}$ into W is the vector $\hat{\mathbf{v}} \in W$ such that $\mathbf{v} - \hat{\mathbf{v}}$ is orthogonal to every element in W.[8] Note that while the notations are similar, the projection $\hat{\mathbf{v}}$ should not be confused with the vector $\widehat{f}$ of Fourier coefficients.

17. Prove that the word "the" in the definition of projection is justified. In other words, prove that if both $\mathbf{u}$ and $\mathbf{w}$ are in the subspace W and both $\mathbf{v} - \mathbf{u}$ and $\mathbf{v} - \mathbf{w}$ are orthogonal to every element of W, then $\mathbf{u} = \mathbf{w}$. (Use exercise 10 to show that $\mathbf{u} - \mathbf{w}$ is orthogonal to itself and hence must be $\mathbf{0}$.)

18. Let W be a subspace of an inner product space V and let $\mathbf{v} \in V$. Prove that the projection $\hat{\mathbf{v}}$ of $\mathbf{v}$ into W satisfies $\|\mathbf{v} - \hat{\mathbf{v}}\| \leq \|\mathbf{v} - \mathbf{w}\|$ for all $\mathbf{w} \in W$. (Use exercise 12 on $\mathbf{v} - \mathbf{w} = (\mathbf{v} - \hat{\mathbf{v}}) + (\hat{\mathbf{v}} - \mathbf{w})$.)

19. Suppose that $\{\mathbf{v}_1, \mathbf{v}_2, \ldots, \mathbf{v}_n\}$ is a set of non-zero orthogonal vectors in an inner product space V and that $\mathbf{v} \in V$. Prove that

$$\hat{\mathbf{v}} = \frac{\langle \mathbf{v}, \mathbf{v}_1 \rangle}{\langle \mathbf{v}_1, \mathbf{v}_1 \rangle} \mathbf{v}_1 + \frac{\langle \mathbf{v}, \mathbf{v}_2 \rangle}{\langle \mathbf{v}_2, \mathbf{v}_2 \rangle} \mathbf{v}_2 + \cdots + \frac{\langle \mathbf{v}, \mathbf{v}_n \rangle}{\langle \mathbf{v}_n, \mathbf{v}_n \rangle} \mathbf{v}_n$$

is the projection of $\mathbf{v}$ onto $\mathrm{Span}\{\mathbf{v}_1, \mathbf{v}_2, \ldots, \mathbf{v}_n\}$. (Use exercise 9 to prove that $\mathbf{v} - \hat{\mathbf{v}}$ is orthogonal to all the vectors in $\mathrm{Span}\{\mathbf{v}_1, \mathbf{v}_2, \ldots, \mathbf{v}_n\}$.)

10.1 L_2 -spaces: filling the gaps

20. Prove that if $f \in L_2([a, b])$ and c is a scalar, then $cf \in L_2([a, b])$.

---

[8] Such projections always exist when W is finite dimensional (spanned by a finite set of vectors). In general, W must be topologically closed to guarantee the existence of projections.

21. Use the properties of the Lebesgue integral to verify that $\langle \cdot, \cdot \rangle$ satisfies the algebraic properties of an inner product on $L_2 ([a, b])$. In other words, prove the remaining parts of theorem 102. (For (2) and (5), write $f = u_1 + iv_1$ and $g = u_2 + iv_2$.)

22. Prove that a measurable, complex-valued function $f = u + iv$ is square integrable if and only if both u and v are square integrable.

23. Prove part (a) of theorem 103.

24. Prove that $l_2 (\mathcal{I})$ is closed under scalar multiplication and vector addition. In other words, prove that if $\mathbf{x}, \mathbf{y} \in l_2 (\mathcal{I})$ and c is a scalar, then $\mathbf{x} + \mathbf{y} \in l_2 (\mathcal{I})$ and $c\mathbf{x} \in l_2 (\mathcal{I})$. (Mirror the proof of theorem 101.)

25. In proving part (7) of theorem 104, why can we assume that $\mathbf{y} \neq \mathbf{0}$?

10.1 L_2 -spaces: deeper reflections

Let X be a compact subset of $\mathbb{R}$ and let μ be a finite measure on X. (See Sections 8.5 and 8.6 of Chapter 8.) Define $L_2 (X, \mu)$ to be the set of all μ-measurable, complex-valued functions f on X such that $\int_X |f|^2 \, d\mu < \infty$.

26. Prove that $L_2 (X, \mu)$ is a subspace of the vector space of all complex-valued functions on X. (This only requires that you verify that $L_2 (X, \mu)$ contains the zero function and is closed under scalar multiplication and the addition of functions.)

27. Prove that $L_2 (X, \mu)$ is an inner product space under the inner product $\langle f, g \rangle = \int_X f \overline{g} \, d\mu$. (See definition after exercise 5.)

28. Prove that $L_p ([a, b]) \subset L_q ([a, b])$ for $1 \leq q \leq p$. In particular, any square integrable function is integrable.

29. **Jensen's inequality.** Let $f : [0, +\infty) \to [0, +\infty)$ be a continuous increasing function that is also onto and let $g = f^{-1}$.

   (a) Suppose that $f (a) = b$. Explain why $ab = \int_0^a f \, d\lambda + \int_0^b g \, d\lambda$. (Geometrically, place $\int_0^a f \, d\lambda$ on the x-axis and $\int_0^b g \, d\lambda$ on the y-axis.)

   (b) Suppose that $b > f (a)$. Prove that $ab < \int_0^a f \, d\lambda + \int_0^b g \, d\lambda$. (First explain why $g (y) > a$ for any y in the interval $(f (a), b)$.)

   (c) Use symmetry to prove that $ab < \int_0^a f \, d\lambda + \int_0^b g \, d\lambda$ for $0 \leq b < f (a)$.

30. Prove that for any $p, q > 1$ with $\frac{1}{p} + \frac{1}{q} = 1$ and $a, b \geq 0$,

$$ab \leq \frac{a^p}{p} + \frac{b^q}{q}.$$

(Apply Jensen's inequality from exercise 29 to the function $f(x) = x^{p-1}$.)

If $f \in L_p([a, b])$ for some $1 \leq p < \infty$, then $\|f\|_p$ is defined to be $\|f\|_p = (\int_0^a |f|^p \, d\lambda)^{1/p}$. $L_\infty([a, b])$ is defined to be the set of all measurable functions f on $[a, b]$ for which $|f|$ is bounded except on a set of measure zero. $\|f\|_\infty$ is defined to be the smallest such bound.

31. Prove that if $f \in L_1([a, b])$ and $g \in L_\infty([a, b])$ then $fg \in L_1([a, b])$ with $\int_a^b |fg| \, d\lambda \leq \|f\|_1 \|g\|_\infty$.

32. Prove **Hölder's inequality**, a generalization of both exercise 31 and the Cauchy-Schwarz inequality. Suppose that $p, q \geq 1$ with $\frac{1}{p} + \frac{1}{q} = 1$, $f \in L_p([a, b])$, and $g \in L_q([a, b])$. Then $fg \in L_1([a, b])$ with $\int_a^b |fg| \, d\lambda \leq \|f\|_p \|g\|_q$. (Apply exercise 30 with $a = \frac{|f|}{\|f\|_p}$ and $b = \frac{|g|}{\|g\|_q}$ and integrate the resulting inequality.)

33. Even though $L_p([a, b])$ is not an inner product space, prove that $\|f\|_p = (\int_a^b |f|^p \, d\lambda)^{1/p}$ is a norm on $L_p([a, b])$ for $1 \leq p < \infty$. In other words, prove that, for $f, g \in L_p([a, b])$ and any scalar α,

(a) $\|\alpha f\|_p = |\alpha| \|f\|_p$ and
(b) $\|f + g\|_p \leq \|f\|_p + \|g\|_p$ (**Minkowski's inequality**).

(For part (b), modify the proof of theorem 101 to show that $f + g \in L_p([a, b])$ and so $|f + g|^{p-1} \in L_{\frac{p}{p-1}}([a, b])$. Then apply Hölder's inequality to the right side of $|f + g|^p \leq (|f| + |g|) |f + g|^{p-1} = |f| |f + g|^{p-1} + |g| |f + g|^{p-1}$ and relate $\| |f + g|^{p-1} \|_{\frac{p}{p-1}}$ to $\|f + g\|_p$.)

10.2 Completeness: filling the gaps

34. In theorem 105

(a) Explain why g is finite a.e. (Let $E = \{x \in [a, b] : g(x) = +\infty\}$. Show that for all $n \in \mathbb{N}$, $\lambda(E) \leq \frac{1}{n^2} \|g\|_2^2$.)
(b) Explain why $\{f_{n_k}\}$ converges to f a.e.

35. Prove that if $\{f_n\}$ converges to f in the norm of $L_2\left([a,b]\right)$ and $\{f_{n_k}\}$ converges a.e. to some function g, then $f = g$ a.e.

10.2 Completeness: deeper reflections

36. Modify the proof of theorem 105 to show that $L_2\left(X,\mu\right)$ is complete. (See exercises 25 to 27.)

37. Modify the proof of theorem 105 to show that $L_p\left([a,b]\right)$ is complete.

10.3 Density: filling the gaps

Take $\mathcal{B} = \{\ldots, \cos 2\pi 3x, \cos 2\pi 2x, \cos 2\pi x, 1, \sin 2\pi x, \sin 2\pi 2x,$
$\sin 2\pi 3x, \ldots\}$ in exercises 38 through 49.

38. Prove that Span $(\mathcal{B})$ is dense in $L_2\left([a,b]\right)$ if and only if for any $f \in L_2\left([a,b]\right)$, the sequence $\{f_n\}$ defined by

$$f_n(x) = a_0 + 2 \sum_{k=1}^{n} \left(a_k \sin\left(2\pi k x\right) + b_k \cos\left(2\pi k x\right)\right)$$

converges in norm to f. (Use exercises 18 and 19.)

39. Let $V = $ Span $(\mathcal{B})$. Prove that V is an algebra. In other words, prove that V is closed under scalar multiplication, addition, and multiplication. (Use the identities $2\cos\alpha\cos\beta = \cos\left(\alpha-\beta\right) + \cos\left(\alpha+\beta\right)$, $2\sin\alpha\sin\beta = \cos\left(\alpha-\beta\right) - \cos\left(\alpha+\beta\right)$, and $2\sin\alpha\cos\beta = \sin\left(\alpha+\beta\right) + \sin\left(\alpha-\beta\right)$.)

40. Prove that if $f = u + iv \in L_2\left([a,b]\right)$, then u^+, u^-, v^+, and u^- also belong to $L_2\left([a,b]\right)$. (Begin by explaining why $(u^+ - u^-)^2 = (u^+)^2 + (u^-)^2$.)

41. Suppose that F, E, and U are measurable sets satisfying $F \subseteq E \subseteq U$, $\lambda\left(U\backslash E\right) < \varepsilon/2$, and $\lambda\left(E\backslash F\right) < \varepsilon/2$. Explain why $\lambda\left(U\backslash F\right) < \varepsilon$.

42. Prove that if $\{f_n\}$ is a sequence of functions on $[a,b]$ that converges uniformly to the function f, then $\{f_n\}$ also converges to f in the norm of $L_2\left([a,b]\right)$.

43. Prove that if an algebra is uniformly dense in a set $\mathcal{F}$ of functions on $[a,b]$ then the algebra is also dense in $\mathcal{F}$ using the $L_2\left([a,b]\right)$ norm. (Show that any sequence that converges uniformly to f also converges to f in the L_2 norm.)

44. In the proof of the Stone-Weierstrass theorem (theorem 108)

    (a) Why does $\mathcal{C}$ contain all constant functions?
    (b) Why is it sufficient to prove that any continuous function f with a maximum of 1 and a minimum of -1 can be uniformly approximated by an element of $\mathcal{C}$?
    (c) Why are E and F closed and non-empty?

45. Prove that if $\mathcal{A}$ is an algebra of functions and $\overline{\mathcal{A}}$ is the closure of $\mathcal{A}$, then $\overline{\mathcal{A}}$ is also an algebra as long as the operations of scalar multiplication, addition, and multiplication are continuous in the sense that $\lim c f_n = c \lim f_n$, $\lim (f_n + g_n) = \lim f_n + \lim g_n$, and $\lim (f_n g_n) = \lim f_n \lim g_n$. (The limit need not be a uniform limit. It may be pointwise, L_2, or some other type of limit.)

46. Let S be a set of functions and let $\bar{S}$ be the uniform closure of S. Prove that the uniform closure of $\bar{S}$ is $\bar{S}$. In other words, prove that any function that can be uniformly approximated by functions in $\bar{S}$ can be uniformly approximated by functions from S. (Use the triangle inequality.)

47. In theorem 109

    (a) Supply the details explaining why it is sufficient to assume that $|f| \le 1$.
    (b) Supply the details to prove that if $\{p_n\}$ converges to $\sqrt{x}$ uniformly on $[0, 1]$ and $|f| \le 1$, then $\{p_n \circ f^2\}$ converges to $|f|$ uniformly on the domain of f.

48. Prove that if f is a continuous function, $f(x_k) > \varepsilon$ for $k \in \mathbb{N}$, and x^* is a cluster point of $\{x_k\}$, then $f(x^*) \ge \varepsilon$. (If f is continuous and $f(x^*) < \varepsilon$, then $f < \varepsilon$ on some neighborhood of x^*.)

49. Let $\mathcal{C}$ be an algebra of complex-valued functions and let $\mathcal{C}_\mathbb{R}$ be the set of real-valued functions in $\mathcal{C}$. Prove that $\mathcal{C}_\mathbb{R}$ is an algebra over $\mathbb{R}$.

10.3 Density: deeper reflections
50. The recursion used to approximate $\sqrt{t}$ in the proof of theorem 109 is related to Newton's method for finding a zero of a function.

    (a) Derive the recursion of Newton's method.
    (b) Explain why one would choose to employ the alternate recursion in the proof of theorem 109 instead of using Newton's recursion.
    (c) Rework the proof of theorem 109 using the recursion you found in part (a) but starting with $p_0 = 1$.

51. Define $E_n = [\frac{n-2^k}{2^k}, \frac{n+1-2^k}{2^k}]$ where $2^k \le n < 2^{k+1}$.

   (a) Prove that the sequence $\{1_{E_n}\}$ converges in norm to 0 but does not converge to 0 at any $x \in [0, 1]$.

   (b) Find a subsequence of $\{1_{E_n}\}$ that converges to 0 everywhere.

   In exercises 52 and 53, take $\mathcal{B}^* = \{1, \cos 2\pi x, \cos 2\pi 2x, \cos 2\pi 3x, \ldots\}$.

52. Suppose that $f : [0, 1] \to \mathbb{C}$ satisfies $f(x) = f(1-x)$. Further suppose that f can be uniformly approximated on $[0, \frac{1}{2}]$ by functions in $\mathrm{Span}(\mathcal{B}^*)$. Prove that f is also uniformly approximated on $[0, 1]$ by functions in $\mathrm{Span}(\mathcal{B}^*)$. (If $x \in [\frac{1}{2}, 1]$, then $1 - x \in [0, \frac{1}{2}]$.)

53. Suppose that $f : [0, 1] \to \mathbb{C}$ fails to satisfy $f(x) = f(1-x)$. Show that f cannot be uniformly approximated on $[0, 1]$ by functions in $\mathrm{Span}(\mathcal{B}^*)$. (Find an $a \in [0, 1]$ and an $\varepsilon > 0$ so that $|f(a) - f(1-a)| > 2\varepsilon$. Use these to show that if $g \in \mathrm{Span}(\mathcal{B}^*)$ and $|f - g| < \varepsilon$ on $[0, \frac{1}{2}]$, then $|f(x) - g(x)| > \varepsilon$ for some $x \in [\frac{1}{2}, 1]$.)

54. Why does the following approach fail when trying to prove that $\mathrm{Span}(\mathcal{B})$ is dense in $L_2([0, 1])$?

   (a) Given $f \in L_2([0, 1])$, find a continuous function g that uniformly approximates f on $[0, 1]$.

   (b) Use the fact that the functions in $\mathcal{B}$ separate points in $[\frac{1}{n}, 1]$ to find a function $\varphi_n \in \mathrm{Span}(\mathcal{B})$ that uniformly approximates g on $[\frac{1}{n}, 1]$.

   (c) Show that $\lim_n \|f - \varphi_n\|_2 = 0$.

10.6   References

Bartle, R.G. (1966). *The Elements of Integration*. John Wiley & Sons.

Bressoud, D.M. (2006). *A Radical Approach to Real Analysis* (2nd ed.). Mathematical Association of America.

Hewitt, E. and K. Stromberg (1975). *Real and Abstract Analysis*. Springer.

Lay, D. (2012). *Linear Algebra and Its Applications* (4th ed.). Pearson.

Stoock, D.W. (1999). *A Concise Introduction to the Theory of Integration* (3rd ed.). Birkhauser.

Young, N. (1988). *An Introduction to Hilbert Space*. Cambridge University Press.

Appendices: A Compendium of Definitions and Results

A.1   Sets of real numbers

While nearly all of the definitions that follow apply in a much broader context, you may assume here that all the sets consist of real numbers.

Boundary point: A point x is a **boundary point** of a set E if every open set containing x includes at least one point in E and one point in the complement of E.

Bounded set: A set E is **bounded** if there is a real number B so that $|x| < B$ for all $x \in E$.

Closed set: A set F is **closed** if it is the complement of an open set. Alternatively, a set F is closed if it contains all of its boundary points.

The complement of any open set is closed and the complement of any closed set is open.

The intersection of an arbitrary collection of closed sets is closed.

The union of a finite number of closed sets is closed.

Closure: The **closure** of a set S, denoted by $\bar{S}$, is the smallest closed set containing S. Alternatively, $\bar{S}$ consists of S together with all of its boundary points. In a space with a norm, $\bar{S}$ consists of all points that are a limit of a sequence of points from S.

Compact set: A set C is **compact** if, given any cover of C by open sets $\{G_\alpha\}$, there is a finite subset $\left\{G_{\alpha_i}\right\}_{i=1}^{n}$ of $\{G_\alpha\}$ that is also a cover of C.

A set of real numbers is compact if and only if it is both closed and bounded.

Complement of a set: The **complement** E^c of a set E is the set of real numbers that are not elements of E. $E^c = \{x \in \mathbb{R} : x \notin E\}$. (See relative complement.)

Contained: A set A is **contained** in a set B, denoted by $A \subseteq B$, if every element of A is an element of B. If, in addition, there is at least one element of B that is not an element of A, then we say that A is strictly or properly contained in B and write $A \subset B$.

Cover: A collection of sets $\{A_\alpha\}$ is a **cover** of a set E if E is contained in the union of the $\{A_\alpha\}$. ($E \subseteq \cup_\alpha A_\alpha$.)

De Morgan's Laws: For any collection of sets $\{A_\alpha\}$, $\cup_\alpha A_\alpha^c = (\cap_\alpha A_\alpha)^c$ and $\cap_\alpha A_\alpha^c = (\cup_\alpha A_\alpha)^c$. In the special case of two sets A and B, $A^c \cup B^c = (A \cap B)^c$ and $A^c \cap B^c = (A \cup B)^c$.

Dense: Given two sets $A \subset B$ in a space X with norm $\|\cdot\|$, the set A is **dense** in B if for any $x \in B$ there is a sequence $\{x_i\}$ from A such that $\lim \|x - x_n\| = 0$. In a more general context, A is **dense** in B if given any $x \in B$ and any neighborhood U of x, U contains points of A.

Extended real numbers: The **extended real numbers** consist of the real numbers together with $-\infty$ and $+\infty$ which are understood to be respectively less than and greater than any real number. We denote the extended real numbers by $\overline{\mathbb{R}}$.

Integers: The set of **integers**, denoted by $\mathbb{Z}$, is the set $\mathbb{Z} = \{\ldots, -3, -2, -1, 0, 1, 2, 3, \ldots\}$.

Intersection: The **intersection** of a collection of sets $\{A_\alpha\}$, written as $\cap_\alpha A_\alpha$, is the set of elements individually belonging to all the sets A_α. The intersection of a pair of sets A and B is written as $A \cap B$.

Natural numbers: The set of **natural numbers**, denoted by $\mathbb{N}$, is the set $\mathbb{N} = \{1, 2, 3, 4, \ldots\}$.

Neighborhood: Given a point x, a **neighborhood** of x is a open set containing x.

Nondegenerate: An interval I is **nondegenerate** if it contains an open interval. Alternatively, I contains more than a single point.

Nonoverlapping: Two intervals are **nonoverlapping** if their intersection consists of at most a single point. The common point will of necessity be an end point of both intervals.

Open set: A set E is **open** if, given any element x of E, there is an $\varepsilon > 0$ so that any point y satisfying $|y - x| < \varepsilon$ is also an element of E. Alternatively, a set E is open if it contains none of its boundary points. (See relatively open.)

The complement of any open set is closed and the complement of any closed set is open.

The union of an arbitrary collection of open sets is open.

The intersection of a finite number of open sets is open.

Positive rational numbers: $\mathbb{Q}^+ = \{x \in \mathbb{Q} : x > 0\}$.

Rational numbers: A real number x is called **rational** if x can be expressed as the ratio of integers, $x = \frac{p}{q}$ where $p, q \in \mathbb{Z}, q \neq 0$. The set of rational numbers is denoted by $\mathbb{Q}$.

Real numbers: The set of **real numbers** is denoted by $\mathbb{R}$.

Relative complement of a set: Given a pair of sets A and B, the **relative complement** of A with respect to B is $B \backslash A = \{x \in B : x \notin A\} = B \cap A^c$, the set of points belonging to B but not A. (See complement.)

Relatively open: A set E is **relatively open** in X if $E = X \cap U$ for some open set U. Alternatively, given any $x \in U$, there is $\varepsilon > 0$ so that any point $y \in X$ satisfying $|x - y| < \varepsilon$ is also an element of E. (See open.)

Union: The **union** of a collection of sets $\{A_\alpha\}$, written as $\cup_\alpha A_\alpha$, is the set of elements belonging to one or more of the sets A_α. The union of a pair of sets A and B is written as $A \cup B$.

A.2 Infimums and supremums

A nonempty, bounded set of real numbers will have an infimum and a supremeum.

Infimum: The **infimum** of a set S of real numbers, denoted by $\inf S$, is the largest real number α satisfying $\alpha \leq x$ for all $x \in S$.

The existence of $\inf S$ as a real number for all nonempty and bounded S is equivalent to the completeness of the real numbers.

In the extended real numbers, $\overline{\mathbb{R}}$, we write $\inf S = -\infty$ when S is not bounded below and $\inf S = +\infty$ when S is empty.

For any nonempty, bounded set S,

$$\inf S = -\sup(-S) = -\sup\{-x : x \in S\}.$$

If A and B are sets with $A \subseteq B$, then $\inf A \geq \inf B$.

Sometimes, a set that determines the range of values to be considered will appear under the inf. For example, $\inf_A f = \inf\{f(x) : x \in A\}$ where f is a real-valued function whose domain includes A.

If $\{f_j\}_{j=1}^{\infty}$ is a sequence of functions with a common domain, then $g_n = \inf_{j \geq n} f_j$ is the function defined by $g_n(x) = \inf\{f_j(x) : n \leq j < \infty\}$.

Supremum: The **supremum** of a set S of real numbers, denoted by $\sup S$, is the smallest real number α satisfying $\alpha \geq x$ for all $x \in S$.

The existence of $\sup S$ as a real number for all nonempty and bounded S is equivalent to the completeness of the real numbers.

In the extended real numbers, $\overline{\mathbb{R}}$, we write $\sup S = +\infty$ when S is not bounded above and $\sup S = -\infty$ when S is empty.

For any nonempty, bounded set S,

$$\sup S = -\inf(-S) = -\inf\{-x : x \in S\}.$$

If A and B are sets with $A \subseteq B$, then $\sup A \leq \sup B$.

Sometimes, a set that determines the range of values to be considered will appear under the sup. For example, $\sup_A f = \sup\{f(x) : x \in A\}$ where f is a real-valued function whose domain includes A.

If $\{f_j\}_{j=1}^{\infty}$ is a sequence of functions with a common domain, then $g_n = \sup_{j \geq n} f_j$ is the function defined by $g_n(x) = \sup\{f_j(x) : n \leq j < \infty\}$. The sequence $\{g_n\}$ will be a decreasing sequence of functions.

A.3   Sequences of real numbers

A sequence of real numbers is a function from $\mathbb{N}$ to $\mathbb{R}$. The sequence is typically expressed using the form $\{a_n\}_{n=1}^{\infty}$ or $\{a_n\}$ where a_n is the value of the function when the value n is input.

Bounded: A sequence $\{a_n\}$ is **bounded** if there is a real number B such that $|a_n| \leq B$ for all $n \in \mathbb{N}$. The sequence is said to be **bounded above** if $a_n \leq B$ for all $n \in \mathbb{N}$ and **bounded below** if $a_n \geq B$ for all $n \in \mathbb{N}$.

Every bounded sequence has at least one cluster point. If the cluster point is unique, then the cluster point is the limit of the sequence.

Cauchy: A sequence $\{a_n\}$ is called a **Cauchy sequence** if, given any $\varepsilon > 0$, there is a natural number n such that $|a_j - a_i| < \varepsilon$ whenever $i, j \geq n$.

Every Cauchy sequence converges.

Cluster point: A point x is a **cluster point** (also known as an accumulation point) of a sequence $\{a_n\}$ if, given any $\varepsilon > 0$ and any natural number n, there is another natural number j with $j \geq n$ for which $|a_j - x| < \varepsilon$.

Every bounded sequence has at least one cluster point.

If a bounded sequence has a unique cluster point, then the sequence converges to that cluster point.

Complete: A set with a norm (like the absolute value) is **complete** if every Cauchy sequence converges. The set $\mathbb{R}$ of real numbers is complete as is any closed subset of $\mathbb{R}$. The axiom that $\mathbb{R}$ is complete is equivalent to the axiom that every bounded nonempty subset of $\mathbb{R}$ has a supremum.

Converge: A sequence $\{a_n\}$ is said to **converge** if there is a real value x for which, given any $\varepsilon > 0$, there is a natural number n such that $|a_j - x| < \varepsilon$ whenever $j \geq n$. In this case, $\{a_n\}$ is called a **convergent** sequence and x is its **limit**.

Any convergent sequence is bounded.

A monotone sequence converges if and only if it is bounded.

A sequence converges if and only if it is a Cauchy sequence.

Decreasing: A sequence $\{a_n\}$ is **decreasing** if $a_n \geq a_{n+1}$ for all $n \in \mathbb{N}$. If $a_n > a_{n+1}$ for all $n \in \mathbb{N}$, the sequence is said to be **strictly decreasing**.

Any bounded, decreasing sequence converges.

Increasing: A sequence $\{a_n\}$ is **increasing** if $a_n \leq a_{n+1}$ for all $n \in \mathbb{N}$. If $a_n < a_{n+1}$ for all $n \in \mathbb{N}$, the sequence is said to be **strictly increasing**.

Any bounded, increasing sequence converges.

Liminf: Let $\{a_n\}$ be a sequence that is bounded below. The **limit infimum** of $\{a_n\}$ is $\underline{\lim}_n a_n = \liminf_n a_n = \lim_n \inf_{j \geq n} a_j = \lim_n \inf \{a_j : j \geq n\}$.

If $\{a_n\}$ is bounded below, $\{\inf_{j \geq n} a_j\}$ is an increasing sequence so the liminf always exists. If $\{a_n\}$ is not bounded below, we say that $\underline{\lim}_n a_n = -\infty$.

Limit: A point x is the **limit** of a sequence $\{a_n\}$ if, given any $\varepsilon > 0$, there is a natural number n for which $|a_j - x| < \varepsilon$ whenever $j \geq n$.

The limit of a sequence, when it exists, is unique.

If $\{a_n\}$ is monotone and bounded, it has a limit. (See converges.)

A sequence converges to a limit if and only if it is a Cauchy sequence.

Limsup: Let $\{a_n\}$ be a sequence that is bounded above. The **limit supremum** of $\{a_n\}$ is $\overline{\lim}_n a_n = \limsup_n a_n = \lim_n \sup_{j \geq n} a_j = \lim_n \sup \{a_j : j \geq n\}$.

If $\{a_n\}$ is bounded above, $\{\sup_{j \geq n} a_j\}$ is a decreasing sequence so the limsup always exists. If $\{a_n\}$ is not bounded above, we say that $\overline{\lim}_n a_n = +\infty$.

Monotone: A sequence is **monotone** if it is either an increasing sequence or a decreasing sequence.

A.4   Real-valued functions

Given a subset X of $\mathbb{R}$, a real-valued function f on X is a rule that assigns to each element of X a single value from $\mathbb{R}$. The set X is called the **domain** of f. In the following, the set X is assumed to be the domain of f unless otherwise specified.

Absolutely continuous: A function f is **absolutely continuous** on a set A if given any $\varepsilon > 0$ there is a $\delta > 0$ so that $\sum_k |f(x_k) - f(y_k)| < \varepsilon$ for any finite choice of disjoint intervals $\{(x_k, y_k)\}$ from A satisfying $\sum_k |x_k - y_k| < \delta$.

Bounded: A function f is **bounded** if there is a real number B such that $|f| \le B$. By $|f| \le B$ we mean that $|f(x)| \le B$ for all $x \in X$. Alternatively, f is bounded if the range of f, $f(X) = \{f(x) : x \in X\}$, is a bounded set.

If f is continuous on a compact domain D, then $f(D)$ is compact.

If f is continuous on a compact domain, then f is bounded.

Continuous at x: A function f is **continuous at a point** $x \in X$ if any of the following equivalent statements is true. (See continuous.)

1. Given any $\varepsilon > 0$ there is a $\delta > 0$ so that $|f(x) - f(y)| < \varepsilon$ for all $y \in X$ satisfying $|x - y| < \delta$.

2. Given any open set V containing $f(x)$, there is a open set U containing x such that $f(X \cap U) = \{f(x) : x \in X \cap U\} \subseteq V$.

3. Given any sequence $\{x_n\}$ from X that converges to x, $\lim_n f(x_n) = f(x)$.

Continuous: A function f is **continuous** if any of the following equivalent statements is true.

1. The function f is continuous at each point $x \in X$. (See continuous at a point and uniformly continuous.)

2. Given any $x \in X$ and $\varepsilon > 0$ there is a $\delta > 0$ so that $|f(x) - f(y)| < \varepsilon$ for all $y \in X$ satisfying $|x - y| < \delta$.

3. Given any open set V in $\mathbb{R}$, its preimage, $f^{-1}(V) = \{x \in X : f(x) \in V\}$, is an open set relative to X.

4. Given any $x \in X$ and any sequence $\{x_n\}$ from X that converges to x, $\lim_n f(x_n) = f(x)$.

If f is continuous on a compact domain, then $f(D)$ is compact.
If f is continuous on a compact domain, then f is bounded.
If f is continuous on a compact domain, then f is uniformly continuous.
(See also intermediate value theorem.)

Decreasing: A function f is **decreasing** if $f(x) \geq f(y)$ whenever $x, y \in X$ with $x < y$. If $f(x) > f(y)$, then we say that f is **strictly decreasing.**

Differentiable at a point: A function f is **differentiable at the point** $a \in X$ if $\lim_{x \to a} \frac{f(x) - f(a)}{x-a}$ exists. In this case, $f'(a) = \lim_{x \to a} \frac{f(x) - f(a)}{x-a}$ is the derivative of f at a.

Differentiable: A function f is **differentiable** on a set if it is differentiable at every point in the set.

Increasing: A function f is **increasing** if $f(x) \leq f(y)$ whenever $x, y \in X$ with $x < y$. If $f(x) < f(y)$, then we say that f is **strictly increasing.**

Intermediate value theorem: If f is continuous on $[a, b]$ and v is a value between $f(a)$ and $f(b)$ then there is a $c \in (a, b)$ such that $f(c) = v$.

Monotone: A function is **monotone** if it is either increasing or decreasing.

Mean value theorem: If f is continuous on $[a, b]$ and differentiable on (a, b), then there is a point $c \in (a, b)$ satisfying $f'(c) = \frac{f(b) - f(a)}{b - a}$.

Preimage: The **preimage** of a set E under a function f is $f^{-1}(E) = \{x \in X : f(x) \in E\}$.

Uniformly continuous: A function f is **uniformly continuous** if for any $\varepsilon > 0$ there is a $\delta > 0$ such that $|f(x) - f(y)| < \varepsilon$ for all $x, y \in X$ satisfying $|x - y| < \delta$.

In contrast to the definition of continuous, the δ in the definition of uniformly continuous cannot depend on the value of x. Given $\varepsilon > 0$, one value of δ must serve for all choices of x. (See continuous.)

If f is continuous on a compact set X, then f is uniformly continuous.

A.5 Sequences of functions

In the following, we assume that $\{f_n\}$ is a sequence of functions with a common domain X.

Convergence: A sequence of functions $\{f_n\}$ **converges (pointwise)** to a function f if the sequence $\{f_n(x)\}$ of real numbers converges to $f(x)$ for all $x \in X$. Alternatively, given any $x \in X$ and any $\varepsilon > 0$, there is a natural number n so that $\left| f_j(x) - f(x) \right| < \varepsilon$ for all $j \geq n$. (See uniform convergence.)

Decreasing: A sequence of functions $\{f_n\}$ is **decreasing** if $f_n \geq f_{n+1}$ for all $n \in \mathbb{N}$. By $f_n \geq f_{n+1}$ we mean that $f_n(x) \geq f_{n+1}(x)$ for all $x \in X$.

Increasing: A sequence of functions $\{f_n\}$ is **increasing** if $f_n \leq f_{n+1}$ for all $n \in \mathbb{N}$. By $f_n \leq f_{n+1}$ we mean that $f_n(x) \leq f_{n+1}(x)$ for all $x \in X$.

Infimum: The **infimum** of a sequence of functions $\{f_n\}$ is the function $g = \inf_n f_n$ defined by $g(x) = \inf_n f_n(x) = \inf\{f_n(x) : n \in \mathbb{N}\}$. The function $\inf_n f_n$ is only defined as a real-valued function if $\{f_n(x)\}$ is bounded below for all $x \in X$. We can always define $\inf_n f_n$ as a function taking on extended real values.

Limit: A sequence of functions $\{f_n\}$ has a **limit** if there is a real-valued function f such that, given any $x \in X$ and any $\varepsilon > 0$, there is a natural number n so that $\left| f_j(x) - f(x) \right| < \varepsilon$ for all $j \geq n$.

Monotone: A sequence of functions $\{f_n\}$ is **monotone** if it is either increasing or decreasing.

Supremum: The **supremum** of a sequence of functions $\{f_n\}$ is the function $g = \sup_n f_n$ defined by $g(x) = \sup_n f_n(x) = \sup\{f_n(x) : n \in \mathbb{N}\}$. The function $\sup_n f_n$ is only defined as a real-valued function if $\{f_n(x)\}$ is bounded above for all $x \in X$. We can always define $\sup_n f_n$ as a function taking on extended real values.

Uniform convergence: A sequence of functions $\{f_n\}$ **converges uniformly** to a function f given any $\varepsilon > 0$, there is a natural number n so that $\left| f_j(x) - f(x) \right| < \varepsilon$ for all $j \geq n$ and for all $x \in X$. (See convergence.)

Uniformly bounded: A sequence of functions $\{f_n\}$ is **uniformly bounded** if there is a real number B such that $|f_n| \leq B$ for all $n \in \mathbb{N}$. Equivalently, $|f_n(x)| \leq B$ for all $n \in \mathbb{N}$ and all $x \in X$.

A.6 Complex numbers

The set $\mathbb{C}$ of complex numbers is the set of all numbers of the form $x = a + ib$ where $a, b \in \mathbb{R}$ and $i^2 = -1$.

Complex conjugate: The **complex conjugate** of a complex number $x = a + ib$ is $\overline{x} = a - ib$. The **complex conjugate** of a complex-valued function f is the function $\overline{f}$ where $\overline{f}(x) = \overline{f(x)}$. For complex numbers x and y

$$\overline{x + y} = \overline{x} + \overline{y},$$
$$\overline{xy} = \overline{x}\,\overline{y}, \text{ and}$$
$$x\overline{x} \geq 0 \text{ with } x\overline{x} = 0 \text{ only if } x = 0.$$

Imaginary part: Let $x = a + ib$ be a complex number. The **imaginary part** of x, denoted $\operatorname{Im} x$, is $\operatorname{Im} x = b = \frac{1}{2i}(x - \overline{x})$.

Modulus: We define the **modulus** of a complex number $x = a + ib$ to be $|x| = \sqrt{x\overline{x}} = \sqrt{a^2 + b^2}$. The modulus of x is the same as the Eucliean distance from the point (a, b) to the origin in the plane. For complex numbers x and y,

$$|\overline{x}| = |x|,$$
$$|xy| = |x|\,|y|, \text{ and}$$
$$|x + y| \leq |x| + |y|.$$

Real part: Let $x = a + ib$ be a complex number. The **real part** of x, denoted by $\operatorname{Re} x$, is $\operatorname{Re} x = a = \frac{1}{2}(x + \overline{x})$.

A.7   Inner product spaces and projections

The following definitions and results are stated for complex vector spaces. By dropping any complex conjugates, the statements apply to real vector spaces.

Inner product space: An **inner product space** is a vector space V together with an inner product $\langle \cdot, \cdot \rangle : V \times V \to \mathbb{C}$ that for any $\mathbf{x}, \mathbf{y}, \mathbf{z} \in V$ and any scalar α, satisfies

1. $\langle \mathbf{y}, \mathbf{x} \rangle = \overline{\langle \mathbf{x}, \mathbf{y} \rangle}$,

2. $\langle \alpha \mathbf{x}, \mathbf{y} \rangle = \alpha \langle \mathbf{x}, \mathbf{y} \rangle$,

3. $\langle \mathbf{x} + \mathbf{y}, \mathbf{z} \rangle = \langle \mathbf{x}, \mathbf{z} \rangle + \langle \mathbf{y}, \mathbf{z} \rangle$, and

4. $\langle \mathbf{x}, \mathbf{x} \rangle \geq 0$ with $\langle \mathbf{x}, \mathbf{x} \rangle = 0$ if and only if $\mathbf{x} = 0$.

Linear combination: Given a finite set of vectors $\{\mathbf{v}_1, \mathbf{v}_2, \mathbf{v}_3, \ldots, \mathbf{v}_n\}$ any vector $\mathbf{v}$ that can be expressed as $\mathbf{v} = \alpha_1 \mathbf{v}_1 + \alpha_2 \mathbf{v}_2 + \cdots + \alpha_n \mathbf{v}_n$ where $\alpha_1, \alpha_2, \ldots, \alpha_n$ are scalars is a **linear combination** of the vectors in $\{\mathbf{v}_1, \mathbf{v}_2, \mathbf{v}_3, \ldots, \mathbf{v}_n\}$.

Norm: A norm on a vector space V is a mapping $\|\cdot\| : V \to \mathbb{R}$ that for any vectors $\mathbf{u}$ and $\mathbf{v}$ from V and any scalar α satisfies

1. $\|\mathbf{u}\| \geq 0$ with $\|\mathbf{u}\| = 0$ if and only if $\mathbf{u} = \mathbf{0}$.

2. $\|\alpha\mathbf{u}\| = |\alpha|\,\|\mathbf{u}\|$, and

3. $\|\mathbf{u} + \mathbf{v}\| \leq \|\mathbf{u}\| + \|\mathbf{v}\|$ (triangle inequality).

Any inner product space has an associated norm defined by $\|\mathbf{u}\| = \sqrt{\langle \mathbf{u}, \mathbf{u} \rangle}$.

Orthogonal set: A set of vectors $\{\mathbf{v}_\alpha\}_{\alpha \in I}$ in an inner product space is **orthogonal** if $\langle \mathbf{v}_\alpha, \mathbf{v}_\beta \rangle = 0$ for $\alpha \neq \beta$. If vectors $\mathbf{u}$ and $\mathbf{v}$ are orthogonal then $\|\mathbf{u} + \mathbf{v}\|^2 = \|\mathbf{u}\|^2 + \|\mathbf{v}\|^2$.

Projection: Given a subspace W of an inner product space V and a vector $\mathbf{v} \in V$, the **projection** of $\mathbf{v}$ into W is the unique vector $\hat{\mathbf{v}}$ satisfying $\hat{\mathbf{v}} \in W$ and $\mathbf{v} - \hat{\mathbf{v}} \in W^\perp$ where $W^\perp$ is the vector space of all vectors in V that are orthogonal to every vector in W. Note that, while the notation is similar, $\hat{\mathbf{v}}$ should not be confused with sequence of Fourier coefficients.

The projection $\hat{\mathbf{v}}$ satisfies $\|\mathbf{v} - \hat{\mathbf{v}}\| \leq \|\mathbf{v} - \mathbf{w}\|$ for all $w \in W$.

If $\{\mathbf{v}_1, \mathbf{v}_2, \ldots, \mathbf{v}_n\}$ is a set of non-zero orthogonal vectors in an inner product space V and $\mathbf{v} \in V$, then the projection of $\mathbf{v}$ into $\text{Span}\{\mathbf{v}_1, \mathbf{v}_2, \ldots, \mathbf{v}_n\}$ is

$$\hat{\mathbf{v}} = \frac{\langle \mathbf{v}, \mathbf{v}_1 \rangle}{\langle \mathbf{v}_1, \mathbf{v}_1 \rangle} \mathbf{v}_1 + \frac{\langle \mathbf{v}, \mathbf{v}_2 \rangle}{\langle \mathbf{v}_2, \mathbf{v}_2 \rangle} \mathbf{v}_2 + \cdots + \frac{\langle \mathbf{v}, \mathbf{v}_n \rangle}{\langle \mathbf{v}_n, \mathbf{v}_n \rangle} \mathbf{v}_n.$$

Span: The **span**, $\text{Span}(S)$ of a set S of vectors is the set of all vectors that can be expressed as linear combinations of vectors from S.

Subspace: Let V be a vector space. A subset W of V is a **subspace** of V if W, together with the operations of V, satisfies the vector space axioms.

To prove that a subset W is a subspace of V, it is sufficient to verify that W contains the vector $\mathbf{0}$ and is closed under addition and scalar multiplication.

Index

About the Author

C. Ray Rosentrater is a Professor of Mathematics at Westmont College where he has also served as department chair and Associate Dean for Curriculum. He has been recognized as Westmont's Teacher of the Year in the Natural and Behavioral Sciences and has received the Faculty Research Award. He earned a PhD in mathematics from Indiana University and an MSc in computer science from the University of Toronto. Awarded a Fulbright Fellowship in 1995, he now serves as Westmont's Fulbright Program Advisor. He served multiple terms on the ACMS board including terms as Vice President and President. His other publications include papers in operator theory and articles connecting analysis to computer science and linear algebra to statistics. He co-wrote two chapters in *Mathematics through the Eyes of Faith*.